# Fracture Resistance of Aluminum Alloys

## Notch Toughness, Tear Resistance, and Fracture Toughness

J. Gilbert Kaufman

**The Aluminum Association**
Incorporated
900 19th Street, N.W., Washington, D.C. 20006

**The Materials Information Society**

Materials Park, Ohio 44073-0002
www.asminternational.org

*ASM International staff who worked on this project included Veronica Flint, Manager of Book Acquisitions; Bonnie Sanders, Manager of Production; Nancy Hrivnak, Copy Editor; Kathleen Dragolich, Production Editor; and Scott Henry, Assistant Director of Reference Publications.*

Library of Congress Cataloging-in-Publication Data

Kaufman, J.G. (John Gilbert), 1931-
Fracture resistance of aluminum alloys/J. Gilbert Kaufman.
p.cm.
1. Aluminum alloys—Mechanical properties. 2. Fracture mechanics I. Title.
TA480.A6 K355 2000        620.1'866—dc21        2001022228

ISBN: 0-87170-732-2
SAN: 204-7586

ASM International®
Materials Park, OH 44073-0002
www.asminternational.org

Printed in the United States of America

# Preface

On behalf of the Aluminum Association, Inc., Alcoa, Inc., and ASM International, we are pleased to provide this summary of data on the fracture characteristics of aluminum alloys. It is broadly based on a publication produced by Alcoa in 1964, called *Fracture Characteristics of Aluminum Alloys*, and we want to acknowledge the support of Alcoa, Inc., notably Dr. Robert J. Bucci and Dr. William G. Truckner, in arranging to have the copyright to that publication transferred to the Aluminum Association, Inc. Further, we acknowledge the support of Dr. John A.S. Green of the Aluminum Association, Inc. in making it available for a joint publication with ASM International.

In particular, we note the contributions of the members of the Aluminum Association Engineering and Design Task Force, Dr. Andrew J. Hinkle, Chair, through their review of and input to the organization and content of the book.

This book is unique in the degree to which it presents individual test results for many individual lots of a wide range of aluminum alloys, tempers, and products, rather than simply broad summaries of data; it is also unique for the breadth of types of fracture parameters presented. This combination provides not only the ability to dig out specific data needed to evaluate alloy and temper selections for individual applications, but also the ability to check the degree to which the various fracture parameters provide consistent relative ratings for specific alloys and tempers. We believe these capabilities will benefit a wide range of needs, from alloy evaluation and selections to design.

A word is needed about the inclusion in the book of data for a number of alloys and tempers that are considered obsolete today. Such alloys are included because they may have been used in fracture-critical structures in years past, and specialists dealing with maintenance and retrofit of those structures may be looking for data on the old alloys, even though it is unlikely that new structures will be made of them.

An explanation is also needed about the treatment of units in this book. Because all of these data were generated in an environment of the usage of English/engineering units, and because of the mass of data involved, almost the entire book is presented in those units. While this is contrary to the normal ASM International and Aluminum Association, Inc. policies to present engineering and scientific data in both Standard International (SI) and English/engineering units, it saves a prodigious amount of expense related to both time for conversion and to the space required for dual presentation. Further, it avoids the inevitable compromises surrounding rounding techniques for such conversions in a multitude of units. Additional help for those interested in SI conversion is provided in Appendix 2.

J. Gilbert Kaufman

# ASM International Technical Books Committee (2000–2001)

# Contents

# Figures

viii

# Tables

# CHAPTER 1

# Introduction

## 1.1 Synopsis

THE TEST METHODS and criteria used to evaluate the fracture characteristics of aluminum alloys are reviewed, and a substantial amount of representative test data for individual lots of aluminum sheet, plate, forgings, extrusions, and castings are shown for a wide variety of aluminum alloys, tempers, and products at room, subzero, and elevated temperatures. The significance and use of various measures of toughness are discussed, and the more valuable fracture indexes are identified.

From the tensile test, elongation and reduction in area provide a measure of the behavior of materials in very simple stress fields but offer only a broad indication of fracture behavior. Notch toughness, as measured by the notch-yield ratio, is a useful relative measure of the capabilities of materials to deform plastically in the presence of stress-raisers. Tear resistance, as measured by unit propagation energy from the tear test, provides a meaningful measure of relative resistance to either slow or unstable crack growth. Fracture toughness, based upon fracture-mechanics concepts, defines the conditions for unstable crack growth in an elastic-stress field; it is a direct measure of toughness in that it provides structural designers with specific guidance as to how to avoid "brittle" catastrophic fractures. The fracture mechanics approach is most useful for high-strength aluminum alloys but has restricted applicability to many broadly used commercial alloys, most of which have great ability to deform plastically at crack tips and absorb energy. Unstable crack growth in elastic-stress fields is rarely encountered for high-strength aluminum alloys.

Of the structural aluminum alloys, the 5xxx series provide the most attractive combination of strength and toughness for critical applications such as liquefied natural gas storage and transportation tankage. Among the higher strength alloys, premium toughness alloys such as 2124, 2524, 7050, 7175, and 7475 provide excellent toughness at high-yield-strength levels and so are attractive for fracture-critical aerospace and transportation applications.

The 5xxx series are also outstanding for high toughness at subzero temperatures, providing both strength and toughness well above room temperature values at temperatures down to –320 °F; even at temperatures as low as –452 °F (near absolute zero), the toughness levels for many of these alloys and tempers are quite high.

For welded structures, 5xxx filler alloys are recommended over aluminum alloy 4043 where high toughness is important at any service temperature.


# 1.2 Introduction

With the continued development of high-strength aluminum alloys and tempers and their use in very critical components in aerospace, automotive, marine, and cryogenic applications, the ability to adequately describe and predict their fracture resistance remains important. These needs range from (a) in alloy development, determining which alloys and tempers of a given group have the greatest fracture resistance, (b) in alloy selection, making decisions on alloy choices for specific applications, and (c) in design, establishing safe design stresses and/or predicting critical crack or discontinuity sizes under specific service conditions.

Most commercial aluminum alloys and tempers are so tough that "brittle" or "low ductility fracture" (i.e., unstable or self-propagating crack growth in elastically stressed material) rarely occurs under any conditions. For these alloys, the merit-rating approach is generally sufficient, and measures of notch toughness or tear resistance providing relative qualitative ratings may be sufficient. However, there are a number of higher-strength aluminum alloys and tempers that are used principally in aerospace applications, where strength must be used to the maximum advantage and the consequences of unexpected low-ductility failure must be considered. For these particular alloys and tempers, more precise evaluations of toughness by methods such as fracture toughness testing are required for quantitative evaluation of fracture behavior under specific service conditions and, subsequently, the design of fracture resistance into the structure.

It is the purpose of this publication to build on an earlier work (Ref 1) to (a) describe various criteria for evaluating the toughness or fracture resistance of aluminum alloys, how they are determined, and their usefulness and limitations; (b) provide a background of representative data from various types of toughness tests for a wide range of aluminum alloys and tempers, and (c) provide some general guidance as to which alloys may be most useful for applications where high toughness is vital.

It is not the intent of this book to describe and provide extensive performance data for other types of fracture mechanisms such as fatigue and corrosion beyond showing the logical interfaces. For comprehensive coverage of these subjects and more in-depth design approaches, readers are referred to the work of Bucci, Nordmark, and Stark (Ref 2) in *Fatigue and Fracture*, Volume 19 of *ASM Handbook*. For readers interested in a broader range and depth of discussion on applications for aluminum alloys, as well as other aspects of the aluminum industry, reference is made to Altenpohl's book *Aluminum: Technology, Applications, and Environment* (Ref 3).

Much of the data provided herein are from the highly respected Alcoa Laboratories research organization of Alcoa, Inc., which has been active for more than 40 years in the fracture-testing field. Included are data obtained using consistent and well-documented methods from many papers published by Alcoa scientists, as well as data from several previously unpublished reports. Also presented are representative data from the Aluminum Association fracture toughness database, ALFRAC, put together under contract from the Metals Properties Council and subsequently made available through a grant from the National Institute of Standards and Technology and the National Materials Property Data Network.

The data included herein are not intended to be exhaustive but to provide a good representation of a wide range of types of toughness indexes for a broad spectrum of aluminum alloys, including both wrought and cast alloys. The data are presented for their value in understanding the fracture behavior of aluminum alloys but are not intended for design.

A word of explanation is needed about the inclusion in the book of data for a number of aluminum alloys and tempers that are no longer considered useful for various reasons and that are now designated as obsolete by the Aluminum Association, Inc. Such alloys are included because they may have been used in fracture-critical structures in years past, and so specialists dealing with maintenance and retrofit for such structures may be looking for data on the old alloys. Their inclusion herein provides a useful source and potentially valuable comparisons with data for alloys currently recommended for comparable applications. All obsolete alloys are identified by appropriate footnotes in the tables in which they appear.

It is also appropriate to note that all the data presented and discussed in this book were generated in accordance with the ASTM Standard Test Methods (Ref 4–11) applicable at the time. While there has been some evolution in those standards over the years, especially in the field of fracture toughness testing, the results presented are believed to have been determined by procedures reasonably, if not exactly, consistent with current standards.

Finally, some explanation is needed about the treatment of units in this book. Because all of these data were generated in an environment using

English/engineering units, and because of the mass of data involved, the entire book is presented in English units. While the normal ASM International and Aluminum Association, Inc. policies (Ref 12) are to present engineering and scientific data in both International Standard (SI) and English/engineering units, an appreciable amount of time and expense would be required for the complete conversion and for the dual presentation of all the tables included herein. In addition, the inevitable compromises surrounding rounding techniques for such conversions with a multitude of complex units have been avoided. Those readers interested in SI conversion are referred to Appendix 2 for some guidance.

Some additional valuable sources on aluminum alloy products, standards, and properties are included for the reader (Ref 12–18).

# CHAPTER 2

# Definition of Terms Related to Fracture Behavior

IN THE DISCUSSION that follows, a number of general and specific terms are used to describe the various aspects of the fracture behavior of aluminum alloys. It is desirable to define a number of these terms at the outset; many are discussed in greater detail subsequently.

**ductility.** A general term describing the ability of a material to deform plastically, before fracture, usually measured by the elongation or reduction in area in a tensile test. For purposes of this discussion, it is not considered to encompass notch toughness, tear resistance, or fracture toughness.

**toughness.** A general term describing the resistance of a material to low ductility fracture under stress, without reference to the specific conditions or mode of fracture. Generally it is considered to encompass notch toughness, tear resistance, and fracture toughness.

**notch toughness.** A general term describing the ability of a material to deform plastically and locally in the presence of stress-raisers (either cracks, flaws, or design discontinuities) without cracking and thus to redistribute loads to adjacent material or components. It is the inverse of notch sensitivity in the sense that as the notch toughness of a material increases, notch sensitivity decreases. While notch toughness is associated more closely with the resistance of a material to the initiation of cracking and fracture than with its resistance to crack propagation, it correlates well with other indexes of resistance to unstable crack growth (see "Notch Toughness and Notch Sensitivity," Chapter 5, and ASTM Standards E 338 and E 602).

**notch-tensile strength.** The net fracture strength of a notched tensile specimen, that is, maximum load supported by the notched specimen divided by its net cross-sectional area. It has little direct value since the notch geometry rarely mirrors service conditions; its principal usefulness is in its relationship to the tensile and yield strengths of the material.

**notch-yield ratio.** The ratio of notch-tensile strength to tensile yield strength of the material. This provides a measure of notch toughness and, hence, of the inverse of notch sensitivity. Notch-yield ratio is considered by many engineers to be a more useful measure of notch toughness than notch-strength ratio (defined next) because it provides a relative measure of the ability of a material to plastically deform locally in the presence of a stress-raiser and thus to redistribute the stress.

**notch-strength ratio.** The ratio of notch-tensile strength to tensile strength of the material. This provides a measure of tensile efficiency for the specific design of notch. It is not consistently reliable as a measure of notch toughness.

**tear resistance.** A general term describing the resistance of a material to crack propagation under static loading, in either an elastic stress field (brittle fracture) or a plastic stress field (tearing). Like fracture toughness, it is generally used in connection with crack growth, not crack initiation. The term *tear resistance* is generally applied to data obtained from tear tests, usually as measured specifically by unit propagation energy. (See Chapter 6 and ASTM Standard Methods B 871).

**unit propagation energy.** A specific term expressed in in.-lb/in.$^2$ describing the amount of energy required to propagate a crack across a unit area in a tear specimen in terms of the total energy to propagate the crack divided by the nominal crack area (i.e., the original net area of the specimen). It provides a measure of tear resistance and, indirectly, a measure of fracture toughness.

**tear strength.** A specific term, expressed in psi, describing the maximum nominal direct-and-bending stress developed by a tear specimen. Its significance is similar to that of notch-tensile strength, and its primary usefulness is in its relationship to the tensile yield strength of the material. The ratio of tear strength to yield strength (tear-yield ratio) is a measure of the relative resistance of a material to the development of fracture in the presence of a stress-raiser.

**tear-yield ratio.** The ratio of tear strength to the tensile yield strength. Similar to notch-yield ratio, it is a relative index of notch toughness.

**fracture toughness.** A general term describing the resistance of a material to unstable crack propagation at elastic stresses or to low-ductility or brittle fracture of any kind. As used in this book, it does not involve resistance to crack initiation but only to the unstable propagation of a crack already present. The term fracture toughness is sometimes used to denote specifically the critical strain energy release rate, but this is not the literal definition (see "Fracture Toughness," Chapter 7, and ASTM Standard Methods E 399, E 561, B 645, and B 646).

**strain-energy release rate, G.** A specific term, expressed in in.-lb/in., defining the rate of release of elastic strain energy during crack growth in an elastic stress field. The "critical" value of strain-energy release rate is measured at the onset of unstable crack growth and is one measure of fracture toughness.

**stress-intensity factor, K.** A specific term, expressed in ksi $\sqrt{\text{in.}}$, relating the gross stress in a material and the size of a crack or discontinuity present in the stress field. It also describes the stress field local to the crack tip. Stress-intensity factor is proportional to the square root of the strain-energy release rate, and so the critical value is a measure of the conditions for unstable crack growth.

**crack or discontinuity size.** A specific term, expressed in inches, defining the overall length of an opening in the stress field from which unstable crack growth might develop. It may represent a material flaw or crack growing out of a design detail (rivet hole, port hole, etc.); in this latter case, the discontinuity size includes the size of the design discontinuity and the crack length.

**unstable crack growth.** A general term describing a situation in which the elastic strain energy released by an increment of crack growth by any mechanism (i.e., static load, fatigue, creep, or corrosion) is sufficient to cause the crack to grow another increment in length; in other words, for the crack to become self-propagating.

**crack resistance curve.** A plot of resistance of a material to slow, stable crack extension, expressed in the same units as the stress intensity factor, $K$, or the crack extension force, $G$, as a function of the amount (length) of slow, stable crack extension. Comparison of the crack driving forces with this curve provides an estimate of the conditions for crack growth instability.

**stress condition.** A descriptor of the nature of the stress configuration in a component or at a specific location in a component or test specimen in terms of directionality and multiaxiality, thus indicating the degree of constraint on elastic and plastic deformation in the component or specimen.

**plane stress.** The condition in which all the stresses act in a single plane so that the stress in the third principal direction (normal to the plane) and the associated shear stresses are essentially zero. The strains in all three directions may be significant, so that the cross section may not remain uniform or plane. This is the condition in a thin, wide sheet under axial tension, where the stress in the short-transverse direction (normal to the surfaces of the sheet) is zero and local deformation takes place in the short-transverse direction.

**plane strain.** The condition in which the stresses in all three directions may be significant (i.e., a triaxial stress condition may prevail) and the strains in one principal direction are essentially uniform or zero, usually through the thickness. This condition is approximated at the tip of a crack in thick plate, where the strain in the short-transverse direction along the crack front is zero.

**specimen orientation.** Refers to the orientation of a specimen with respect to the major axes of the component from which it is taken.

For cylindrical tensile and notch tensile specimens, specimen orientation is generally defined in terms of the relation of the axis of the specimen to the major grain flow pattern, as follows:

- *Longitudinal, or L:* axis of specimen parallel to the major direction of grain flow
- *Long transverse, or LT (or simply transverse, or T) for thin components:* axis of specimen perpendicular to the axis of major grain flow, in the plane of the component
- *Short transverse, or ST:* axis of specimen normal to the axis of major grain flow and to the plane of the component
- For tear specimens, specimen orientation is generally defined in terms of the relation of the direction of applied stress to the major grain flow pattern and the plane of the component, as shown in Fig. A1.1 and as follows:
- *Longitudinal, or L:* applied stress parallel to the major direction of grain flow, in the plane of the component
- *Long transverse, or LT (or simply transverse, T, for thin components):* applied stress perpendicular to the axis of major grain flow, in the plane of the component
- *Short transverse, or ST:* applied normal to the axis of major grain flow and to the plane of the component

For fracture toughness specimens, specimen orientation is defined in terms of the relationship of the direction of applied stress and also the direction of crack growth to the grain flow and to the major plane of the component, as shown in Fig. A1.2(a) and as follows:

- *L-T:* applied stress in the major direction of grain flow and crack growth across the width or major plane of the component
- *L-S:* applied stress in the major direction of grain flow and crack growth through or normal to the major plane of the component
- *T-L:* applied stress normal to the major direction of grain flow and crack growth along the direction of major grain flow
- *T-S:* applied stress normal to the major direction of grain flow and crack growth through or normal to the major plane of the component
- *S-L:* applied stress normal to the major plane of the component and crack growth in the major direction of grain flow
- *S-T:* applied stress normal to the major plane of the component and crack growth normal to the major direction of grain flow

For most fracture toughness testing programs, specimens representing only the L-T, T-L, and S-L are used.

The orientations and positions of specimens from welded components are included in Appendix A1, Fig. A1.2(b) (compact tension specimens) and A1.2(c) (notch bend specimens).

# CHAPTER 3

# Tensile Properties as Indicators of Fracture Behavior

ELONGATION AND REDUCTION in area from the tensile test are measures of ductility and might be considered the simplest indicators of fracture behavior. As generally measured, elongation is a combination of uniform and nonuniform local deformation in a specific gage length. Because neither elongation nor reduction in area from the tensile test incorporates any measure of stress-sustaining capability in the presence of these types of deformation, however, neither is sufficiently descriptive of the fracture behavior to be very useful to the materials engineer or to the designer concerned with design to resist unstable crack growth (Ref 19).

On the other hand, it is fair to say that elongation and reduction in area do provide very broad indications of fracture behavior, so that one material having appreciably greater elongation and/or reduction in area than another is likely to have greater toughness as well. Elongation and reduction in area may also be somewhat useful indicators for comparing different lots of a given alloy, temper, and product if the data under consideration are all from one test direction and if the specimens are all of a single type and size. The correlations among various indicators of fracture resistance are discussed in "Interrelation of Fracture Characteristics," Chapter 8, and both the advantages and limitations of these properties as indicators of toughness are illustrated in greater detail.

The ratio of yield strength to tensile strength and the area under the tensile stress-strain curve have also been suggested as useful indications of toughness. Although they may be useful for some purposes, they are completely unreliable as indexes of resistance to low-ductility fracture. Alloys 2020-T6 and 6061-T6 both have similar ratios of yield strength to tensile strength (Table 6.1) and similar areas under their stress-strain curves but significantly different toughness levels by any measure. A comparison of both alloys is sufficient to demonstrate the inadequacy of these properties, as is shown by the average values in the following table:

| Parameter | 2020-T6, T651 | | 6061-T6, T651 | | Source |
|---|---|---|---|---|---|
| | L | LT | L | LT | |
| Ratio yield strength/ tensile strength | | | | | |
| Sheet | 0.94 | 0.93 | 0.91 | 0.88 | Tables 5.2, 5.5, |
| Plate | 0.95 | 0.93 | 0.93 | 0.90 | 6.1(b), 6.2 |
| Stress-strain curve area, in.-lb | | | | | |
| Sheet | $7.8 \times 10^3$ | $5.3 \times 10^3$ | $4.7 \times 10^3$ | $4.7 \times 10^3$ | Estimated |
| Plate | $6.3 \times 10^3$ | $3.3 \times 10^3$ | $6.5 \times 10^3$ | $6.6 \times 10^3$ | as elongation $\times$ (TS + YS)/2 |
| Notch-yield ratio, sheet | 0.76 | 0.70 | 1.18 | 1.06 | Table 5.1 |
| Notch-yield ratio, plate | 0.50 | 0.47 | 1.08 | 1.01 | Table 5.5 |
| Unit propagation energy, sheet, in.-lb/in.$^2$ | 30 | 15 | 900 | 740 | Table 6.1(b) |
| Unit propagation energy, plate, in.-lb/in.$^2$ | 100 | 50 | 905 | 775 | Table 6.2 |
| Plane-strain fracture toughness, $K_{Ic}$, plate, psi-in.$^{0.5}$ | 17,600 | 16,800 | 26,200 | 26,900 | Tables 7.1, 7.3 |

L, longitudinal; LT, long transverse; TS, tensile strength; YS, yield strength

The net result is that relying on any measurements from tensile tests for any more than very broad qualitative indicators of notch toughness, tear resistance, or fracture toughness is not recommended.

# CHAPTER 4

# Notched-Bar Impact and Related Tests for Toughness

THE TEMPERATURE SENSITIVITY of the fracture behavior of ferritic steels, that is, the transition over a relatively narrow range of temperatures from a high resistance to fracture to a very low resistance to fracture, brought about the development of various tests to determine their "transition temperature." Charpy and Izod notched-bar impact tests (Ref 5) are among those widely used. The U.S. Navy tear test (Ref 20) has served the same purpose. Drop-weight tests of various types have also been developed for that purpose (Ref 21–22) and are reported to be the most reliable.

The significant feature of all these tests is that their sole purpose is to establish a limiting temperature below which special precautions must be exercised in using materials displaying such a sudden transition in fracture behavior. The significance of the numbers obtained from the tests is that they define the critical temperature range of a fracture transition. The failure-analysis diagrams developed by Pellini and associates at the U.S. Naval Research Laboratories represented a significant refinement in the handling of transition-temperature data (Ref 21), and this approach has had an important impact on the steel industry.

Aluminum alloys, like other face-centered cubic materials, do not exhibit any sudden changes in fracture behavior with a decrease or other change in temperature. Therefore, transition-temperature tests, such as the Charpy and Izod impact tests, have little merit for aluminum alloys except to show the absence of a transition, as the data in Fig. 4.1 illustrate. In addition, many aluminum alloys are so tough that they do not fracture completely in Charpy and Izod tests, so that the numbers obtained in the tests are of no interpretive usefulness. In fact, they usually include the energy required to throw the bent specimen across the room. This is often

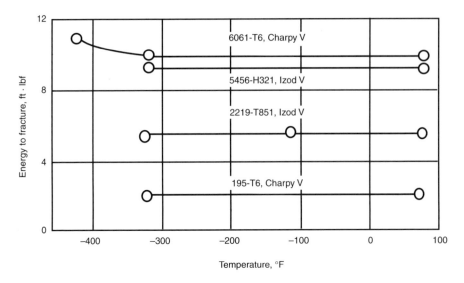

**Fig. 4.1** Notched-bar impact data for aluminum alloys, transverse direction

overlooked in the reporting and analysis of impact test data, and, as a result, there is a considerable amount of meaningless impact data in the literature for aluminum alloys.

The net result is that notched-bar impact tests have never been considered useful indicators of the fracture characteristics of aluminum alloys and are not discussed further herein.

# CHAPTER 5

# Notch Toughness and Notch Sensitivity

ONE OF THE EARLIEST APPROACHES to the evaluation of the fracture characteristics of aluminum alloys was via tensile tests of specimens containing various types of stress raisers (Fig. A1.3–A1.7). The results from these tests were analyzed in terms of the theoretical stress concentration factors (Ref 23) of the stress raisers. However, this approach has not always been very useful in design because the same theoretical stress concentration factors can be obtained with a great variety of different geometrical notch and specimen configurations, each of which has a unique influence on the numerical results of the tests; if design is the goal, the notched specimen must mirror the stress conditions in the component, including its stress raisers.

Therefore, the results of tensile tests of notched specimens have been used primarily to qualitatively merit-rate aluminum alloys with respect to their notch toughness; that is, their ability to plastically deform locally in the presence of stress raisers, and thus redistribute the stress. The notch tensile strength itself is of little value for this rating, but the relationship of the notch tensile strength to the tensile yield strength is much more meaningful.

For many years, the criterion most often used from notch tensile test results was the notch-strength ratio, the ratio of the notch tensile strength to the tensile strength of the material. However, this ratio tells little about the relative abilities of alloys to deform plastically in the presence of stress raisers. In fact, for different notch geometries it can indicate contradictory ratings (Ref 24). There are instances, of course, when the notch-strength ratio is useful; for example, when a measure of tensile efficiency of a specific structural member is required, or when the ultimate strength is the primary data taken for the smooth specimens, as in fatigue tests or stress-rupture tests.

A more meaningful indication of the inherent ability of a material to plastically deform locally in the presence of a severe stress raiser is provided by the notch-yield ratio, which is the ratio of the notch tensile strength to the tensile yield strength (Ref 24). The yield strength, although arbitrarily defined, is a measure of the lowest stress at which appreciable plastic deformation occurs in a tensile test. Therefore, the relationship of the notch tensile strength to the yield strength tells more about the behavior of the material in the presence of a stress raiser than the ratio of the notch tensile strength to the tensile strength. If the notch tensile strength is appreciably above the yield strength (regardless of its relation to the tensile strength), the material has exhibited an ability to deform locally in the presence of the stress raiser.

If the notch tensile strength is appreciably below the yield strength, the fracture must have taken place without very much plastic deformation. For a specific notch design this may or may not provide much specific design information, but it is quite useful as a relative measure of how several alloys behave in that situation. Further indication of this fact is the experimental result that the notch-yield ratio provides rather consistent ratings for many alloys and tempers for a wide variety of notch geometries (Ref 24), and the ratings are consistent with those from other fracture parameters, as described later.

While a number of different designs of notch have been used by different investigators, very sharp, 60 degree V-notches provide the greatest discrimination among the different alloys. In addition, such notches come

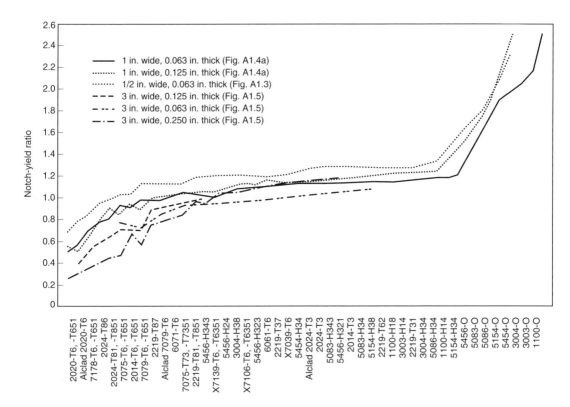

**Fig. 5.1** Similarity of ratings of alloys with respect to notch sensitivity from notch-yield ratio with different types of notched sheet-type specimens

close to representing the most severe unintentional stress raiser that is likely to exist in a structure: a crack. ASTM standards for notch-tensile testing (Ref 6, 7) call for notch-tip radii equal to or less than 0.0005 in., easily maintained in machining aluminum specimens, though quality assurance measurements are recommended.

The specific designs of notches for which data for a wide variety of alloys are available are shown in Fig. A1.3 through A1.7. Representative data for various aluminum alloys with each of the notches are shown in Tables 5.1 through 5.5, primarily from Ref 25 to 35. The types of notches and the dimensions of the various specimens, except those of the 0.5 in. wide, edge-notched sheet-type specimen (Table 5.1) first used many years ago, are consistent with the early (Ref 12) or more recent (Ref 13) recommendations of the ASTM Fracture Committee E-24 (now Committee E9). The data were obtained by the ASTM recommended practices applicable at the time.

Data are presented for wrought aluminum alloys, cast aluminum alloys, and welds in aluminum alloys, as follows (tables are at the end of this Chapter):

| | | |
|---|---|---|
| Wrought alloys | Table 5.1 | 0.5 in. wide, edge-notched specimens (Fig. A1.3) |
| | Table 5.2 | 1 in. wide, edge-notched specimens (Fig. A1.4a) |
| | Table 5.3 | 3 in. wide, edge-notched specimens (Fig. A1.5, ASTM E 338) |
| | Table 5.4 | 3 in. wide, center-notched specimens (Fig A1.6, ASTM E 338) |
| | Table 5.5 | 0.5 in. diameter, circumferentially notched specimens (Fig A1.7(a), ASTM E 602) |
| Cast alloys | Table 5.6 | 0.5 in. diameter circumferentially notched specimens (Fig. A1.7(a), ASTM E 602) |
| Welds in wrought alloys | Table 5.7 | 1 in. wide, edge-notched specimens (Fig. A1.4b) |
| | Table 5.8 | 0.5 in. diameter, circumferentially notched specimens (Fig. A1.7b) |
| Welds in cast alloys | Table 5.9 | 0.5 in. diameter, circumferentially notched specimens (Fig A1.7) |

The relative ratings of various alloys and tempers and also the similarity of the ratings based on a variety of different designs of sheet-type specimens are illustrated in Fig. 5.1, where the alloys and tempers are shown from left to right in order of increasing notch toughness as indicated by notch-yield ratio for several designs of notched specimen. The order in which alloys and tempers are shown was selected from the average ratings with the different designs of specimen. Although there are isolated discrepancies because of the differences in the numbers of lots tested, the overall ratings are quite consistent.

## 5.1 Wrought Alloys

It is clear from Fig. 5.1 that the annealed (-O temper) non-heat-treatable alloys that have the lowest yield strengths rate highest as a group. The very high-strength 2xxx and 7xxx series of alloys rate lowest. It is not sufficient, however, to conclude that notch toughness increases as yield strength decreases. Additional information may be gained by plotting the notch-yield ratios as a function of tensile yield strength, as in Fig. 5.2 and 5.3, where the notch-yield ratios associated with 0.250 in. thick, 3 in. wide, edge-notched specimens (Fig. A1.5) and 0.5 in. diam, cylindrically notched specimens (Fig. A1.7a), respectively, are plotted against yield strength.

From Fig. 5.2, the general trend for decreasing notch toughness with increasing strength is obvious, but it is also clear that the 7xxx (Al-Zn-Mg) series of alloys provides a better combination of notch toughness and yield strength than alloys of the other series represented. In Fig. 5.3 for cylindrically notched specimens, that same trend is apparent, as is a broader indication of the rather closely defined relationship between notch-yield ratio and tensile yield strength.

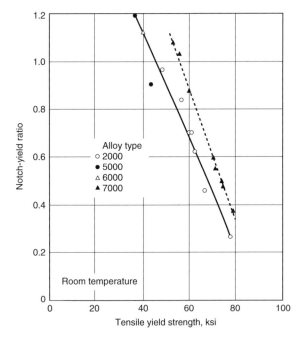

**Fig. 5.2** Notch-yield ratio vs. tensile yield strength of 0.250 in. plate. Transverse direction. Edge-notched specimen per Fig. A1.5

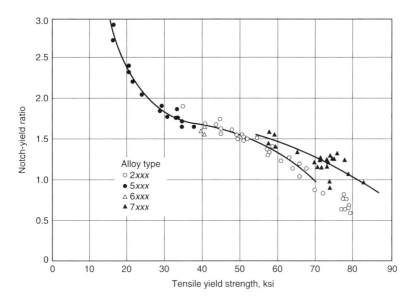

**Fig. 5.3** Notch-yield ratio vs. tensile yield strength for wrought aluminum alloys. Transverse direction (Table 5.5). Specimens per Fig. A1.7(a)

## 5.2 Cast Alloys

Relative rankings of the cast alloys are presented in Fig. 5.4. Alloy A444.0-F, the lowest-strength cast alloy, ranks highest, but B535.0-F also stands out for its high notch-yield ratio. The poorest performance is for sand cast alloys 240.0-F and 356.0-T6. Among the higher-strength casting alloys, the premium-quality castings (that is, sand castings made with special care to provide high metal chill rates in highly stressed regions) rate well, and A356.0-T6 consistently has higher toughness than does 356.0-T6, the positive effect of its higher purity (i.e., lower content of impurities such as iron and silicon).

Looking at the relationship between notch-yield ratio (NYR) and tensile yield strength (TYS) also provides interesting information for castings (Fig. 5.5), most notably the relationship of their performance to that of wrought alloys. Alloys A444.0-F and B535.0-F fall in the band for wrought alloy data, but the other alloys fall at least slightly below the band. The premium-quality castings show the best performance in this respect, and the sand cast alloys the poorest; permanent mold castings generally fall in the middle of the range.

## 5.3 Welds

Relative rankings for welds are shown in Fig. 5.6. In general, welds made with 5356 and 5556 filler alloys have higher notch-yield ratios and, therefore, higher toughness than welds made with 4043 filler alloy. This

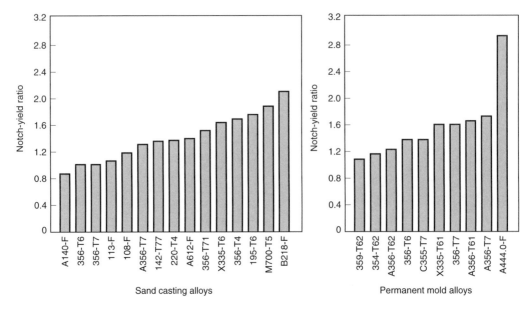

**Fig. 5.4** Notch-yield ratios (notch tensile strength/tensile yield strength) for cast slabs and separately cast tensile bars of aluminum sand and permanent mold cast slabs. Specimens per Fig. A1.7(a), $K_t \geq 16$

is not entirely consistent for reasons not clear from the data, but it is reasonable based upon the higher toughness of aluminum-magnesium (5xxx) alloys in general compared to the limited data for aluminum-silicon (4xxx) alloys.

Once again, looking at the data on the basis of NYR versus TYS (Fig. 5.7) reveals additional information. The notch toughness of welds as measured by NYR is generally somewhat less than for parent metal of the

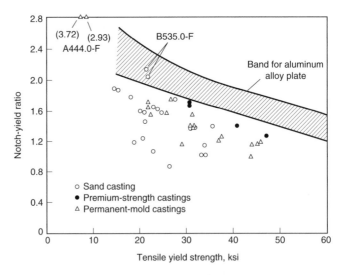

**Fig. 5.5** Notch-yield ratio vs. tensile yield strength for aluminum alloy castings from notched round specimens (Fig. A1.7a)

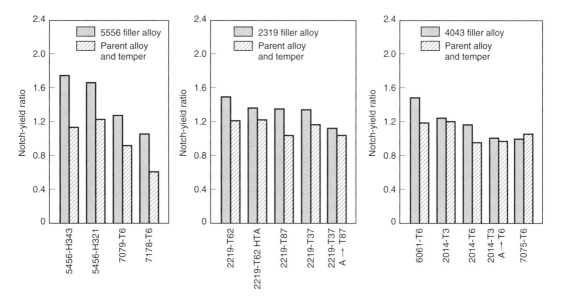

**Fig. 5.6** Ratings of aluminum alloy welds based on notch-yield ratio (notched tensile strength/yield strength) from sheet-type specimens (Fig. A1.4b). HTA, heat treated and artificially aged after welding; A, artificially aged after welding (to indicate temper)

same strength, the principal exceptions being welds made with the 5*xxx* series filler alloys. Many data for 4043 welds fall well below the band for wrought alloys, a notable exception being when the 4043 weld in 6061-T6 was heat treated and aged after welding.

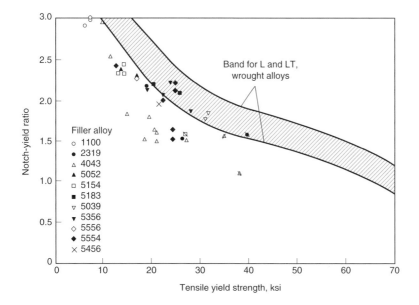

**Fig. 5.7** Notch-yield ratio (notch tensile strength/tensile yield strength) vs. tensile yield strength for welds in wrought and cast alloys (Tables 5.8 and 5.9). Specimens per Fig. A1.7(b)

## 5.4 ASTM Standard Notch Tensile Test Methods

Emphasizing a point made previously, while data have been generated in the past and are presented herein for a number of geometries of notched specimens, the recommended approach for the future is to use those designs covered by the current ASTM standards, namely ASTM E 338 (Ref 6) for materials up to about 0.500 in. in thickness using sheet-type specimens (Fig. A1.5 and A1.6), and ASTM E 602 (Ref 7) for thicker materials using cylindrical specimens (Fig. A1.7a).

**Table 5.1(a)   Results of tensile tests of smooth and 0.5 in. wide, edge-notched sheet-type specimens of aluminum alloy sheet, longitudinal**

| Alloy and temper | Ultimate tensile strength (UTS), ksi | Tensile yield strength (TYS), ksi | Elongation in 2 in., % | Notch tensile strength (NTS), ksi | NTS/TS | NTS/YS |
|---|---|---|---|---|---|---|
| 2014-T6 | 73.5 | 68.3 | 10.0 | 70.4 | 0.96 | 1.03 |
| 2020-T6(a) | 81.5 | 77.4 | 7.0 | 64.4 | 0.79 | 0.83 |
| | 80.2 | 75.4 | 8.0 | 51.3 | 0.64 | 0.68 |
| Alclad 2020-T6 | 74.0 | 70.8 | 7.5 | 57.2 | 0.77 | 0.81 |
| 2219-T31 | 56.5 | 44.9 | 17.2 | 55.2 | 0.98 | 1.23 |
| 2219-T37 | 62.9 | 54.0 | 9.0 | 62.5 | 1.00 | 1.16 |
| 2219-T62 | 59.1 | 39.2 | 9.0 | 49.3 | 0.84 | 1.26 |
| 2219-T81 | 69.4 | 54.2 | 9.2 | 63.8 | 0.92 | 1.18 |
| 2219-T87 | 72.1 | 59.8 | 9.2 | 68.0 | 0.94 | 1.14 |
| 2024-T3 | 67.8 | 51.4 | 18.2 | 61.8 | 0.92 | 1.25 |
| Alclad 2024-T3 | 68.3 | 52.7 | 18.2 | 63.4 | 0.93 | 1.19 |
| 2024-T86 | 76.9 | 72.9 | 6.0 | 71.4 | 0.93 | 0.98 |
| 5083-O | 44.6 | 21.9 | 22.0 | 40.0 | 0.90 | 1.83 |
| 5083-H34 | 53.2 | 43.2 | 10.5 | 53.1 | 1.00 | 1.23 |
| 5086-O | 41.2 | 19.6 | 22.2 | 38.0 | 0.93 | 1.95 |
| 5086-H34 | 48.7 | 38.8 | 11.0 | 50.2 | 1.03 | 1.30 |
| 5154-H38 | 49.6 | 42.7 | 9.8 | 52.2 | 1.06 | 1.22 |
| 5454-O | 37.6 | 15.8 | 21.5 | 36.8 | 0.98 | 2.33 |
| 5454-H34 | 46.0 | 39.9 | 10.8 | 47.6 | 1.03 | 1.19 |
| 5456-O | 47.4 | 24.3 | 21.8 | 41.4 | 0.87 | 1.70 |
| 5456-H24 | 54.9 | 43.0 | 12.0 | 48.8 | 0.89 | 1.14 |
| 6061-T6 | 44.6 | 41.4 | 11.0 | 48.8 | 1.10 | 1.18 |
| 7075-T6 | 83.0 | 75.7 | 10.5 | 79.7 | 0.96 | 1.06 |
| 7178-T6 | 88.6 | 81.6 | 11.5 | 72.6 | 0.82 | 0.89 |

Specimens per Fig. A1.3; each line is the average of duplicate or triplicate tests of an individual lot of material. For yield strengths, offset is 0.2%. (a) Obsolete alloy

**Table 5.1(b)   Results of tensile tests of smooth and 0.5 in. wide, edge-notched sheet-type specimens of aluminum alloy sheet, transverse**

| Alloy and temper | Ultimate tensile strength (UTS), ksi | Tensile yield strength (TYS), ksi | Elongation in 2 in., % | Notch tensile strength (NTS), ksi | NTS/TS | NTS/YS |
|---|---|---|---|---|---|---|
| **Nominal sheet thickness, 0.063 in.** | | | | | | |
| 2014-T6 | 72.1 | 56.4 | 9.50 | 67.3 | 0.94 | 1.03 |
| 2020-T6(a) | 81.8 | 75.9 | 7.00 | 61.0 | 0.75 | 0.81 |
| | 80.3 | 75.0 | 6.50 | 51.1 | 0.64 | 0.68 |
| Alclad 2020-T6 | 74.8 | 69.8 | 6.50 | 53.8 | 0.72 | 0.77 |
| 2219-T31 | 56.6 | 40.9 | 17.00 | 54.9 | 0.97 | 1.34 |
| 2219-T37 | 63.1 | 50.9 | 11.20 | 62.9 | 1.00 | 1.24 |
| 2219-T62 | 58.7 | 38.9 | 9.50 | 49.9 | 0.85 | 1.28 |
| 2219-T81 | 68.7 | 52.8 | 9.50 | 62.6 | 0.91 | 1.19 |
| 2219-T87 | 72.9 | 60.6 | 9.20 | 66.2 | 0.91 | 1.09 |
| 2024-T3 | 66.6 | 50.0 | 20.00 | 57.8 | 0.89 | 1.30 |
| Alclad 2024-T3 | 66.2 | 46.6 | 19.80 | 61.3 | 0.92 | 1.32 |
| 2024-T86 | 75.8 | 71.3 | 5.20 | 64.4 | 0.86 | 0.91 |
| 5083-O | 43.8 | 22.0 | 23.20 | 37.6 | 0.86 | 1.71 |
| 5083-H34 | 51.9 | 38.2 | 11.80 | 50.6 | 0.98 | 1.33 |
| 5086-O | 40.4 | 19.8 | 24.00 | 36.7 | 0.91 | 1.86 |
| 5086-H34 | 49.0 | 37.2 | 12.80 | 51.2 | 1.00 | 1.37 |
| 5154-H38 | 49.9 | 42.5 | 14.20 | 56.3 | 1.13 | 1.32 |
| 5454-O | 36.0 | 15.7 | 20.50 | 35.3 | 0.98 | 2.25 |
| 5454-H34 | 47.8 | 38.3 | 10.20 | 49.8 | 1.04 | 1.30 |
| 5456-O | 46.9 | 25.7 | 24.20 | 41.6 | 0.88 | 1.62 |
| 5456-H24 | 55.2 | 38.7 | 14.50 | 49.7 | 0.90 | 1.29 |
| 6061-T6 | 45.2 | 40.7 | 11.00 | 48.2 | 1.06 | 1.18 |
| 7075-T6 | 82.2 | 72.9 | 10.50 | 74.8 | 0.91 | 1.02 |
| 7178-T6 | 80.5 | 78.5 | 11.20 | 61.6 | 0.70 | 0.78 |

Specimens per Fig. A1.3; each line is the average of duplicate or triplicate tests of an individual lot of material. For yield strengths, offset is 0.2%. (a) Obsolete alloy

**Table 5.2(a)  Results of tensile tests of smooth and notched 1 in. wide, edge-notched sheet-type tensile specimens of aluminum alloy sheet, longitudinal**

| Alloy and temper | Ultimate tensile strength (UTS), ksi | Tensile yield strength (TYS), ksi | Elongation in 2 in., % | Notch tensile strength (NTS), ksi | NTS/TS | NTS/YS |
|---|---|---|---|---|---|---|
| **Nominal sheet thickness, 0.063 in.** | | | | | | |
| 1100-O | 14.2 | 4.9 | 35.1 | 13.0 | 0.91 | 2.65 |
| 1100-HI4 | 17.9 | 16.8 | 13.0 | 19.2 | 1.08 | 1.14 |
| 1100-HI8 | 27.7 | 26.3 | 5.5 | 28.9 | 1.04 | 1.10 |
| 2014-T6 | 73.5 | 68.3 | 10.0 | 63.4 | 0.86 | 0.93 |
| | 72.5 | 65.6 | 11.5 | 67.9 | 0.83 | 0.92 |
| 2020-T6(a) | 80.2 | 75.4 | 8.0 | 37.4 | 0.47 | 0.50 |
| | 80.2 | 75.9 | 7.8 | 42.2 | 0.53 | 0.56 |
| Alclad 2020-T6(a) | 74.2 | 70.8 | 7.5 | 45.9 | 0.62 | 0.65 |
| 2024-T3 | 71.5 | 55.5 | 19.2 | 60.2 | 0.84 | 1.09 |
| | 67.8 | 51.4 | 18.2 | 57.1 | 0.84 | 1.11 |
| Alclad 2024-T3 | 68.3 | 52.7 | 18.2 | 57.6 | 0.84 | 1.09 |
| 2024-T81 | 72.6 | 68.0 | 6.2 | 55.5 | 0.76 | 0.82 |
| 2024-T86 | 76.9 | 72.9 | 6.0 | 59.0 | 0.77 | 0.81 |
| 3003-O | 17.0 | 7.1 | 34.5 | 15.8 | 0.93 | 2.23 |
| 3003-H14 | 22.2 | 21.4 | 10.0 | 24.2 | 1.09 | 1.13 |
| 3004-O | 27.1 | 11.4 | 22.2 | 23.8 | 0.88 | 2.09 |
| 3004-H34 | 35.6 | 31.7 | 8.8 | 35.6 | 1.00 | 1.12 |
| 3004-H38 | 44.0 | 41.0 | 8.0 | 43.8 | 1.00 | 1.07 |
| 5083-H34 | 54.1 | 44.8 | 10.7 | 49.2 | 0.91 | 1.10 |
| 5154-O | 35.4 | 16.4 | 24.5 | 31.4 | 0.89 | 1.92 |
| 5154-H34 | 45.8 | 37.8 | 9.8 | 44.0 | 0.96 | 1.16 |
| 5154-H39 | 49.4 | 42.8 | 9.8 | 46.8 | 0.95 | 1.09 |
| 5454-O | 37.6 | 15.8 | 21.5 | 31.7 | 0.84 | 2.00 |
| 5454-H34 | 46.0 | 39.9 | 10.8 | 43.6 | 0.95 | 1.09 |
| 5456-O | 47.4 | 24.3 | 21.8 | 34.3 | 0.72 | 1.41 |
| 5456-H24 | 54.9 | 43.0 | 12.0 | 42.2 | 0.77 | 0.98 |
| 6061-T6 | 46.6 | 43.4 | 11.5 | 47.0 | 1.01 | 1.08 |
| 6071-T6(a) | 54.6 | 52.1 | 9.5 | 54.4 | 1.00 | 1.04 |
| 7075-T6 | 82.3 | 76.1 | 11.5 | 68.1 | 0.83 | 0.90 |
| | 80.4 | 71.9 | 10.0 | 71.0 | 0.88 | 0.99 |
| 7075-T73 | 72.0 | 61.0 | 10.2 | 67.2 | 0.93 | 1.10 |
| 7079-T6(a) | 77.0 | 70.2 | 10.8 | 71.4 | 0.93 | 1.02 |
| | 72.8 | 64.0 | 11.0 | 67.3 | 0.92 | 1.05 |
| | 78.0 | 71.1 | 10.8 | 66.3 | 0.77 | 0.86 |
| Alclad 7079-T6(a) | 68.9 | 61.2 | 11.4 | 59.0 | 0.87 | 0.97 |
| X7106-T6(a) | 61.2 | 53.9 | 11.0 | 59.7 | 0.98 | 1.11 |
| X7139-T6(a) | 65.2 | 56.1 | 11.0 | 60.6 | 0.92 | 1.08 |
| 7178-T6 | 89.4 | 82.4 | 11.5 | 57.8 | 0.65 | 0.70 |
| **Nominal sheet thickness, 0.125 in.** | | | | | | |
| 2014-T3 | 66.0 | 46.7 | 20.4 | 53.2 | 0.81 | 1.14 |
| 2014-T6 | 70.2 | 65.8 | 10.3 | 65.3 | 0.93 | 0.99 |
| 2020-T6(a) | 80.5 | 76.8 | 7.8 | 48.4 | 0.60 | 0.63 |
| Alclad 2020-T6(a) | 75.2 | 70.2 | 8.0 | 38.8 | 0.52 | 0.55 |
| | 71.0 | 67.0 | 8.0 | 35.4 | 0.50 | 0.53 |
| | 72.6 | 68.6 | 8.5 | 37.2 | 0.51 | 0.54 |
| 2024-T81 | 70.9 | 64.8 | 8.5 | 62.2 | 0.88 | 0.96 |
| 2219-T37 | 57.1 | 48.3 | 15.9 | 53.6 | 0.94 | 1.11 |
| 2219-T62 | 58.3 | 39.0 | 11.0 | 47.9 | 0.82 | 1.23 |
| 2219-T87 | 70.9 | 59.2 | 11.0 | 62.7 | 0.88 | 1.06 |
| | 70.1 | 58.4 | 11.5 | 55.6 | 0.79 | 0.95 |
| | 68.2 | 57.3 | 11.6 | 61.0 | 0.90 | 1.06 |
| 5083-O | 44.6 | 20.2 | 22.5 | 35.1 | 0.79 | 1.74 |
| 5083-H343 | 52.6 | 42.3 | 10.5 | 48.0 | 0.91 | 1.13 |
| 5086-O | 40.8 | 19.4 | 23.5 | 35.3 | 0.87 | 1.82 |
| 5086-H34 | 49.6 | 38.0 | 12.8 | 47.0 | 0.95 | 1.24 |
| 5454-O | 33.8 | 12.6 | 22.5 | 32.0 | 0.95 | 2.54 |
| 5456-O | 49.8 | 23.8 | 20.5 | 38.5 | 0.77 | 1.62 |
| 5456-H24 | 54.7 | 37.8 | 12.8 | 41.8 | 0.76 | 1.11 |
| | 55.2 | 40.8 | 12.2 | 44.2 | 0.80 | 1.08 |

(continued)

Specimens per Fig. A1.4. Each line is the average of duplicate or triplicate tests of an individual lot of material. For yield strengths, offset is 0.2%. (a) Obsolete alloy

**Table 5.2(a)    (continued)**

| Alloy and temper | Ultimate tensile strength (UTS), ksi | Tensile yield strength (TYS), ksi | Elongation in 2 in., % | Notch tensile strength (NTS), ksi | NTS/TS | NTS/YS |
|---|---|---|---|---|---|---|
| Nominal sheet thickness, 0.125 in. (continued) | | | | | | |
| 5456-H321 | 55.3 | 40.4 | 13.0 | 45.3 | 0.82 | 1.12 |
|  | 57.4 | 39.5 | 14.5 | 47.6 | 0.83 | 1.11 |
| 5456-H323 | 56.0 | 42.1 | 11.2 | 46.9 | 0.83 | 1.11 |
| 5456-H343 | 58.2 | 46.1 | 8.3 | 48.7 | 0.84 | 1.06 |
|  | 59.3 | 46.2 | 8.5 | 47.8 | 0.81 | 1.04 |
| 6061-T6 | 44.9 | 40.9 | 13.8 | 46.2 | 1.03 | 1.12 |
| 7075-T6 | 82.1 | 74.4 | 11.2 | 68.4 | 0.83 | 0.92 |
|  | 82.8 | 76.6 | 11.2 | 60.8 | 0.73 | 0.79 |
| 7075-T73 | 73.0 | 61.8 | 12.8 | 65.4 | 0.89 | 1.03 |
| 7079-T6(a) | 80.9 | 75.6 | 11.5 | 67.8 | 0.84 | 0.90 |
|  | 77.5 | 72.3 | 11.5 | 70.8 | 0.91 | 0.98 |
| X7106-T6(a) | 58.8 | 52.5 | 12.5 | 59.6 | 1.01 | 1.13 |
| X7139-T6(a) | 64.6 | 56.2 | 11.0 | 62.5 | 0.97 | 1.11 |
| 7178-T6 | 84.2 | 75.3 | 11.0 | 56.3 | 0.67 | 0.75 |
|  | 90.0 | 83.6 | 12.2 | 51.8 | 0.58 | 0.62 |

Specimens per Fig. A1.4. Each line is the average of duplicate or triplicate tests of an individual lot of material. For yield strengths, offset is 0.2%. (a) Obsolete alloy

**Table 5.2(b)    Results of tensile tests of smooth and notched 1 in. wide, edge-notched sheet-type tensile specimens of aluminum alloy sheet, transverse**

| Alloy and temper | Ultimate tensile strength (UTS), ksi | Tensile yield strength (TYS), ksi | Elongation in 2 in., % | Notch tensile strength (NTS), ksi | NTS/TS | NTS/YS |
|---|---|---|---|---|---|---|
| Nominal sheet thickness, 0.063 in. | | | | | | |
| 1100-O | 14.0 | 5.3 | 41.3 | 12.8 | 0.91 | 2.41 |
| 1100-HI4 | 18.0 | 16.4 | 9.2 | 19.8 | 110 | 1.21 |
| 1100-HI8 | 27.7 | 26.3 | 5.50 | 28.9 | 1.04 | 1.10 |
| 2014-T6 | 72.1 | 65.4 | 9.5 | 57.2 | 0.79 | 0.88 |
|  | 72.3 | 62.6 | 11.8 | 56.4 | 0.78 | 0.90 |
| 2020-T6 | 80.3 | 75.0 | 6.5 | 34.7 | 0.43 | 0.46 |
|  | 81.1 | 75.8 | 7.5 | 36.5 | 0.45 | 0.48 |
| Alclad 2020-T6 | 74.8 | 69.8 | 6.5 | 35.6 | 0.48 | 0.51 |
| 2024-T3 | 68.9 | 47.8 | 19.5 | 54.0 | 0.78 | 1.13 |
|  | 66.6 | 44.8 | 20.0 | 51.2 | 0.77 | 1.14 |
| Alclad 2024-T3 | 66.2 | 46.6 | 19.8 | 53.2 | 0.81 | 1.14 |
| 2024-T81 | 71.6 | 66.7 | 6.2 | 53.7 | 0.75 | 0.80 |
| 2024-T86 | 75.8 | 71.3 | 5.2 | 51.7 | 0.68 | 0.72 |
| 3003-O | 16.5 | 7.3 | 30.8 | 15.6 | 0.94 | 2.14 |
| 3003-H14 | 22.2 | 20.8 | 6.9 | 24.6 | 1.11 | 1.18 |
| 3004-O | 26.6 | 11.5 | 22.8 | 23.6 | 0.89 | 2.05 |
| 3004-H34 | 36.6 | 30.6 | 8.0 | 36.8 | 1.01 | 1.20 |
| 3004-H38 | 44.4 | 9.6 | 7.8 | 43.4 | 0.98 | 1.10 |
| 5083-H34 | 55.2 | 42.9 | 11.8 | 49.8 | 0.90 | 1.16 |
| 5154-O | 34.6 | 16.0 | 25.0 | 30.2 | 0.87 | 1.89 |
| 5154-H34 | 46.2 | 35.4 | 13.0 | 44.8 | 0.97 | 1.27 |
| 5154-H39 | 49.9 | 42.6 | 14.2 | 50.1 | 1.00 | 1.18 |

(continued)

Specimens per Fig. A1.4. Each line is the average of duplicate or triplicate tests of an individual lot of material. For yield strengths, offset is 0.2%. (a) Obsolete alloy

**Table 5.2(b)** **(continued)**

| Alloy and temper | Ultimate tensile strength (UTS), ksi | Tensile yield strength (TYS), ksi | Elongation in 2 in., % | Notch tensile strength (NTS), ksi | NTS/TS | NTS/YS |
|---|---|---|---|---|---|---|
| **Nominal sheet thickness, 0.063 in. (continued)** | | | | | | |
| 5454-O | 36.0 | 15.7 | 20.5 | 31.3 | 0.87 | 1.99 |
| 5454-H34 | 47.8 | 38.3 | 10.2 | 45.6 | 0.95 | 1.19 |
| 5456-O | 46.9 | 25.7 | 24.2 | 34.6 | 0.74 | 1.35 |
| 5456-H24 | 55.2 | 38.7 | 14.5 | 41.7 | 0.75 | 1.08 |
| 6061-T6 | 46.1 | 41.8 | 11.5 | 46.7 | 1.01 | 1.12 |
| 6071-T6 | 53.4 | 50.2 | 9.0 | 50.0 | 0.94 | 1.00 |
| 7075-T6 | 82.6 | 73.5 | 11.0 | 63.3 | 0.77 | 0.86 |
| | 79.7 | 70.6 | 10.0 | 65.4 | 0.82 | 0.93 |
| 7075-T73 | 74.0 | 62.9 | 10.5 | 61.4 | 0.83 | 0.98 |
| 7079-T6 | 76.1 | 67.3 | 10.5 | 65.3 | 0.86 | 0.97 |
| | 72.8 | 62.4 | 11.0 | 67.5 | 0.93 | 1.08 |
| | 77.1 | 68.7 | 10.2 | 59.7 | 0.77 | 0.87 |
| Alclad 7079-T6 | 67.7 | 58.7 | 11.0 | 58.4 | 0.86 | 0.99 |
| X7106-T6 | 62.4 | 54.0 | 11.0 | 60.0 | 0.96 | 1.12 |
| X7139-T6(a) | 65.7 | 56.1 | 10.0 | 60.0 | 0.92 | 1.07 |
| 7178-T6 | 88.0 | 77.6 | 11.0 | 54.6 | 0.62 | 0.70 |
| **Nominal sheet thickness, 0.125 in.** | | | | | | |
| 2014-T3 | 66.0 | 40.8 | 20.5 | 50.3 | 0.76 | 1.23 |
| 2014-T6 | 73.1 | 66.2 | 11.8 | 58.5 | 0.80 | 0.88 |
| 2020-T6(a) | 82.8 | 77.2 | 6.0 | 36.6 | 0.44 | 0.47 |
| Alclad 2020-T6(a) | 75.8 | 69.1 | 7.0 | 31.6 | 0.42 | 0.46 |
| | 71.7 | 66.4 | 8.0 | 33.2 | 0.46 | 0.50 |
| | 74.4 | 68.4 | 6.5 | 33.2 | 0.45 | 0.49 |
| 2024-T81 | 70.8 | 64.6 | 7.5 | 56.6 | 0.80 | 0.88 |
| 2219-T37 | 60.3 | 46.7 | 13.3 | 54.9 | 0.91 | 1.11 |
| 2219-T62 | 57.4 | 38.2 | 12.0 | 46.1 | 0.80 | 1.21 |
| 2219-T87 | 71.3 | 59.2 | 10.0 | 58.0 | 0.81 | 0.98 |
| | 72.6 | 59.8 | 10.5 | 57.5 | 0.79 | 0.96 |
| | 70.1 | 58.4 | 10.8 | 58.1 | 0.82 | 0.98 |
| 5083-O | 43.9 | 20.2 | 23.5 | 33.6 | 0.77 | 1.66 |
| 5083-H343 | 52.6 | 38.0 | 11.5 | 45.8 | 0.87 | 1.18 |
| 5086-O | 40.0 | 19.4 | 24.0 | 33.2 | 0.83 | 1.71 |
| 5086-H34 | 48.8 | 38.9 | 17.8 | ... | ... | ... |
| 5454-O | 34.0 | 12.9 | 21.5 | 31.6 | 0.93 | 2.45 |
| 5456-O | 49.4 | 24.0 | 20.0 | 36.0 | 0.73 | 1.50 |
| 5456-H24 | 54.4 | 36.8 | 15.0 | 39.4 | 0.72 | 1.07 |
| | 54.9 | 39.4 | 15.8 | 45.0 | 0.82 | 1.14 |
| 5456-H321 | 52.8 | 31.8 | 16.5 | 40.9 | 0.77 | 1.29 |
| | 57.9 | 39.6 | 17.5 | 48.2 | 0.83 | 1.22 |
| 5456-H323 | 56.6 | 40.1 | 13.5 | 46.7 | 0.83 | 1.12 |
| 5456-H343 | 59.7 | 43.0 | 12.6 | 47.0 | 0.79 | 1.09 |
| | 58.8 | 43.2 | 11.8 | 47.0 | 0.80 | 1.09 |
| 6061-T6 | 44.3 | 38.0 | 14.0 | 45.6 | 1.03 | 1.20 |
| 7075-T6 | 83.7 | 71.8 | 13.0 | 60.5 | 0.72 | 0.84 |
| | 82.8 | 74.1 | 11.5 | 57.0 | 0.69 | 0.78 |
| 7075-T73 | 74.6 | 61.1 | 10.8 | 62.4 | 0.84 | 1.02 |
| 7079-T6(a) | 83.0 | 73.8 | 10.8 | 58.8 | 0.71 | 0.80 |
| | 79.3 | 71.4 | 11.3 | 61.5 | 0.78 | 0.86 |
| X7106-T6(a) | 61.0 | 52.7 | 11.0 | 59.1 | 0.98 | 1.13 |
| X7139-T6(a) | 66.5 | 57.1 | 10.0 | 61.4 | 0.93 | 1.08 |
| 7178-T6 | 85.8 | 75.3 | 10.0 | 49.8 | 0.58 | 0.66 |
| | 91.4 | 79.2 | 12.5 | 46.9 | 0.51 | 0.59 |

Specimens per Fig. A1.4. Each line is the average of duplicate or triplicate tests of an individual lot of material. For yield strengths, offset is 0.2%. (a) Obsolete alloy

**Table 5.3(a)   Results of tensile tests of 3 in. wide, edge-notched sheet-type specimens of aluminum alloy sheet, longitudinal**

| Alloy and temper | Ultimate tensile strength (UTS), ksi | Tensile yield strength (TYS), ksi | Elongation in 2 in., % | Notch tensile strength (NTS), ksi | NTS/TS | NTS/YS |
|---|---|---|---|---|---|---|
| **Nominal sheet thickness, 0.063 in.** | | | | | | |
| 5154-H38 | 49.4 | 42.8 | 9.8 | 44.2 | 0.89 | 1.03 |
| 6071-T69(a) | 56.8 | 54.2 | 9.5 | 45.6 | 0.80 | 0.84 |
| 7075-T6 | 81.5 | 73.0 | 10.2 | 59.2 | 0.73 | 0.81 |
| | 80.4 | 71.9 | 10.0 | 54.0 | 0.67 | 0.75 |
| 7075-T73 | 72.8 | 61.3 | 9.9 | 57.0 | 0.78 | 0.93 |
| 7079-T6(a) | 77.0 | 70.2 | 10.8 | 50.6 | 0.66 | 0.72 |
| Alclad 7079-T6(a) | 68.9 | 61.2 | 11.4 | 52.6 | 0.76 | 0.86 |
| | 78.0 | 71.1 | 10.8 | 51.5 | 0.66 | 0.72 |
| **Nominal sheet thickness, 0.125 in.** | | | | | | |
| 2014-T6 | 70.2 | 64.0 | 10.5 | 52.7 | 0.75 | 0.82 |
| Alclad 2020-T6(a) | 71.0 | 67.0 | 8.0 | 31.2 | 0.44 | 0.47 |
| | 72.6 | 68.6 | 8.5 | 32.1 | 0.44 | 0.47 |
| 2219-T87 | 68.2 | 57.3 | 11.6 | 53.1 | 0.78 | 0.93 |
| | 66.2 | 54.7 | 10.2 | 50.6 | 0.76 | 0.92 |
| 5456-H343 | 59.3 | 46.2 | 8.5 | 44.8 | 0.76 | 0.97 |
| 7075-T6 | 82.8 | 76.6 | 11.2 | 56.8 | 0.69 | 0.74 |
| | 84.7 | 78.2 | 11.0 | 53.0 | 0.63 | 0.68 |
| | 80.7 | 73.2 | 11.5 | ... | ... | ... |
| 7079-T6 | 80.9 | 75.6 | 11.5 | 52.1 | 0.64 | 0.69 |
| 7178-T6 | 90.0 | 83.6 | 12.2 | 45.0 | 0.50 | 0.54 |
| **Nominal sheet thickness, 0.250 in.** | | | | | | |
| 2014-T651 | 70.3 | 65.0 | 11.0 | 46.4 | 0.66 | 0.71 |
| 2020-T651(a) | 81.6 | 77.4 | 8.5 | 22.6 | 0.28 | 0.29 |
| 2219-T87 | 69.3 | 57.6 | 10.5 | 50.6 | 0.73 | 0.88 |
| 7075-T651 | 84.3 | 78.0 | 14.0 | 44.7 | 0.53 | 0.57 |
| 7079-T651(a) | 79.0 | 73.8 | 14.0 | 55.2 | 0.70 | 0.75 |
| | 79.4 | 74.0 | 12.8 | 53.2 | 0.67 | 0.72 |

Specimens per Fig. A1.5. Each line is the average of duplicate or triplicate tests of one lot of material. Yield strength offset is 0.2.
(a) Obsolete alloy

**Table 5.3(b)  Results of tensile tests of 3 in. wide, edge-notch sheet-type specimens of aluminum alloy sheet, transverse**

| Alloy and temper | Ultimate tensile strength (UTS), ksi | Tensile yield strength (TYS), ksi | Elongation in 2 in., % | Notch tensile strength (NTS), ksi | NTS/TS | NTS/YS |
|---|---|---|---|---|---|---|
| **Nominal sheet thickness, 0.063 in.** | | | | | | |
| 5154-H38 | 49.9 | 42.6 | 14.2 | 48.1 | 0.96 | 1.13 |
| 6071-T6(a) | 56.2 | 52.2 | 10.0 | 41.6 | 0.74 | 0.80 |
| 7075-T6 | 81.0 | 71.8 | 10.2 | 55.5 | 0.69 | 0.80 |
|  | 79.7 | 70.6 | 10.0 | 54.0 | 0.68 | 0.76 |
| 7075-T73 | 72.2 | 61.8 | 10.1 | 54.4 | 0.75 | 0.88 |
| 7079-T6(a) | 76.1 | 67.3 | 10.5 | 52.7 | 0.69 | 0.78 |
| Alclad 7079-T6(a) | 67.7 | 67.7 | 11.0 | 48.8 | 0.72 | 0.83 |
|  | 77.1 | 68.7 | 10.2 | 47.0 | 0.61 | 0.68 |
| **Nominal sheet thickness, 0.125 in.** | | | | | | |
| 2014-T6 | 70.0 | 62.2 | 10.5 | 47.0 | 0.67 | 0.76 |
| Alclad 2020-T6(a) | 71.7 | 66.4 | 8.0 | 22.4 | 0.31 | 0.34 |
|  | 74.4 | 68.4 | 6.5 | 22.3 | 0.30 | 0.34 |
| 2219-T87 | 70.1 | 58.4 | 10.8 | 50.0 | 0.71 | 0.86 |
|  | 68.2 | 55.9 | 10.5 | 46.2 | 0.68 | 0.83 |
| 5456-H343 | 58.8 | 43.2 | 11.8 | 42.0 | 0.71 | 0.97 |
| 7075-T6 | 82.8 | 74.1 | 11.5 | 48.8 | 0.59 | 0.66 |
|  | 86.9 | 77.0 | 10.5 | 39.9 | 0.46 | 0.52 |
|  | 82.7 | 72.9 | 11.0 | 49.7 | 0.60 | 0.68 |
| 7079-T6 | 83.0 | 73.8 | 10.8 | 48.4 | 0.58 | 0.66 |
| 7178-T6 | 91.4 | 79.2 | 12.5 | 33.0 | 0.36 | 0.42 |
| **Nominal sheet thickness, 0.250 in.** | | | | | | |
| 2014-T651 | 68.5 | 61.5 | 11.0 | 42.2 | 0.62 | 0.69 |
|  | 69.6 | 62.8 | 10.5 | 38.2 | 0.55 | 0.61 |
| 2020-T651(a) | 83.1 | 78.0 | 6.0 | 20.0 | 0.24 | 0.26 |
| 2024-T851 | 72.6 | 67.2 | 7.0 | 30.0 | 0.41 | 0.45 |
| 2219-T851 | 66.2 | 49.2 | 11.0 | 47.2 | 0.71 | 0.96 |
| 2219-T87 | 73.4 | 60.8 | 10.5 | 42.2 | 0.58 | 0.69 |
|  | 70.2 | 57.2 | 10.5 | 47.5 | 0.68 | 0.83 |
| 5456-H321 | 53.7 | 37.2 | 19.5 | 44.0 | 0.82 | 1.18 |
| 5456-H343 | 61.4 | 43.8 | 11.2 | 39.2 | 0.64 | 0.90 |
| 6061-T651 | 44.8 | 40.7 | 14.8 | 45.2 | 1.01 | 1.11 |
| 7075-T651 | 86.1 | 74.8 | 12.5 | 34.6 | 0.40 | 0.46 |
|  | 84.8 | 74.2 | 13.0 | 34.9 | 0.41 | 0.47 |
| 7075-T7351 | 71.8 | 59.4 | 12.0 | 50.2 | 0.70 | 0.85 |
| 7079-T651(a) | 79.6 | 71.0 | 12.5 | 41.2 | 0.52 | 0.58 |
|  | 80.2 | 71.2 | 12.0 | 38.1 | 0.48 | 0.54 |
|  | 81.0 | 72.6 | 11.5 | 29.8(b) | 0.37(b) | 0.41(b) |
| X7106-T6351 | 62.0 | 54.0 | 12.2 | 57.8 | 0.93 | 1.07 |
| X7139-T6351 | 65.8 | 56.8 | 12.5 | 58.0 | 0.88 | 1.02 |
| 7178-T651 | 87.6 | 79.2 | 11.0 | 28.8 | 0.33 | 0.36 |

Specimens per Fig. A1.5. Each line is the average of duplicate or triplicate tests of one lot of material. Yield strength offset is 0.2%. (a) Obsolete alloy. (b) Value is unreasonably low and could not be checked; omitted from all comparisons

**Table 5.4(a)   Results of tensile tests of smooth and center-notched sheet-type specimens of aluminum alloy sheet and plate, longitudinal**

| Alloy and temper | Ultimate tensile strength (UTS), ksi | Tensile yield strength (TYS), ksi | Elongation in 2 in., % | Notch tensile strength (NTS), ksi | NTS/TS | NTS/YS |
|---|---|---|---|---|---|---|
| **Nominal sheet thickness, 0.125 in.** | | | | | | |
| 2014-T6 | 70.2 | 64.0 | 10.5 | 54.0 | 0.77 | 0.84 |
| 2024-T81 | 71.3 | 65.2 | 9.0 | 51.9 | 0.73 | 0.80 |
| | 71.0 | 64.9 | 9.0 | 51.4 | 0.72 | 0.79 |
| | 69.6 | 62.2 | 8.1 | 51.8 | 0.74 | 0.83 |
| 2219-T87 | 66.2 | 54.7 | 10.2 | 51.0 | 0.77 | 0.43 |
| 7075-T6 | 80.7 | 73.2 | 11.5 | 57.9 | 0.72 | 0.74 |
| | 94.7 | 78.2 | 11.0 | 54.9 | 0.65 | 0.70 |
| 7178-T6 | 89.6 | 83.5 | 12.6 | 43.0 | 0.48 | 0.51 |
| **Nominal sheet thickness, 0.125 in.** | | | | | | |
| 2014-T651 | 70.3 | 65.0 | 11.0 | 49.6 | 0.77 | 0.77 |
| 2020-T651(a) | 81.6 | 77.4 | 8.5 | 24.4 | 0.32 | 0.32 |
| 2219-T87 | 69.3 | 57.6 | 10.5 | 51.4 | 0.89 | 0.89 |
| 7075-T651 | 83.5 | 77.3 | 14.5 | 41.2 | 0.49 | 0.53 |
| 7075-T7351 | 70.2 | 54.2 | 13.5 | ... | ... | ... |
| 7079-T651(a) | 80.2 | 74.7 | 11.0 | 44.2 | 0.59 | 0.54 |

Specimens per Fig. A1.6. Each line is the average of duplicate or triplicate tests of an individual lot of material. For yield strengths, offset is 0.2%. (a) Obsolete alloy

**Table 5.4(b)   Results of tensile tests of smooth and center-notched sheet-type specimens of aluminum alloy sheet and plate, transverse**

| Alloy and temper | Ultimate tensile strength (UTS), ksi | Tensile yield strength (TYS), ksi | Elongation in 2 in., % | Notch tensile strength (NTS), ksi | NTS/TS | NTS/YS |
|---|---|---|---|---|---|---|
| **Nominal sheet thickness, 0.063 in.** | | | | | | |
| 7075-T6 | 81.0 | 71.8 | 10.0 | 55.0 | 0.68 | 0.77 |
| 7075-T73 | 72.2 | 61.8 | 10.1 | 55.2 | 0.76 | 0.89 |
| **Nominal sheet thickness, 0.125 in.** | | | | | | |
| 2014-T6 | 70.0 | 62.2 | 10.5 | 47.7 | 0.68 | 0.77 |
| 2024-T81 | 72.5 | 66.4 | 8.0 | 47.4 | 0.65 | 0.71 |
| | 71.7 | 66.0 | 8.2 | 46.6 | 0.65 | 0.71 |
| | 70.5 | 64.1 | 7.4 | 46.2 | 0.66 | 0.72 |
| 2219-T87 | 68.2 | 55.9 | 10.5 | 47.6 | 0.70 | 0.85 |
| 7075-T6 | 82.7 | 72.9 | 11.0 | 50.7 | 0.61 | 0.70 |
| | 86.9 | 77.0 | 10.5 | 44.0 | 0.51 | 0.57 |
| 7178-T6 | 89.8 | 77.4 | 12.8 | 36.3 | 0.40 | 0.47 |
| **Nominal sheet thickness, 0.250 in.** | | | | | | |
| 2014-T651 | 69.6 | 62.8 | 10.5 | 39.3 | 0.56 | 0.63 |
| 2020-T651(a) | 83.1 | 78.0 | 6.0 | 17.1 | 0.21 | 0.22 |
| 2219-T87 | 70.2 | 57.2 | 10.5 | 48.0 | 0.68 | 0.84 |
| 7075-T651 | 94.8 | 74.2 | 13.0 | 37.0 | 0.00 | 0.50 |
| 7075-T7351 | 71.8 | 59.4 | 12.0 | 50.5 | 0.70 | 0.85 |
| 7079-T651(a) | 81.0 | 72.6 | 11.5 | 34.2 | 0.42 | 0.47 |

Specimens per Fig. A1.6. Each line is the average of duplicate or triplicate tests of an individual lot of material. For yield strengths, offset is 0.2%. (a) Obsolete alloy

**Table 5.5(a) Results of tensile tests of smooth and 0.5 in. diameter, notched round specimens from aluminum alloy plate, longitudinal**

| Alloy and temper | Nominal thickness, in. | Ultimate tensile strength (UTS), ksi | Tensile yield strength (TYS), ksi | Elongation in 2 in., % | Reduction of area, % | Notch tensile strength (NTS), ksi | Notch reduction of area, % | NTS/TS | NTS/YS |
|---|---|---|---|---|---|---|---|---|---|
| 2014-T651 | 1.000 | 69.0 | 63.5 | 10.2 | 24 | 82.8 | 2 | 1.20 | 1.30 |
| 2020-T651(a) | 0.875 | 81.8 | 77.0 | 2.9 | 4 | 51.4 | ... | 0.63 | 0.67 |
| | | 83.0 | 77.4 | 4.4 | 7 | 55.5 | | 0.67 | 0.72 |
| | | 81.4 | 76.6 | 5.5 | 10 | 61.6 | ... | 0.76 | 0.80 |
| | 0.900 | 82.6 | 78.3 | 5.0 | 7 | 63.4 | 1 | 0.77 | 0.81 |
| | 1.250 | 79.0 | 74.7 | 3.8 | 7 | 51.6 | 0 | 0.65 | 0.69 |
| | 1.375 | 83.2 | 77.5 | 6.0 | 7 | 66.7 | ... | 0.80 | 0.86 |
| | | 81.6 | 76.1 | 5.8 | 8 | 74.9 | | 0.92 | 0.98 |
| | | 81.9 | 76.3 | 6.0 | 9 | 73.3 | | 0.89 | 0.96 |
| 2024-T351 | 1.000 | 70.0 | 56.2 | 17.5 | 22 | 81.6 | 3 | 1.17 | 1.45 |
| | 1.500 | 69.4 | 53.0 | 19.5 | 28 | 80.9 | 4 | 1.16 | 1.52 |
| 2024-T851 | 0.875 | 71.9 | 67.9 | 8.0 | 21 | 86.1 | 1 | 1.20 | 1.27 |
| | 1.375 | 72.0 | 65.8 | 7.8 | 19 | 85.1 | ... | 1.18 | 1.29 |
| | | 72.0 | 66.1 | 8.0 | 20 | 81.8 | | 1.14 | 1.24 |
| | | 71.8 | 65.6 | 8.5 | 22 | 85.2 | | 1.19 | 1.30 |
| 2024-T86 | 0.875 | 76.5 | 72.8 | 8.5 | 22 | 87.0 | 0 | 1.14 | 1.19 |
| 2219-T31 | 0.500 | 54.5 | 37.7 | 28.0 | ... | 68.3 | 6 | 1.25 | 1.81 |
| 2219-T37 | 0.500 | 54.5 | 45.2 | 21.0 | ... | 78.3 | 7 | 1.44 | 1.73 |
| 2219-T62 | 1.000 | 64.1 | 48.1 | 11.0 | ... | 74.7 | 3 | 1.16 | 1.55 |
| | | 60.1 | 41.4 | 13.0 | ... | 72.2 | 3 | 1.20 | 1.74 |
| 2219-T851 | 0.500 | 67.8 | 52.2 | 11.4 | ... | 80.5 | 3 | 1.19 | 1.59 |
| | 1.000 | 65.7 | 52.8 | 12.0 | ... | 79.4 | ... | 1.21 | 1.50 |
| | 1.250 | 65.8 | 50.8 | 11.0 | 23 | 73.8 | 2 | 1.12 | 1.45 |
| | 1.375 | 66.8 | 51.1 | 10.2 | 22 | 80.6 | ... | 1.21 | 1.58 |
| | | 66.4 | 50.6 | 10.2 | 24 | 81.2 | ... | 1.22 | 1.60 |
| | | 66.6 | 52.0 | 11.0 | 25 | 79.9 | ... | 1.20 | 1.54 |
| 2219-T87 | 0.500 | 69.3 | 56.8 | 12.6 | ... | 85.0 | 3 | 1.26 | 1.50 |
| | 1.000 | 67.9 | 56.9 | 12.5 | ... | 82.3 | ... | 1.21 | 1.45 |
| | | 68.4 | 57.1 | 11.5 | 26 | 83.1 | 3 | 122 | 1.46 |
| 2618-T651 | 1.356 | 62.4 | 57.6 | 10.8 | ... | 81.2 | ... | 1.30 | 1.41 |
| 5083-O | 0.750 | 45.5 | 20.4 | 20.5 | 32 | 51.6 | 7 | 1.13 | 2.53 |
| 5083-H113 | 0.750 | 49.9 | 34.5 | 14.5 | 25 | 58.7 | 4 | 1.17 | 1.70 |
| 5086-O | 0.750 | 41.2 | 20.5 | 25.0 | 35 | 48.7 | 8 | 1.18 | 2.37 |
| 5086-H32 | 0.750 | 45.0 | 30.4 | 16.0 | 24 | 55.4 | 6 | 1.23 | 1.82 |
| 5086-H34 | 0.750 | 50.7 | 38.2 | 12.5 | 18 | 65.5 | 5 | 1.29 | 1.72 |
| 5154-O | 0.750 | 35.1 | 16.1 | 30.7 | 51 | 46.0 | 10 | 1.31 | 2.88 |
| 5356-O | 0.750 | 43.5 | 20.9 | 28.8 | 39 | 50.0 | 8 | 1.15 | 2.39 |
| 5356-H321 | 0.750 | 53.3 | 34.7 | 16.0 | 18 | 60.4 | 6 | 1.13 | 1.74 |
| 5454-O | 0.750 | 35.9 | 16.6 | 25.0 | 52 | 49.0 | 16 | 1.37 | 2.95 |
| 5454-H32 | 0.750 | 39.2 | 38.2 | 16.5 | 36 | 51.3 | 13 | 1.28 | 1.82 |
| | 0.750 | 40.9 | 28.9 | 15.7 | 32 | 56.2 | 12 | 1.37 | 1.94 |
| 5456-O | 0.750 | 49.9 | 23.2 | 21.8 | 31 | 50.9 | 5 | 1.04 | 2.19 |
| 5456-H321 | 0.750 | 56.3 | 34.5 | 13.5 | 16 | 59.7 | 8 | 1.06 | 1.73 |
| | | 52.9 | 34.2 | 16.0 | 18 | 60.8 | 5 | 1.15 | 1.78 |
| | | 55.4 | 35.8 | 13.2 | is | 62.6 | 5 | 1.13 | 1.75 |
| 6061-T651 | 0.750 | 42.2 | 39.2 | 16.0 | 41 | 64.5 | 4 | 1.53 | 1.65 |
| | 1.250 | 44.9 | 42.2 | 16.5 | 50 | 69.2 | 5 | 1.54 | 1.64 |
| 7001-T75(a) | 1.000 | 81.9 | 74.8 | 11.0 | ... | 80.0 | ... | 0.98 | 1.07 |
| | 1.000 | 81.4 | 74.4 | 10.2 | ... | 65.0 | ... | 0.80 | 0.87 |
| | | 81.8 | 72.2 | 9.5 | 17 | 93.1 | ... | 1.14 | 1.29 |
| | | 80.6 | 70.6 | 9.5 | 18 | 91.3 | ... | 1.13 | 1.29 |
| | | 80.6 | 70.6 | 9.5 | 17 | 91.4 | ... | 1.13 | 1.29 |
| 7075-T651 | 1.000 | 85.2 | 76.4 | 10.0 | ... | 99.3 | 2 | 1.16 | 1.29 |
| | 1.250 | 90.4 | 81.6 | 10.0 | 16 | 97.3 | 2 | 1.08 | 1.19 |
| | 1.375 | 84.0 | 75.4 | 11.4 | 17 | 97.8 | ... | 1.16 | 1.30 |
| | | 87.3 | 79.1 | 10.9 | 16 | 100.8 | ... | 1.15 | 1.27 |
| | | 88.9 | 80.6 | 11.2 | 14 | 101.7 | ... | 1.14 | 1.26 |
| 7075-T7351 | 1.375 | 76.7 | 66.3 | 12.0 | 29 | 93.2 | ... | 1.22 | 1.41 |
| | | 69.8 | 58.3 | 12.5 | 29 | 87.6 | ... | 1.26 | 1.50 |
| | | 70.8 | 59.1 | 12.5 | 29 | 89.7 | ... | 1.27 | 1.52 |
| 7079-T651(a) | 1.500 | 84.2 | 76.8 | 10.0 | 20 | 103.9 | 2 | 1.23 | 1.35 |
| | 1.375 | 84.0 | 77.6 | 11.5 | 17 | 103.0 | ... | 1.23 | 1.33 |
| | | 82.8 | 76.0 | 11.0 | 17 | 100.0 | ... | 1.21 | 1.32 |
| | | 82.2 | 75.2 | 11.2 | 20 | 100.0 | ... | 1.22 | 1.33 |
| X7106-T6351(a) | 0.500 | 65.6 | 57.7 | 14.8 | ... | 90.8 | ... | 1.38 | 1.57 |
| | 1.250 | 67.5 | 60.0 | 12.5 | ... | 92.0 | ... | 1.36 | 1.54 |
| 7178-T651 | 1.250 | 93.6 | 84.2 | 9.0 | 12 | 87.9 | 1 | 0.94 | 1.04 |
| 7178-T7651 | 1.000 | 80.6 | 71.7 | 11.0 | ... | 95.6 | ... | 1.18 | 1.33 |

Specimens per Fig. A1.7(a). Each line is the average of duplicate or triplicate tests of an individual lot of material. For yield strengths, offset is 0.2%. (a) Obsolete alloy

**Table 5.5(b)  Results of tensile tests of smooth and 0.5 in. diameter, notch round specimens from aluminum alloy plate, transverse**

| Alloy and temper | Nominal thickness, in. | Ultimate tensile strength (UTS), ksi | Tensile yield strength (TYS), ksi | Elongation in 2 in., % | Reduction of area, % | Notch tensile strength (NTS), ksi | Notch reduction of area, % | NTS/TS | NTS/YS |
|---|---|---|---|---|---|---|---|---|---|
| 2014-T651 | 1.000 | 69.5 | 62.7 | 8.8 | 16 | 79.8 | 0 | 1.15 | 1.27 |
| 2020-T651(a) | 0.875 | 82.3 | 77.7 | 4.5 | 8 | 47.8 | ... | 0.58 | 0.62 |
| | | 83.6 | 78.2 | 2.2 | 4 | 52.0 | ... | 0.62 | 0.66 |
| | | 82.8 | 77.6 | 3.2 | 6 | 58.3 | ... | 0.70 | 0.75 |
| | 0.900 | 83.8 | 78.6 | 3.7 | 5 | 54.3 | 0 | 0.65 | 0.69 |
| | 1.250 | 81.5 | 76.5 | 2.2 | 5 | 49.2 | 0 | 0.60 | 0.64 |
| | 1.375 | 82.4 | 78.4 | 1.8 | 2 | 53.6 | ... | 0.65 | 0.68 |
| | | 82.2 | 77.5 | 2.6 | 4 | 59.0 | ... | 0.72 | 0.76 |
| | | 82.2 | 77.4 | 2.4 | 4 | 62.0 | ... | 0.75 | 0.80 |
| 2024-T4 | 0.625 | 67.2 | 43.4 | 18.0 | 22 | 73.2 | 2 | 1.09 | 1.69 |
| 2024-T351 | 0.625 | 67.2 | 44.7 | 17.0 | 22 | 72.8 | 2 | 1.08 | 1.63 |
| | 1.000 | 70.0 | 50.3 | 14.0 | 22 | 78.8 | 2 | 1.13 | 1.57 |
| | 1.500 | 68.4 | 47.4 | 17.2 | ... | 77.0 | 3 | 1.13 | 1.62 |
| 2024-T36 | 0.625 | 73.0 | 60.5 | 10.0 | 16 | 74.6 | 1 | 1.02 | 1.23 |
| 2024-T6 | 0.625 | 68.6 | 57.2 | 9.1 | 16 | 74.2 | 0 | 1.01 | 1.30 |
| 2024-T81 | 0.625 | 68.8 | 63.4 | 6.6 | 16 | 70.9 | 2 | 1.03 | 1.12 |
| 2024-T851 | 0.875 | 71.5 | 66.9 | 7.0 | 16 | 73.8 | 0 | 1.03 | 1.10 |
| | 1.375 | 70.9 | 65.0 | 7.0 | 13 | 76.2 | ... | 1.07 | 1.17 |
| | | 71.2 | 65.5 | 7.0 | 14 | 68.2 | ... | 0.96 | 1.04 |
| | | 70.8 | 64.4 | 7.2 | 14 | 76.7 | ... | 1.08 | 1.19 |
| 2024-T86 | 0.625 | 75.1 | 71.6 | 4.8 | 12 | 58.4 | 0 | 0.78 | 0.82 |
| | 0.875 | 73.9 | 69.9 | 6.0 | 13 | 67.2 | 0 | 0.91 | 0.96 |
| 2219-T31 | 0.500 | 56.3 | 34.8 | 24.0 | ... | 68.3 | 13 | 1.21 | 1.96 |
| 2219-T37 | 0.500 | 57.4 | 44.2 | 16.0 | ... | 77.7 | 5 | 1.35 | 1.76 |
| 2219-T62 | 1.000 | 64.2 | 44.3 | 9.0 | ... | 69.9 | 7 | 1.09 | 1.57 |
| | | 59.9 | 40.6 | 12.0 | ... | 68.8 | 4 | 1.15 | 1.69 |
| 2219-T851 | 0.500 | 68.5 | 50.9 | 9.7 | ... | 75.3 | 2 | 1.10 | 1.48 |
| | 1.000 | 66.1 | 51.3 | 10.5 | ... | 77.0 | 2 | 1.16 | 1.50 |
| | 1.250 | 65.8 | 49.1 | 11.0 | 21 | 73.8 | 2 | 1.12 | 1.50 |
| | 1.375 | 66.0 | 50.8 | 10.2 | 20 | 77.2 | ... | 1.17 | 1.52 |
| | | 65.6 | 51.2 | 11.0 | 18 | 76.9 | ... | 1.17 | 1.50 |
| | | 65.8 | 49.3 | 10.0 | 20 | 76.1 | ... | 1.16 | 1.54 |
| 2219-T87 | 0.500 | 70.8 | 58.1 | 9.1 | ... | 81.2 | 1 | 1.15 | 1.40 |
| | 1.000 | 69.3 | 57.0 | 10.0 | ... | 76.2 | ... | 1.10 | 1.34 |
| | | 69.1 | 57.1 | 9.0 | 16 | 78.8 | 1 | 1.14 | 1.38 |
| 2618-T651 | 1.356 | 61.1 | 54.6 | 8.8 | ... | 83.2 | ... | 1.36 | 1.52 |
| 5083-O | 0.750 | 45.9 | 20.5 | 25.0 | 33 | 48.3 | 6 | 1.05 | 2.36 |
| 5083-H113 | 0.750 | 50.1 | 34.3 | 16.1 | 24 | 56.6 | 3 | 1.13 | 1.65 |
| 5086-O | 0.750 | 41.1 | 20.6 | 27.8 | 36 | 49.7 | 7 | 1.09 | 2.41 |
| 5086-H32 | 0.750 | 44.8 | 30.4 | 18.6 | 38 | 54.0 | 4 | 1.20 | 1.78 |
| 5086-H34 | 0.750 | 52.4 | 37.4 | 15.7 | 33 | 62.1 | 3 | 1.19 | 1.66 |
| 5154-O | 0.750 | 36.1 | 16.2 | 29.6 | 47 | 44.3 | 11 | 1.23 | 2.74 |
| 5356-O | 0.750 | 44.7 | 21.7 | 27.7 | 32 | 48.0 | 4 | 1.07 | 2.21 |
| 5356-H321 | 0.750 | 51.8 | 33.2 | 21.0 | 31 | 59.2 | 3 | 1.14 | 1.78 |
| 5454-O | 0.750 | 35.6 | 16.8 | 24.0 | 45 | 47.2 | 12 | 1.33 | 2.92 |
| 5454-H32 | 0.750 | 40.0 | 28.8 | 19.2 | 46 | 53.4 | 8 | 1.33 | 1.85 |
| | 0.750 | 41.4 | 29.4 | 18.8 | 46 | 56.5 | 10 | 1.36 | 1.92 |
| 5456-O | 0.750 | 48.8 | 24.0 | 21.2 | 26 | 49.5 | 6 | 1.01 | 2.06 |
| 5456-H321 | 0.750 | 55.9 | 33.6 | 19.0 | 27 | 58.5 | 4 | 1.05 | 1.79 |
| | | 52.7 | 34.4 | 17.2 | 22 | 58.3 | 4 | 1.11 | 1.70 |
| | | 55.4 | 33.3 | 16.5 | 26 | 61.0 | 4 | 1.10 | 1.83 |
| 6061-T651 | 0.750 | 42.0 | 39.1 | 16.5 | ... | 62.6 | 4 | 1.49 | 1.60 |
| | 0.625 | 45.1 | 40.2 | 15.8 | 42 | 62.6 | 2 | 1.39 | 1.55 |
| | 1.250 | 44.9 | 40.4 | 15.2 | 42 | 67.8 | 5 | 1.51 | 1.68 |
| 7001-T75(a) | 1.000 | 81.8 | 73.7 | 8.5 | ... | 68.9 | ... | 0.84 | 0.93 |
| | 1.000 | 81.8 | 73.4 | 9.2 | ... | 64.6 | ... | 0.79 | 0.88 |
| | | 80.8 | 71.3 | 8.8 | 14 | 81.2 | ... | 1.00 | 1.14 |
| | | 79.9 | 69.6 | 9.0 | 14 | 83.6 | ... | 1.05 | 1.20 |
| | | 80.5 | 70.6 | 8.8 | 14 | 80.7 | ... | 1.00 | 1.14 |

(continued)

Specimens per Fig. A1.7(a). Each line is the average of duplicate or triplicate tests of an individual lot of material. For yield strengths, offset is 0.2%. (a) Obsolete alloy

**Table 5.5(b)** (continued)

| Alloy and temper | Nominal thickness, in. | Ultimate tensile strength (UTS), ksi | Tensile yield strength (TYS), ksi | Elongation in 2 in., % | Reduction of area, % | Notch tensile strength (NTS), ksi | Notch reduction of area, % | NTS/TS | NTS/YS |
|---|---|---|---|---|---|---|---|---|---|
| 7075-T651 | 1.000 | 82.5 | 72.8 | 9.5 | ... | 85.4 | 1 | 1.04 | 1.17 |
| | 0.625 | 82.4 | 71.6 | 11.2 | 18 | 91.6 | 0 | 1.11 | 1.28 |
| | 1.250 | 88.0 | 78.8 | 10.0 | 14 | 83.2 | 1 | 0.95 | 1.06 |
| | 1.375 | 82.4 | 73.4 | 11.2 | 16 | 94.6 | ... | 1.15 | 1.29 |
| | | 86.1 | 77.7 | 10.8 | 15 | 95.8 | ... | 1.11 | 1.23 |
| | | 86.7 | 77.3 | 11.8 | 16 | 102.1 | ... | 1.18 | 1.32 |
| 7075-T7351 | 1.375 | 74.9 | 64.6 | 10.5 | 20 | 85.2 | ... | 1.14 | 1.32 |
| | | 68.2 | 56.8 | 11.8 | 24 | 80.6 | ... | 1.18 | 1.42 |
| | | 70.1 | 58.5 | 11.0 | 23 | 82.1 | ... | 1.17 | 1.40 |
| 7079-T651(a) | 1,500 | 83.4 | 74.4 | 11.2 | 20 | 94.4 | 2 | 1.13 | 1.27 |
| | 1.375 | 83.1 | 74.2 | 11.2 | 17 | 94.4 | ... | 1.14 | 1.27 |
| | | 82.5 | 72.8 | 11.2 | 16 | 91.0 | 2 | 1.10 | 1.25 |
| | | 82.8 | 72.6 | 11.2 | 17 | 86.8 | ... | 1.05 | 1.20 |
| X7106-T6351(a) | 0.500 | 65.4 | 57.7 | 14.0 | ... | 90.9 | ... | 1.39 | 1.57 |
| | 1.250 | 66.4 | 58.9 | 12.5 | ... | 91.2 | ... | 1.38 | 1.56 |
| 7178-T651 | 1.250 | 94.2 | 83.0 | 9.0 | 13 | 79.6 | 1 | 0.84 | 0.96 |
| 7178-T7651 | 1.000 | 79.8 | 70.9 | 10.8 | ... | 88.7 | ... | 1.11 | 1.25 |

Specimens per Fig. A1.7(a). Each line is the average of duplicate or triplicate tests of an individual lot of material. For yield strengths, offset is 0.2%. (a) Obsolete alloy

**Table 5.6** Results of tensile tests of smooth and 0.5 in. diameter, notched round specimens from aluminum alloy castings

| Alloy and temper | Ultimate tensile strength | Tensile yield strength (TYS), ksi | Elongation in 2 in., % | Reduction of area, % | Notch tensile strength (NTS), ksi | NTS/TS | NTS/YS |
|---|---|---|---|---|---|---|---|
| **Sand casting** | | | | | | | |
| 240.0-F | 33.8 | 26.0 | 1.4 | 2 | 22.5 | 0.67 | 0.87 |
| 242.0-T77 | 29.8 | 20.4 | 2.1 | 4 | 27.4 | 0.92 | 1.34 |
| 295.0-T6 | 42.0 | 27.1 | 6.4 | 10 | 47.4 | 1.13 | 1.75 |
| 308.0-F | 25.0 | 18.5 | 1.8 | 2 | 22.1 | 0.88 | 1.19 |
| X335.0-T6 | 37.3 | 23.4 | 8.6 | 12 | 38.2 | 1.02 | 1.63 |
| | 35.3 | 22.6 | ... | ... | 37.4 | 1.06 | 1.65 |
| Average X335.0-T6 | 36.3 | 23.0 | ... | ... | 37.8 | 1.04 | 1.64 |
| 356.0-T4 | 31.1 | 19.0 | 4.4 | 6 | 31.6 | 1.02 | 1.60 |
| | 29.4 | 17.6 | ... | ... | 31.3 | 1.07 | 1.78 |
| Average 356.0-T4 | 30.2 | 18.7 | ... | ... | 31.4 | 1.04 | 1.69 |
| 356.0-T6 | 38.6 | 32.6 | 2.2 | 3 | 37.4 | 0.97 | 1.15 |
| 356.0-T7 | 37.8 | 33.7 | 1.6 | 2 | 34.5 | 0.91 | 1.02 |
| 356.0-T71 | 28.8 | 20.2 | 5.0 | ... | 32.0 | 1.11 | 1.59 |
| | 31.9 | 24.2 | ... | ... | 38.2 | 1.20 | 1.58 |
| | 29.4 | 20.7 | ... | ... | 30.6 | 1.04 | 1.46 |
| Average 356.0-T7 | 30.0 | 21.7 | ... | ... | 33.6 | 1.12 | 1.54 |
| A356.0-T61 | 41.6 | 30.2 | 8.8 | 10 | 51.4 | 1.23 | 1.70 |
| A356.0-T7 | 37.1 | 30.5 | 4.4 | 7 | 44.9 | 1.21 | 1.47 |
| | 37.6 | 33.2 | ... | ... | 38.2 | 1.02 | 1.15 |
| Average A356.0-T7 | 37.4 | 31.8 | ... | ... | 41.6 | 1.12 | 1.31 |
| 520.0-F | 34.2 | 31.6 | 2.1 | 2 | 38.4 | 1.12 | 1.22 |
| B535.0-F | 41.2 | 21.2 | 12.9 | 13 | 43.8 | 1.06 | 2.06 |
| | 42.6 | 21.0 | ... | ... | 44.9 | 1.05 | 2.14 |
| Average B535.0-F | 41.9 | 21.1 | ... | ... | 44.4 | 1.06 | 2.10 |
| A612.0-F | 43.1 | 34.8 | 3.2 | 7 | 45.5 | 1.05 | 1.31 |
| **Permanent-mold casting** | | | | | | | |
| X335.0-T61 | 40.8 | 28.4 | 8.5 | 13 | 45.7 | 1.12 | 1.61 |
| | 35.6 | 25.6 | 3.5 | 4 | 40.4 | 1.14 | 1.58 |
| Average X335.0-T61 | 38.2 | 23.8 | 6.0 | ... | ... | ... | ... |
| 354.0-T62 | 50.1 | 45.5 | 1.1 | 3 | 54.2 | 1.08 | 1.19 |
| | 47.8 | 44.3 | 0.9 | 2 | 51.2 | 1.07 | 1.16 |
| Average 354.0-T62 | 49.0 | 44.9 | 1.0 | 2 | ... | ... | ... |
| C355.0-T7 | 37.0 | 31.0 | 2.1 | 4 | 43.4 | 1.17 | 1.40 |
| | 41.0 | 30.4 | 2.5 | 6 | 41.8 | 1.02 | 1.38 |
| Average C355.0-T7 | 39.0 | 30.7 | 2.3 | 5 | ... | ... | ... |
| 356.0-T6 | 35.8 | 31.1 | 1.4 | 3 | 4.3 | 1.17 | 1.38 |
| 356.0-T7 | 28.4 | 21.4 | 4.3 | 6 | 35.3 | 1.24 | 1.65 |
| | 29.6 | 22.0 | 3.2 | 6 | 34.3 | 1.16 | 1.56 |
| Average 356.0-T7 | 29.8 | 22.0 | 5.0 | 8 | ... | ... | ... |
| A356.0-T61 | 39.4 | 30.8 | 4.3 | 7 | 47.8 | 1.21 | 1.55 |
| | 41.7 | 30.4 | 7.5 | 8 | 45.4 | 1.22 | 1.75 |
| Average A356.0-T61 | 40.6 | 30.6 | 5.9 | 8 | ... | ... | ... |
| A356.0-T62 | 40.9 | 36.7 | 2.1 | 6 | 46.2 | 1.13 | 1.26 |
| | 43.6 | 36.3 | 3.9 | ... | 43.8 | 1.00 | 1.21 |
| Average A356.0-T62 | 42.2 | 36.5 | 3.0 | 6 | ... | ... | ... |
| A356.0-T7 | 28.2 | 21.4 | 5.3 | 9 | 36.9 | 1.31 | 1.72 |
| 359.0-T62 | 46.2 | 43.2 | 1.2 | 3 | 49.7 | 1.08 | 1.15 |
| | 47.4 | 43.1 | 1.6 | 4 | 42.9 | 0.91 | 1.00 |
| Average 359.0-T62 | 34.7 | 43.2 | 1.4 | 4 | ... | ... | ... |
| A444.0-F | 23.2 | 9.7 | 22.2 | 37 | 28.6 | 1.23 | 2.96 |
| | 22.5 | 9.6 | 15.7 | 21 | 27.8 | 1.25 | 2.90 |
| Average A444.0-F | 22.8 | 9.6 | 19.0 | ... | ... | ... | ... |
| A444.0-T4 | 23.0 | 8.0 | 24.4 | 36 | 30.8 | 1.39 | 3.72 |
| **Premium-strength casting** | | | | | | | |
| C355.0-T61 | 43.6 | 30.3 | 6.4 | 9 | 52.6 | 1.21 | 1.74 |
| A356.0-T6 | 41.6 | 30.2 | 8.8 | 10 | 51.4 | 1.23 | 1.7 |
| A357.0-T61 | 51.2 | 40.0 | 11.4 | 13 | 56.2 | 1.1 | 1.41 |
| A357.0-T62 | 53.9 | 46.4 | 5.3 | 7 | 59.4 | 1.1 | 1.28 |

Specimens per Fig. A1.7(a). Each line is the average of two tests of a single lot of material. For tensile yield strength, offset is 0.2%.

**Table 5.7(a)    Results of tensile tests of smooth and notched 1 in. wide, edge-notched sheet-type tensile specimens from welds in 0.125 in. aluminum alloy sheet, longitudinal (transverse weld)**

| Parent alloy and temper | Filler alloy | Post weld heat treatment | Longitudinal (transverse weld) | | | | | |
|---|---|---|---|---|---|---|---|---|
| | | | Ultimate tensile strength (UTS), ksi | Joint yield strength (JYS), ksi | Elongation in 2 in., % | Notch tensile strength (NTS), % | NTS/TS | NTS/YS |
| 2014-T3 | 4043 | None | 50.0 | 41.5 | 2.8 | 47.4 | 0.95 | 1.14 |
| 2014-T3 | 4043 | Aged to T6 | 54.8 | 54.8(a) | (a) | 52.9 | 0.97 | 0.97 |
| 2014-T6 | 4043 | None | 46.3 | 37.8 | 2.8 | 42.9 | 0.93 | 1.13 |
| 2219-T37 | 2319 | None | 41.8 | 28.7 | 4.0 | 36.0 | 0.88 | 1.28 |
| 2219-T37 | 2319 | Aged to T87 | 43.2 | 39.1 | 1.9 | 43.7 | 1.01 | 1.12 |
| 2219-T62 | 2319 | None | 44.0 | 30.5 | 2.7 | 47.3 | 1.08 | 1.55 |
| 2219-T62 | 2319 | RHT to T6 | 60.5 | 43.5 | 7.5 | 59.5 | 0.98 | 1.37 |
| 2219-T87 | 2319 | None | 45.3 | 32.8 | 2.3 | 41.7 | 0.92 | 1.27 |
| 5456-H321 | 5556 | None | 51.9 | 32.6 | 13.0 | 54.4 | 1.05 | 1.67 |
| 5456-H343 | 5556 | None | 51.9 | 30.7 | 7.0 | 53.2 | 1.03 | 1.73 |
| 6061-T6 | 4043 | None | 32.2 | 23.2 | 5.3 | 34.1 | 1.06 | 1.47 |
| 7075-T6 | 5556 | None | 46.5 | 45.0 | 1.0 | 43.6 | 0.94 | 0.97 |
| 7178-T6 | 5556 | None | 54.1 | 51.0 | 1.3 | 53.0 | 0.98 | 1.04 |

Specimens per Fig. A1.4(b). Each line represents the average of three specimens from a single lot of material. For yield strengths, offset is 0.2% in 2 in. gage length. (a) No joint yield strength or elongation identified; failed before reaching 0.2% offset

**Table 5.7(b)    Results of tensile tests of smooth and notched 1 in. wide, edge-notched sheet-type tensile specimens from welds in 0.125 in. aluminum alloy sheet, transverse (longitudinal weld)**

| Parent alloy and temper | Filler alloy | Post weld heat treatment | Transverse (longitudinal weld) | | | | | |
|---|---|---|---|---|---|---|---|---|
| | | | Ultimate tensile strength (UTS), ksi | Joint yield strength (JYS), ksi | Elongation in 2 in., % | Notch tensile strength (NTS), ksi | NTS/TS | NTS/YS |
| 2014-T3 | 4043 | None | 47.6 | 38.6 | 1.0 | 49.5 | 1.04 | 1.28 |
| 2014-T3 | 4043 | Aged to T6 | 49.3 | 49.3(a) | (a) | 48.9 | 0.99 | 0.99 |
| 2219-T37 | 2319 | None | 42.7 | 27.4 | 4.0 | 38.3 | 0.90 | 1.40 |
| 2219-T37 | 2319 | Aged to T87 | 42.7 | 38.2 | 2.3 | 41.1 | 0.96 | 1.08 |
| 2219-T62 | 2319 | None | 45.2 | 30.8 | 3.5 | 44.9 | 0.99 | 1.46 |
| 2219-T62 | 2319 | RHT to T6 | 60.2 | 42.8 | 9.2 | 58.0 | 0.96 | 1.36 |
| 2219-T87 | 2319 | None | 44.6 | 31.0 | 2.2 | 44.2 | 0.99 | 1.43 |
| 5456-H321 | 5556 | None | 51.7 | 30.3 | 8.5 | 51.8 | 1.00 | 1.71 |
| 5456-H343 | 5556 | None | 51.8 | 30.1 | 7.3 | 54.0 | 1.04 | 1.79 |
| 7178-T6 | 5556 | None | 46.4 | 46.4(a) | (a) | 49.6 | 1.07 | 1.07 |

Specimens per Fig. A1.4(b). Each line represents the average of three specimens from a single lot of material. For yield strengths, offset is 0.2% in 2 in. gage length. (a) No joint yield strength or elongation identified; failed before reaching 0.2% offset

**Table 5.8** Results of tensile tests of smooth and 0.5 in. diameter, notched round specimens from welds in aluminum alloy plate

| Base alloy and temper | Filler alloy | Post weld thermal treatment | Ultimate tensile strength (UTS), ksi | Tensile yield strength (TYS), ksi | Elongation in 2 in., % | Reduction of area, % | Joint strength efficiency, % | Location of fracture(a) | Notch tensile strength (NTS), ksi | NTS/TS | NTS/YS |
|---|---|---|---|---|---|---|---|---|---|---|---|
| 1100-H112 | 1100 | None | 11.6 | 6.1 | 26.5 | (b) | (b) | (b) | 17.8 | 1.53 | 2.92 |
| 3003-H112 | 1100 | None | 16.1 | 7.6 | 24.0 | (b) | (b) | (b) | 22.7 | 1.41 | 2.99 |
| 2219-T62 | 2319 | Aged to T62 | 57.3 | 40.2 | 7.5 | 7 | 99 | C | 63.7 | 1.11 | 1.58 |
| 2218-T851 | 2319 | None | 32.7 | 26.8 | 2.0 | 5 | 50 | C | 40.7 | 1.24 | 1.52 |
| 3003-H112 | 1100 | None | 16.1 | 7.6 | 24.0 | 67 | 100 | C | 22.7 | 1.41 | 3.00 |
| 5052-H112 | 5052 | None | 29.1 | 13.9 | 18.0 | (b) | (b) | (b) | 32.8 | 1.13 | 2.36 |
|  | 5154 | None | 29.2 | 13.7 | 15.0 | (b) | (b) | (b) | 32.1 | 1.1 | 2.34 |
| 5083-O | 5183 | None | 42.5 | 20.1 | 21.5 | (b) | (b) | (b) | 44.7 | 1.05 | 2.22 |
| 5083-H321 | 5183 | None | 44.2 | 26.0 | 14.0 | 39 | 96 | C | 54.5 | 1.23 | 2.10 |
|  | 5356 | None | 41.5 | 24.3 | 13.5 | 47 | 90 | A | 53.8 | 1.30 | 2.22 |
|  | 5556 | None | 44.4 | 25.6 | 14.0 | 36 | 97 | A | 53.7 | 1.21 | 2.10 |
| 5086-H32 | 5356 | None | 38.5 | 19.1 | 16.0 | (b) | (b) | (b) | 41.4 | 1.07 | 2.17 |
| 5154-H112 | 5154 | None | 32.6 | 14.5 | 17.0 | (b) | (b) | (b) | 34.1 | 1.05 | 2.35 |
| 5454-H32 | 5554 | None | 33.9 | 17.1 | 18.0 | 42 | 85 | A | 39.3 | 1.16 | 2.30 |
| 5456-O | 5456 | None | 43.9 | 21.7 | 13.0 | (b) | (b) | (b) | 40.7 | 0.93 | 1.87 |
| 5456-H321 | 5556 | None | 44.6 | 22.5 | 13.0 | (b) | (b) | (b) | 45.2 | 1.01 | 2.01 |
| 6061-T6 | 4043 | None | 31.0 | 20.9 | 6.0 | 19 | 69 | C | 34.0 | 1.10 | 1.63 |
|  | 4043 | None | 26.1 | 15.2 | 12.0 | (b) | (b) | (b) | 27.5 | 1.05 | 1.81 |
|  | 4043 | Aged to T6 | 43.3 | 35.9 | 11.0 | 44 | 96 | B | 57.5 | 1.31 | 1.57 |
|  | 4043 | HTA(c) | 43.2 | 38.6 | 2.0 | (b) | (b) | (b) | 42.3 | 0.98 | 1.10 |
|  | 5154 | None | 25.0 | 14.0 | 13.0 | (b) | (b) | (b) | 33.8 | 1.35 | 2.41 |
|  | 5154 | HTA | 35.6 | 27.3 | 5.5 | (b) | (b) | (b) | 42.9 | 1.20 | 1.57 |
|  | 5356 | None | 32.7 | 22.6 | 8.0 | 31 | 73 | A | 46.9 | 1.44 | 2.07 |
|  | 5356 | Aged to T6 | 40.5 | 29.3 | 9.5 | 33 | 90 | B | ... | ... | ... |
| 7005-T53 | 5039 | None | 48.3 | 32.2 | 12.2 | (b) | 78 | (b) | 59.0 | 1.83 | 1.85 |
| 7005-T6351 | 5039 | None | 48.4 | 32.3 | 11.5 | (b) | 85 | (b) | 57.6 | 1.78 | 1.78 |
|  | 5356 | None | 42.1 | 28.2 | 6.8 | (b) | 74 | (b) | 52.7 | 1.87 | 1.87 |

Specimens per Fig. A1.7(b). Each line represents the average of duplicate to triplicate tests of an individual lot of material. For joint yield strength, offset is 0.2%, over a 2 in. gage length. Joint efficiencies based on typical values for parent alloys. (a) Location of fracture of unnotched specimens: A, through weld; B, 0.5 to 2.5 in. from weld; C, edge of weld. (b) Not recorded. (c) HTA, heat treated and artificially aged after welding

**Table 5.9** Results of tensile tests of smooth and 0.5 in. diameter, notched round specimens from welds in aluminum alloy sand castings

| Alloy and temper combination | Filler alloy | Post-weld thermal treatment | Ultimate tensile strength (UTS), ksi | Tensile yield strength (TYS), ksi | Elongation in 2 in., % | Reduction of area, % | Joint strength efficiency, % | Location of fracture(a) | Notch tensile strength (NTS), ksi | NTS/TS | NTS/YS |
|---|---|---|---|---|---|---|---|---|---|---|---|
| A444.0-F to A444.0-F | 4043 | None | 23.8 | 9.5 | 12.1 | 22 | 100 | B | 27.5 | 1.15 | 2.90 |
| A444.0-F to 6061-T6 | 4043 | None | 24.0 | 11.4 | 5.7 | 23 | 100 | B | 29.3 | 1.22 | 2.51 |
| A444.0-F to 5456-H321 | 5556 | None | 24.1 | 12.2 | 12.1 | 27 | 100 | B | 29.5 | 1.22 | 2.42 |
| 354.0-T62 to 354.0-T6 | 4043 | None | 37.8 | 21.5 | 6.4 | 10 | 76 | A | 32.0 | 0.85 | 1.48 |
| 354.0-T62 to 6061-T6 | 4043 | None | 30.8 | 19.0 | 9.3 | 39 | 62 | C | 28.7 | 0.93 | 1.51 |
| 354.0-T62 to 5456-H321 | 5556 | None | 37.7 | 24.6 | 3.6 | 5 | 75 | A | 37.7 | 1.00 | 1.53 |
| C355.0-T61 to 6061-T6 | 4043 | None | 28.9 | 19.3 | 7.1 | 32 | 66 | C | 34.5 | 1.19 | 1.79 |
| C355.0-T61 to 5456-H321 | 5556 | None | 35.4 | 24.4 | 3.6 | 5 | 81 | A | 40.5 | 1.15 | 1.66 |

Specimens per Fig. A1.7(b). Each line represents the average of duplicate tests on one lot of material. For joint yield strength, offset is 0.2%, over a 2 in. gage length. Joint efficiencies based upon typical values for parent alloys. (a) Location of fracture of unnotched specimens: A, through weld; B, 0.5 to 2.5 in. from weld; C, edge of weld

# 6

# Tear Resistance

A TEAR TEST of the type described in ASTM method B 871 was first developed at Alcoa Laboratories in about 1950 to more discriminatively evaluate the fracture characteristics of the aluminum alloys in various tempers (Ref 36, 37). As illustrated schematically in Fig. 6.1, values of the energies required to initiate and propagate cracks in small, sharply edge-notched specimens of the design in Fig. A1.8 are determined from measurements of the appropriate areas under autographic load-deformation curves developed during the tests. The unit propagation energy is equal to the energy required to propagate the crack divided by the initial net area of the specimen, and unit propagation energy is the primary criterion of tear resistance obtained from the tear test.

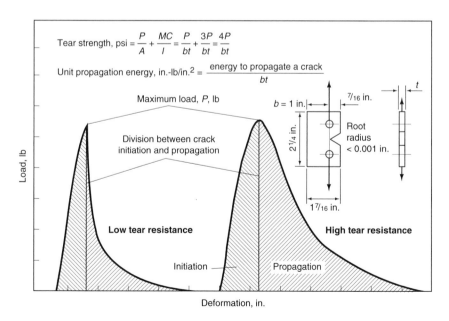

**Fig. 6.1** Tear-test specimen and representation of load-deformation curves. *A*, area; *M*, moment; *C*, moment arm; *I*, moment of inertia

The unit propagation energy, more than data from notch-tensile tests, provides a measure of that combination of strength and ductility that permits a material to resist crack growth under either elastic or plastic stresses. The "tear strength," the maximum nominal direct-and-bending stress developed by the tear specimen, is also calculated, and the ratio of this tear strength to the yield strength provides a measure of notch toughness; it is referred to as the tear-yield ratio.

The usefulness of the data from this test is not dependent upon the development of rapid crack propagation or fracture at elastic stresses. Therefore, the test can be used for all aluminum alloys, even very ductile, tough alloys such as 1100 and 3003, providing a criterion for making direct comparisons of the relative toughness of alloys across the whole range of aluminum alloy types, and directly comparing alloys such as 3003 to the very high-strength alloys.

This test is a modification of the older Navy tear test (Ref 20) but involves a smaller, sharp-notched specimen. The design of the tear-test specimen was selected for several reasons. First, the specimen is small enough to be taken from several orientations within most aluminum alloy products, including forgings, extrusions, castings, sheet, and plate. Second, it can be tested conveniently at different temperatures and in various environments. Third, the very sharp notch, in place of the keyhole notch in the Navy tear specimen, permits crack initiation at relatively low energy levels, thus increasing the accuracy of the measurement of propagation energy. With a relatively blunt notch, the large amount of energy required to initiate a crack overshadows and, on the test record, obscures the energy to propagate the crack.

It should be noted that the numerical results of tear tests are greatly dependent upon specimen size and geometry, although with specimens of the design in Fig. 6.1, thickness variation in the range from about 0.060 to about 0.100 in. generally has an insignificant effect on the values of tear strength and unit propagation energy. It is appropriate to note that the results are also testing-machine dependent, and that relatively stiff machines are preferred; more flexible machines undergo greater extension during testing and contribute greater stored elastic strain energy to fracturing the specimen, potentially obscuring the propagation energy measurements. In any case, it is desirable to use the same machine when developing relative measurements among a group of alloys and tempers.

Representative data for a variety of aluminum alloys and tempers, including welds, taken primarily from Ref 1, 29, and 33–37 plus some unpublished reports, are shown in the following tables at the end of this Chapter:

| | |
|---|---|
| Tables 6.1 and 6.2 | Wrought aluminum alloys in the form of 0.063 in. thick sheet, with 0.063 in. thick specimens |
| Table 6.1(a) and (b) | Non-heat-treated sheet |
| Table 6.2(a) and (b) | Heat treated sheet |
| Tables 6.3, 6.4, and 6.5 | Wrought aluminum alloys in the form of plate, extrusions, and forgings |
| Table 6.3(a) and (b) | Plate |
| Table 6.4(a) and (b) | Extruded shapes |
| Table 6.5(a), (b), and (c) | Forgings |
| Table 6.6(a) and (b) | Cast aluminum alloys, with 0.100 in. thick specimens from cast slabs |
| Table 6.7(a) and (b) | Welds in wrought aluminum alloys, with 0.100 in. thick specimens |
| Table 6.8 | Welds in cast aluminum alloys, with 0.100 in. thick specimens |

Ratings of the alloys and tempers are shown in Fig. 6.2 for wrought alloys based on the tests of sheet; Fig. 6.3 for wrought alloys in the form of plate (Fig. 6.3a), extrusions (Fig. 6.3b), and forgings (Fig. 6.3c); Fig. 6.4 for cast alloys; and Fig. 6.5 for welds in aluminum alloys (Fig. 6.5a for castings welded to other castings and Fig. 6.5b for castings welded to plate).

The ratings based on the values of unit propagation energy for sheet, plate, extrusions, and forgings are generally consistent within the various alloys and tempers where comparisons can be made. There is a general trend for unit propagation energy to decrease with increasing product thickness, and thicker products do show a greater degree of directionality than sheet.

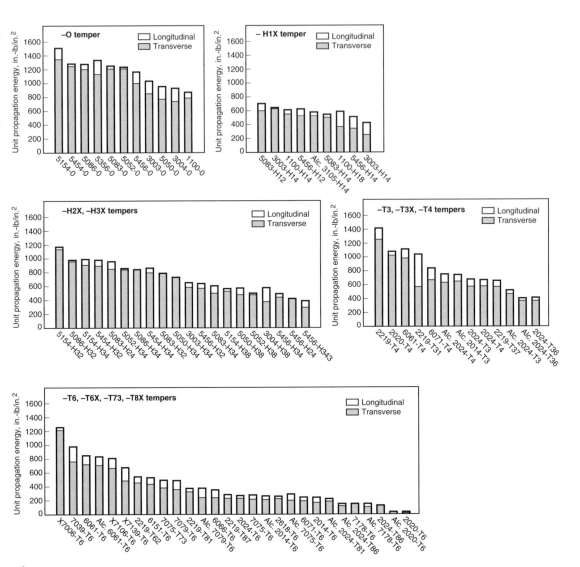

**Fig. 6.2** Ratings of 0.063 in. aluminum alloy sheet based on unit propagation energy

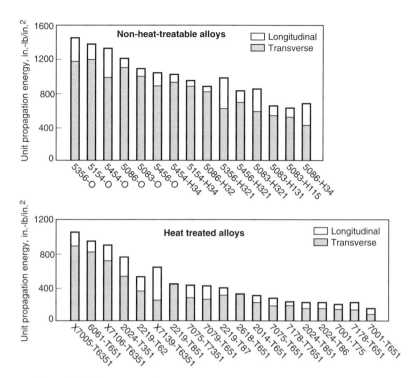

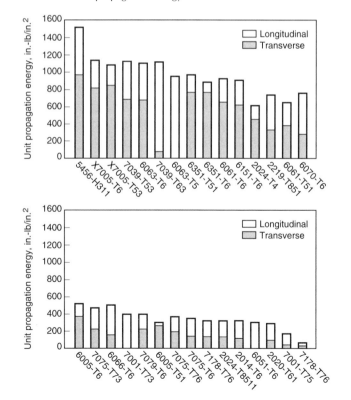

**Fig. 6.3(a)** Ratings of 0.75 to 1.5 in. thick aluminum alloy plate based on unit propagation energy

**Fig. 6.3(b)** Ratings of aluminum alloy extruded shapes based on unit propagation energy

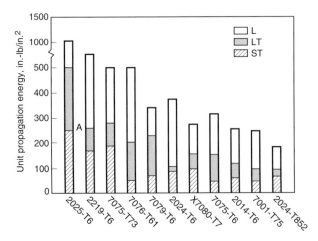

**Fig. 6.3(c)** Ratings of aluminum alloy forgings based on unit propagation energy. Values for 7075-T73, 7079-T6, 7075-T6, and 2014-T6 include stress relieved (TX52) tempers. Value at A is estimated. L, longitudinal; LT, long transverse; ST, short transverse

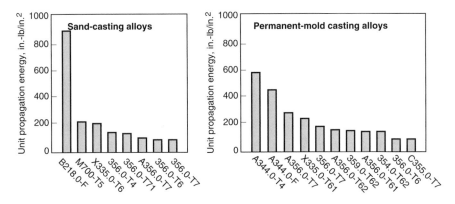

**Fig. 6.4** Ratings of aluminum alloy sand and permanent-mold cast slabs based on unit propagation energy

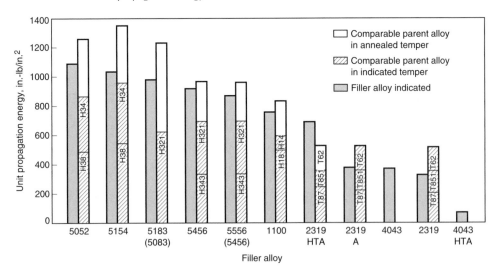

**Fig. 6.5(a)** Ratings of aluminum alloy welds based on unit propagation energy from tear tests. HTA, heat treated and artificially aged after welding. A, artificially aged after welding

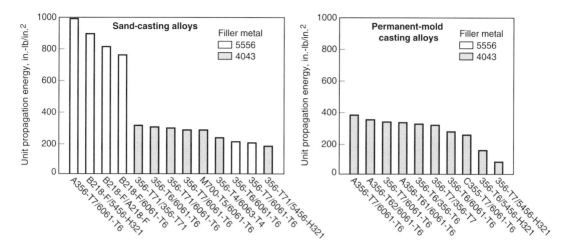

**Fig. 6.5(b)** Ratings of groove welds in cast-to-cast and cast-to-wrought. Based on test unit propagation energy with crack propagation through the weld. No subsequent thermal treatment unless otherwise noted. 356-T4/6063-T4 is aged 4 h at 375 °F after welding.

## 6.1 Wrought Alloys

As with notch toughness, a broader understanding of tear resistance is gained by plotting the unit propagation energy as a function of tensile yield strength, as in Fig. 6.6 based upon the data for 0.063 in. thick sheet (Ref 1). A broad band of data emerges that, if not separated by alloy and temper type, might appear to indicate a lack of association beyond a broad tendency for unit propagation energy to decrease with increase in strength. When separated by alloy type, however, it is clear that individual relationships exist for different types of alloys. The 7*xxx* (Al-Zn-Mg) series provides the superior level of tear resistance for a given level of strength. Of the 2*xxx* (Al-Cu), 5*xxx* (Al-Mg), and 6*xxx* (Al-Mg-Si) series, the 2*xxx* series has a slight advantage. The 1*xxx* and 3*xxx* series fall in the lower portion of the band. Increasing the strength of alloys in any of the series by cold work or thermal treatment reduces the tear resistance.

An important deviation from the general trend is illustrated by data for the annealed (O) temper of the non-heat-treatable alloys (open symbols in Fig. 6.6): for these alloys in the O temper, unit propagation energy increases with increase in yield strength up to approximately 16 ksi and then decreases. This illustrates that there is a contribution of both strength and ductility to tear resistance; the great ductility of 1100-O or 3003-O, for example, is not sufficient to give them exceptionally high tear resistance as measured by the unit energy required to propagate the crack. The 5*xxx* alloys in the annealed temper have about the optimal combination of these properties, yielding the highest unit propagation energies measured. This characteristic was the basis of the selection of these alloys for particularly critical applications (see Chapter 11).

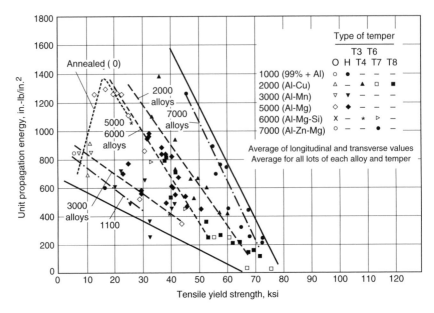

**Fig. 6.6** Unit propagation energy vs. tensile yield strength of 0.063 in. aluminum alloy sheet

A plot relating unit propagation energy to elongation in 2 in. for 0.063 in. sheet is shown in Fig. 6.7. In such a plot, there are few consistent trends and ample evidence that elongation by itself is not a very reliable indicator of resistance to crack growth or tear resistance.

## 6.2 Cast Alloys

Once again, relating unit propagation energy (UPE) to tensile yield strength (TYS), as in Fig. 6.8, reveals more than the bar charts alone (Fig. 6.4). While it is obvious that low-strength alloy A444.0 has relatively high tear resistance as defined by UPE, Fig. 6.8 also reveals that:

- Premium-strength cast alloys (i.e., those produced with high chill rates in key areas of the casting) consistently have among the best combinations of UPE and TYS, especially at relatively high strength levels.
- Sand-cast alloy B535.0-F itself has tear resistance in the same range as wrought alloy plate of the same strength level, and a much better combination of UPE and TYS than most other casting alloys.
- With the exception of B535.0-F, sand castings generally have among the poorest combination of strength and toughness.
- Permanent-mold cast alloys generally fall into the intermediate range, with the notable exceptions that 354.0-T62 and 359.0-T62 essentially match the performance of the premium-strength cast alloys (illustrated by the trend line for the triangular symbols).

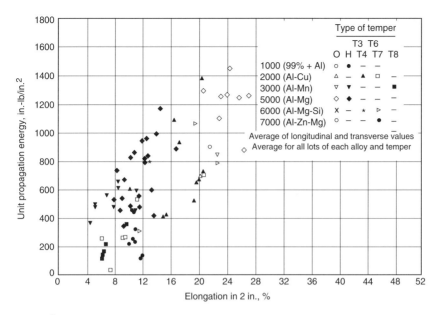

**Fig. 6.7** Unit propagation energy vs. elongation of 0.063 in. aluminum alloy sheet

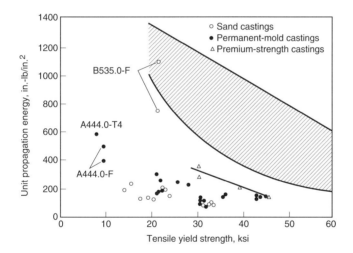

**Fig. 6.8** Unit propagation energy vs. tensile yield strength for aluminum alloy castings. Band is for data for aluminum alloy plate.

## 6.3 Welds

As with notch toughness, the tear resistance of welds made with 5*xxx* filler alloys is generally appreciably higher than that of welds made with high-silicon 4043 filler alloy (Fig. 6.5a and 6.5b). Once again, there are a few exceptions, notably in joints between 6061-T6 plate and 356.0-T6 or T7 sand castings; in these cases, the high silicon in the 3*xx*.0 castings may be overwhelming the inherent high toughness of the 5*xxx* type filler alloys.

When welds in wrought alloys are evaluated on the basis of UPE versus TYS (Fig. 6.9), it is clear that the UPEs of the 1100 and 5*xxx* welds are as high or nearly as high as those of the comparable parent alloys. For filler alloy 2319, welds that have been post-weld aged or heat-treated and aged provide superior combinations of strength and toughness to those of as-welded samples. Filler alloy 4043 consistently provides less desirable combinations of strength and toughness.

A comparable analysis of welds in castings based upon UPE versus TYS is not available because joint yield strengths were not reported and a plot cannot be made. However, a scan of the data in Table 6.5 illustrates that welds made in castings with 4043 filler alloy have lower toughness than those made with 5356 filler alloy. As noted earlier, for the very high-silicon-bearing casting alloys, even 5556 welds have relatively low UPE, the high silicon overwhelming the beneficial effects of the high-magnesium filler alloy.

In general, one can conclude that for applications where high toughness is critical, 5*xxx* filler alloys would be recommended and 4043 filler alloy should be avoided.

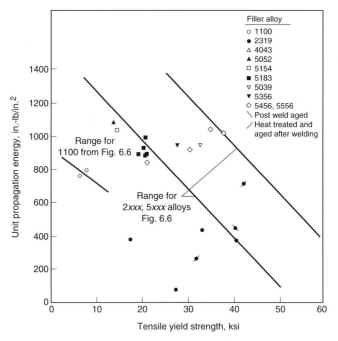

**Fig. 6.9** Unit propagation energy vs. tensile yield strength for welds in wrought aluminum alloys

**Table 6.1(a)  Results of tensile and tear tests of 0.063 in. thick non-heat-treated aluminum alloy sheet, longitudinal**

| Alloy and temper | Tensile tests | | | Tear tests | | | | | |
| | Ultimate tensile strength (UTS), ksi | Tensile yield strength (TYS), ksi | Elongation in 2 in., % | Tear strength, ksi | Ratio tear strength to yield strength, (TYR) | Energy required to: | | Total energy, in.-lb | Unit propagation energy, in.-lb/in.[2] |
| | | | | | | Initiate a crack, in.-lb | Propagate a crack, in.-lb | | |
| 1100-O | 14.6 | 5.2 | 35.0 | 19.5 | 3.75 | 27 | 46 | 73 | 725 |
| | 14.2 | 4.9 | 35.2 | 19.5 | 3.98 | 32 | 57 | 89 | 900 |
| Average 1100-O | 14.4 | 5.0 | 35.1 | 19.5 | 3.86 | 30 | 52 | 81 | 810 |
| 1100-H14 | 17.9 | 16.8 | 13.0 | 31.1 | 1.85 | 20 | 40 | 60 | 635 |
| 1100-H18 | 27.6 | 24.8 | 5.0 | 44.0 | 1.77 | 20 | 35 | 55 | 585 |
| | 27.7 | 26.3 | 5.5 | 43.2 | 1.64 | 17 | 40 | 57 | 630 |
| Average 1100-H18 | 27.6 | 25.6 | 5.2 | 43.6 | 1.70 | 18 | 38 | 56 | 610 |
| 3003-O | 17.0 | 7.1 | 34.5 | 24.6 | 3.47 | 36 | 55 | 91 | 865 |
| 3003-H14 | 21.2 | 20.4 | 11.5 | 35.4 | 1.74 | 19 | 39 | 58 | 600 |
| | 23.2 | 22.4 | 8.5 | 38.1 | 1.70 | 17 | 46 | 63 | 720 |
| Average 3003-H14 | 22.2 | 21.4 | 10.0 | 36.8 | 1.72 | 18 | 42 | 60 | 660 |
| 3003-H18 | 35.6 | 33.4 | 4.0 | 52.3 | 1.57 | 12 | 24 | 36 | 385 |
| | 33.5 | 32.7 | 4.8 | 50.7 | 0.10 | 20 | 34 | 54 | 530 |
| Average 3003-H18 | 34.6 | 33.0 | 4.4 | 51.5 | 1.56 | 16 | 29 | 45 | 460 |
| 3004-O | 27.1 | 11.4 | 22.2 | 33.8 | 2.96 | 33 | 49 | 81 | 750 |
| 3004-H34 | 35.4 | 29.4 | 8.5 | 51.1 | 1.74 | 17 | 39 | 56 | 600 |
| | 36.7 | 31.6 | 8.2 | 54.0 | 1.71 | 19 | 46 | 65 | 700 |
| Average 3004-H34 | 36.0 | 30.5 | 8.4 | 52.6 | 1.72 | 18 | 42 | 60 | 650 |
| 3004-H38 | 44.0 | 41.0 | 8.0 | 58.4 | 1.42 | 11 | 34 | 45 | 530 |
| | 45.0 | 42.4 | 8.0 | 59.1 | 1.39 | 14 | 39 | 53 | 620 |
| Average 3004-H38 | 44.5 | 41.7 | 8.0 | 58.8 | 1.40 | 12 | 36 | 49 | 575 |
| Alclad 3105-H14 | 31.7 | 29.2 | 7.0 | 47.9 | 1.64 | 19 | 38 | 67 | 600 |
| 5050-O | 22.6 | 9.2 | 24.5 | 28.6 | 3.11 | 39 | 48 | 87 | 785 |
| 5050-H34 | 26.4 | 22.8 | 8.8 | 41.9 | 1.84 | 20 | 46 | 66 | 710 |
| | 27.9 | 24.9 | 7.8 | 45.5 | 1.83 | 22 | 47 | 69 | 760 |
| Average 5050-H34 | 27.2 | 23.8 | 8.3 | 43.7 | 1.84 | 21 | 46 | 68 | 735 |
| 5050-H38 | 31.0 | 27.9 | 7.5 | 48.3 | 1.73 | 20 | 37 | 57 | 580 |
| 5052-O | 29.2 | 12.4 | 25.0 | 38.1 | 3.07 | 54 | 80 | 134 | 1240 |
| 5052-H34 | 37.7 | 31.0 | 10.0 | 58.0 | 1.87 | 28 | 55 | 83 | 865 |
| 5052-H38 | 43.6 | 38.3 | 9.5 | 62.8 | 1.64 | 24 | 32 | 56 | 500 |
| 5083-O | 44.3 | 21.9 | 22.7 | 48.5 | 2.21 | 38 | 87 | 125 | 1270 |
| 5083-H12 | 51.7 | 44.3 | 9.8 | 65.7 | 1.48 | 21 | 46 | 67 | 715 |
| 5083-HI4 | 53.4 | 46.7 | 8.0 | 65.7 | 1.41 | 20 | 42 | 62 | 570 |
| 5083-H24 | 49.7 | 32.9 | 16.2 | 60.7 | 1.84 | 34 | 67 | 101 | 970 |
| | 50.4 | 38.6 | 13.8 | 68.0 | 1.76 | 30 | 63 | 93 | 945 |
| Average 5083-H24 | 50.0 | 35.9 | 15.2 | 64.4 | 1.80 | 32 | 65 | 97 | 960 |
| 5083-H32 | 50.5 | 38.4 | 11.5 | 64.3 | 1.67 | 21 | 47 | 68 | 790 |
| 5083-H34 | 54.1 | 44.8 | 10.7 | 69.7 | 1.56 | 22 | 37 | 59 | 610 |
| 5086-O | 37.4 | 17.8 | 23.0 | 45.5 | 2.56 | 52 | 80 | 132 | 1220 |
| | 40.7 | 20.2 | 22.2 | 46.6 | 2.31 | 44 | 92 | 136 | 1385 |
| Average 5086-O | 39.0 | 19.0 | 22.6 | 46.0 | 2.44 | 48 | 86 | 134 | 1300 |
| 5086-H32 | 43.4 | 32.8 | 13.0 | 61.2 | 1.87 | 30 | 63 | 93 | 940 |
| | 46.1 | 34.2 | 12.5 | 61.6 | 1.80 | 29 | 62 | 91 | 1005 |
| Average 5086-H32 | 44.8 | 33.5 | 12.8 | 61.4 | 1.84 | 30 | 62 | 92 | 970 |
| 5086-H34 | 45.6 | 36.7 | 11.0 | 65.8 | 1.79 | 24 | 52 | 76 | 800 |
| | 48.7 | 39.7 | 10.2 | 64.9 | 1.63 | 20 | 48 | 68 | 760 |
| | 47.5 | 36.9 | 11.8 | 64.0 | 1.73 | 24 | 59 | 83 | 915 |
| | 48.8 | 37.6 | 11.8 | 66.0 | 1.76 | 26 | 56 | 82 | 870 |
| Average 5086-H34 | 47.6 | 37.7 | TI-2 | 65.2 | 1.73 | 24 | 54 | 78 | 840 |
| 5154-O | 35.4 | 16.4 | 24.5 | 44.9 | 2.74 | 41 | 106 | 147 | 1630 |
| | 34.0 | 15.5 | 23.0 | 43.4 | 2.80 | 42 | 92 | 134 | 1440 |
| Average 5154-O | 34.7 | 16.0 | 23.8 | 44.2 | 2.77 | 42 | 99 | 140 | 1535 |

(continued)

Each line represents the average of duplicate tear tests (Fig. A1.8) of an individual lot of sheet. For tensile yield strength, offset is 0.2%.

**Table 6.1(a)   (continued)**

| Alloy and temper | Tensile tests | | | Tear tests | | | | | |
|---|---|---|---|---|---|---|---|---|---|
| | Ultimate tensile strength (UTS), ksi | Tensile yield strength (TYS), ksi | Elongation in 2 in., % | Tear strength, ksi | Ratio tear strength to yield strength (TYR) | Energy required to: | | Total energy, in.-lb | Unit propagation energy, in.-lb/in.$^2$ |
| | | | | | | Initiate a crack, in.-lb | Propagate a crack, in.-lb | | |
| 5154-H32 | 40.5 | 30.6 | 13.5 | 59.7 | 1.95 | 25 | 77 | 102 | 1185 |
| 5154-H34 | 41.3 | 32.3 | 12.0 | 60.0 | 1.86 | 28 | 58 | 86 | 915 |
| 5154-H38 | 46.8 | 40.5 | 9.5 | 68.8 | 1.70 | 25 | 35 | 60 | 545 |
| | 49.4 | 42.8 | 9.8 | 66.6 | 1.56 | 21 | 38 | 59 | 595 |
| Average 5154-H38 | 48.1 | 41.6 | 9.6 | 67.7 | 1.63 | 23 | 36 | 60 | 570 |
| 5356-O | 41.4 | 21.0 | 26.5 | 50.0 | 2.38 | 49 | 87 | 136 | 1360 |
| 5454-O | 37.4 | 15.8 | 22.5 | 45.5 | 2.88 | 46 | 80 | 126 | 1210 |
| | 37.5 | 15.6 | 22.0 | 45.8 | 2.94 | 48 | 76 | 124 | 1160 |
| | 36.6 | 16.8 | 19.0 | 44.5 | 2.65 | 45 | 99 | 144 | 1545 |
| Average 5454-O | 37.2 | 16.1 | 21.2 | 45.3 | 2.82 | 46 | 85 | 131 | 1305 |
| 5454-H32 | 40.2 | 32.0 | 11.8 | 57.8 | 1.81 | 27 | 58 | 85 | 905 |
| | 40.6 | 32.6 | 11.0 | 60.6 | 1.86 | 28 | 57 | 85 | 905 |
| Average 5454-H32 | 40.4 | 32.3 | 11.4 | 59.2 | 1.84 | 28 | 58 | 85 | 905 |
| 5454-H34 | 45.6 | 39.9 | 10.5 | 66.2 | 1.66 | 23 | 52 | 75 | 810 |
| | 45.4 | 37.0 | 10.5 | 66.5 | 1.80 | 17 | 59 | 76 | 910 |
| Average 5454-H34 | 45.5 | 38.4 | 10.5 | 66.4 | 1.73 | 20 | 56 | 76 | 860 |
| 5456-O | 47.4 | 24.3 | 21.8 | 51.7 | 2.13 | 33 | 78 | 111 | 1235 |
| | 48.6 | 24.1 | 22.5 | 51.9 | 2.15 | 36 | 77 | 113 | 1155 |
| | 50.0 | 25.6 | 20.8 | 52.5 | 2.05 | 37 | 85 | 122 | 1135 |
| Average 5456-O | 48.7 | 24.7 | 21.7 | 5.2 | 2.11 | 35 | 80 | 115 | 1175 |
| 5456-H12 | 56.2 | 42.9 | 12.2 | 66.2 | 1.54 | 18 | 44 | 62 | 645 |
| 5456-H14 | 60.6 | 52.4 | 8.2 | 64.5 | 1.23 | 13 | 36 | 49 | 540 |
| 5456-H24 | 54.8 | 43.0 | 12.0 | 57.8 | 1.34 | 13 | 26 | 39 | 405 |
| | 53.7 | 39.5 | 13.2 | 55.5 | 1.40 | 12 | 27 | 39 | 425 |
| Average 5456-H24 | 54.2 | 41.2 | 12.6 | 56.6 | 1.37 | 12 | 26 | 39 | 415 |
| 5456-H32 | 56.0 | 42.7 | 12.0 | 68.5 | 1.60 | 22 | 41 | 63 | 620 |
| | 55.9 | 41.6 | 12.2 | 66.1 | 1.59 | 18 | 46 | 64 | 665 |
| Average 5456-H32 | 56.0 | 42.2 | 12.1 | 67.3 | 1.60 | 20 | 44 | 64 | 640 |
| 5456-H34 | 57.7 | 46.0 | 10.5 | 69.8 | 1.52 | 15 | 33 | 48 | 510 |
| | 59.5 | 47.5 | 10.7 | 67.8 | 1.43 | 18 | 34 | 52 | 490 |
| Average 5456-H34 | 58.6 | 46.8 | 10.6 | 68.8 | 1.48 | 16 | 34 | 50 | 500 |
| 5456-H343 | 56.2 | 48.7 | 9.0 | 58.9 | 1.21 | 10 | 26 | 36 | 390 |

Each line represents the average of duplicate tear tests (Fig. A1.8) of an individual lot of sheet. For tensile yield strength, offset is 0.2%.

**Table 6.1(b)  Results of tensile and tear tests of 0.063 in. thick non-heat-treated aluminum alloy sheet, transverse**

| | Tensile tests | | | Tear tests | | | | | | |
|---|---|---|---|---|---|---|---|---|---|---|
| | | | | | Ratio tear strength to yield strength, (TYR) | Energy required to: | | | |
| Alloy and temper | Ultimate tensile strength (UTS), ksi | Tensile yield strength (TYS), ksi | Elongation in 2 in., % | Tear strength, ksi | | Initiate a crack, in.-lb | Propagate a crack, in.-lb | Total energy, in.-lb | Unit propagation energy, in.-lb/in.$^2$ |
| 1100-O | 14.3 | 5.4 | 39.8 | 18.9 | 3.50 | 29 | 47 | 76 | 740 |
| | 14.0 | 5.3 | 41.3 | 20.0 | 3.78 | 38 | 64 | 102 | 1015 |
| Average 1100-O | 14.2 | 5.4 | 40.6 | 19.4 | 3.64 | 34 | 56 | 89 | 875 |
| 1100-H14 | 18.0 | 16.4 | 9.2 | 31.7 | 1.93 | 18 | 36 | 54 | 570 |
| 1100-H18 | 28.5 | 26.3 | 5.0 | 47.3 | 1.80 | 28 | 31 | 59 | 515 |
| | 29.0 | 26.6 | 5.5 | 45.5 | 1.71 | 19 | 16 | 35 | 255 |
| Average 1100-H18 | 28.8 | 26.4 | 5.2 | 46.4 | 1.76 | 24 | 24 | 47 | 385 |
| 3003-O | 16.5 | 7.3 | 30.8 | 24.3 | 3.33 | 35 | 66 | 101 | 1040 |
| 3003-H14 | 21.4 | 20.0 | 8.0 | 36.3 | 1.81 | 20 | 40 | 60 | 615 |
| | 23.1 | 21.6 | 5.8 | 37.4 | 1.73 | 17 | 46 | 63 | 720 |
| Average 3003-H14 | 22.2 | 20.8 | 6.9 | 36.8 | 1.77 | 18 | 43 | 62 | 670 |
| 3003-H18 | 36.0 | 31.6 | 4.2 | 52.3 | 1.66 | 12 | 9 | 21 | 145 |
| | 33.8 | 31.4 | 5.0 | 52.8 | 1.68 | 18 | 26 | 44 | 405 |
| Average 3003-H18 | 34.9 | 31.5 | 4.6 | 52.6 | 1.67 | 15 | 18 | 33 | 275 |
| 3004-O | 26.6 | 11.5 | 22.8 | 35.6 | 3.10 | 36 | 61 | 97 | 950 |
| 3004-H34 | 36.4 | 29.0 | 8.8 | 52.6 | 1.82 | 18 | 37 | 55 | 570 |
| | 36.5 | 29.5 | 9.0 | 54.8 | 1.86 | 22 | 40 | 62 | 610 |
| Average 3004-H34 | 36.4 | 29.2 | 8.9 | 53.7 | 1.84 | 20 | 38 | 58 | 590 |
| 3004-H38 | 44.4 | 39.6 | 7.8 | 57.8 | 1.46 | 12 | 25 | 37 | 390 |
| | 44.5 | 40.6 | 7.5 | 62.9 | 1.55 | 17 | 23 | 40 | 365 |
| Average 3004-H38 | 44.4 | 40.1 | 7.6 | 60.4 | 1.50 | 14 | 24 | 38 | 380 |
| Alclad 3105-H14 | 31.8 | 29.4 | 6.5 | 48.2 | 1.64 | 19 | 34 | 53 | 535 |
| 5050-O | 22.4 | 9 | 28.5 | 28.4 | 3.16 | 41 | 59 | 100 | 970 |
| 5050-H34 | 26.9 | 22.4 | 8.2 | 44.4 | 1.98 | 22 | 45 | 67 | 690 |
| | 29.2 | 24.4 | 7.8 | 47.8 | 1.96 | 33 | 48 | 81 | 775 |
| Average 5050-H34 | 28.0 | 23.4 | 8.0 | 46.1 | 1.97 | 28 | 46 | 74 | 730 |
| 5050-H38 | 33.0 | 29.4 | 8.1 | 51.5 | 1.75 | 22 | 31 | 53 | 485 |
| 5052-O | 28.8 | 12.4 | 26.0 | 37.5 | 3.02 | 53 | 81 | 134 | 1255 |
| 5052-H34 | 38.1 | 30.0 | 11.8 | 58.6 | 1.95 | 29 | 54 | 83 | 850 |
| 5052-H38 | 44.4 | 38.4 | 11.2 | 66.2 | 1.72 | 24 | 31 | 55 | 485 |
| 5083-O | 44.1 | 22.9 | 23.2 | 49.1 | 2.14 | 40 | 84 | 124 | 1230 |
| 5083-H12 | 51.8 | 40.4 | 9.8 | 65.4 | 1.62 | 23 | 40 | 63 | 620 |
| 5083-HI4 | 53.7 | 42.8 | 9.8 | 66.8 | 1.56 | 21 | 38 | 59 | 515 |
| 5083-H24 | 49.5 | 32.8 | 19.2 | 59.8 | 1.82 | 26 | 57 | 83 | 825 |
| | 51.5 | 41.1 | 17.8 | 70.1 | 1.71 | 31 | 53 | 84 | 795 |
| Average 5083-H24 | 50.5 | 37.0 | 15.3 | 65.0 | 1.76 | 28 | 55 | 84 | 810 |
| 5083-H32 | 50.1 | 35.9 | 13.2 | 63.0 | 1.75 | 22 | 47 | 69 | 790 |
| 5083-H34 | 55.2 | 42.9 | 11.8 | 64.8 | 1.51 | 15 | 31 | 46 | 510 |
| 5086-O | 36.3 | 17.6 | 26.0 | 44.6 | 2.53 | 47 | 77 | 124 | 1175 |
| | 40.8 | 20.9 | 24.2 | 48.1 | 2.30 | 52 | 84 | 136 | 1265 |
| Average 5086-O | 38.6 | 19.2 | 25.1 | 46.4 | 2.42 | 50 | 80 | 130 | 1220 |
| 5086-H32 | 45.0 | 29.6 | 14.5 | 59.1 | 2.00 | 30 | 62 | 92 | 925 |
| | 45.1 | 31.2 | 15.5 | 60.9 | 1.95 | 28 | 65 | 93 | 1055 |
| Average 5086-H32 | 45.0 | 30.4 | 15.0 | 60.0 | 1.98 | 29 | 64 | 92 | 990 |
| 5086-H34 | 46.0 | 34.3 | 14.0 | 66.8 | 1.95 | 24 | 53 | 77 | 815 |
| | 49.2 | 37.2 | 12.5 | 64.6 | 1.74 | 21 | 48 | 69 | 760 |
| | 47.5 | 34.6 | 14.8 | 64.0 | 1.85 | 24 | 60 | 84 | 930 |
| | 49.4 | 37.4 | 14.3 | 67.3 | 1.80 | 25 | 55 | 80 | 855 |
| Average 5086-H34 | 48.0 | 35.9 | 13.9 | 65.7 | 1.84 | 24 | 54 | 78 | 840 |
| 5154-O | 34.6 | 16.0 | 25.0 | 43.1 | 2.69 | 56 | 94 | 150 | 1445 |
| | 34.1 | 15.6 | 24.5 | 42:5 | 2.72 | 42 | 83 | 125 | 1300 |
| Average 5154-O | 34.4 | 15.8 | 24.8 | 42.8 | 2.70 | 49 | 88 | 138 | 1370 |
| 5154-H32 | 39.5 | 28.3 | 15.5 | 60.6 | 2.14 | 29 | 75 | 104 | 1150 |
| 5154-H34 | 41.7 | 31.0 | 13.0 | 60.9 | 1.96 | 26 | 64 | 90 | 1000 |
| 5154-H38 | 47.1 | 40.0 | 13.0 | 70.9 | 1.77 | 26 | 29 | 55 | 455 |
| | 49.9 | 42.6 | 14.2 | 70.3 | 1.65 | 23 | 39 | 62 | 610 |

(continued)

Each line represents the average of duplicate tear tests (Fig. A1.8) of an individual lot of sheet. For tensile yield strength, offset is 0.2%.

**Table 6.1(b)** **(continued)**

| Alloy and temper | Tensile tests | | | Tear tests | | | | | |
|---|---|---|---|---|---|---|---|---|---|
| | Ultimate tensile strength (UTS), ksi | Tensile yield strength (TYS), ksi | Elongation in 2 in., % | Tear strength, ksi | Ratio tear strength to yield strength, (TYR) | Energy required to: | | Total energy, in.-lb | Unit propagation energy, in.-lb/in.$^2$ |
| | | | | | | Initiate a crack, in.-lb | Propagate a crack, in.-lb | | |
| Average 5154-H38 | 48.5 | 41.3 | 13.6 | 70.6 | 1.71 | 24 | 34 | 58 | 530 |
| 5356-O | 40.9 | 20.7 | 27.5 | 48.8 | 2.36 | 46 | 74 | 120 | 1155 |
| 5454-O | 35.5 | 15.7 | 19.5 | 45.2 | 2.88 | 45 | 82 | 127 | 1240 |
| | 36.2 | 15.4 | 20.2 | 45.8 | 2.97 | 50 | 74 | 124 | 1130 |
| | 36.2 | 17.6 | 20.5 | 43.6 | 2.48 | 42 | 93 | 135 | 1455 |
| Average 5454-O | 36.0 | 16.2 | 20.1 | 44.9 | 2.78 | 46 | 83 | 129 | 1275 |
| 5454-H32 | 40.7 | 30.4 | 12.8 | 59.4 | 1.95 | 37 | 58 | 95 | 905 |
| | 41.0 | 30.9 | 12.0 | 61.9 | 2.00 | 32 | 67 | 99 | 1065 |
| Average 5454-H32 | 40.8 | 30.6 | 12.4 | 60.6 | 1.98 | 34 | 62 | 97 | 985 |
| 5454-H34 | 47.4 | 38.3 | 9.5 | 68.4 | 1.79 | 26 | 53 | 79 | 830 |
| | 47.1 | 37.3 | 10.5 | 62.8 | 1.68 | 16 | 50 | 66 | 770 |
| Average 5454-H34 | 47.2 | 37.8 | 10.0 | 65.6 | 1.74 | 21 | 52 | 72 | 800 |
| 5456-O | 46.9 | 25.7 | 24.2 | 53.5 | 2.09 | 31 | 76 | 107 | 1195 |
| | 46.5 | 24.7 | 24.0 | 52.2 | 2.11 | 45 | 62 | 107 | 925 |
| | 47.3 | 24.7 | 23.5 | 51.7 | 2.09 | 33 | 71 | 104 | 945 |
| Average 5456-O | 46.9 | 25.0 | 23.9 | 52.5 | 2.10 | 36 | 70 | 106 | 1020 |
| 5456-H12 | 55.5 | 39.1 | 14.2 | 63.5 | 1.62 | 18 | 37 | 55 | 545 |
| 5456-H14 | 60.8 | 48.5 | 9.5 | 63.6 | 1.31 | 13 | 24 | 37 | 360 |
| 5456-H24 | 55.0 | 38.6 | 14.5 | 58.5 | 1.52 | 13 | 29 | 42 | 440 |
| | 54.0 | 38.2 | 15.2 | 60.2 | 1.58 | 10 | 26 | 36 | 410 |
| Average 5456-H24 | 54.5 | 38.4 | 14.8 | 59.4 | 1.55 | 12 | 28 | 39 | 425 |
| 5456-H32 | 54.1 | 37.5 | 13.5 | 65.8 | 1.75 | 18 | 38 | 56 | 575 |
| | 54.8 | 38.7 | 14.2 | 64.6 | 1.67 | 19 | 40 | 59 | 575 |
| Average 5456-H32 | 54.4 | 38.1 | 13.8 | 65.2 | 1.71 | 18 | 39 | 58 | 575 |
| 5456-H34 | 58.5 | 44.1 | 12.5 | 69.8 | 1.58 | 14 | 28 | 42 | 430 |
| | 59.5 | 45.2 | 12.2 | 66.7 | 1.48 | 19 | 31 | 50 | 450 |
| Average 5456-H34 | 59.0 | 44.6 | 12.4 | 682 | 1.53 | 16 | 30 | 46 | 440 |
| 5456-H343 | 56.7 | 42.7 | 9.5 | 58.0 | 1.31 | 8 | 20 | 28 | 300 |

Each line represents the average of duplicate tear tests (Fig. A1.8) of an individual lot of sheet. For tensile yield strength, offset is 0.2%.

**Table 6.2(a)   Results of tensile and tear tests of 0.063 in. thick heat treated aluminum alloy sheet, longitudinal**

| Alloy and temper | Tensile tests | | | Tear tests | | | | | |
|---|---|---|---|---|---|---|---|---|---|
| | Ultimate tensile strength (UTS), ksi | Tensile yield strength (TYS), ksi | Elongation in 2 in., % | Tear strength, ksi | Ratio tear strength to yield strength, (TYR) | Initiate a crack, in.-lb | Propagate a crack, in.-lb | Total energy, in.-lb | Unit propagation energy, in.-lb/in.$^2$ |
| Alclad 2014-T3 | 62.3 | 42.6 | 21.0 | 70.2 | 1.65 | 24 | 50 | 74 | 770 |
| | 65.3 | 42.7 | 19.0 | 70.6 | 1.65 | 20 | 46 | 66 | 725 |
| | 63.4 | 44.9 | 20.5 | 70.7 | 1.57 | 22 | 50 | 72 | 800 |
| Average Alclad 2014-T3 | 63.7 | 43.4 | 20.2 | 70.5 | 1.62 | 22 | 49 | 71 | 765 |
| 2014-T6 | 72.8 | 67.6 | 11.2 | 68.0 | 1.01 | 10 | 13 | 23 | 205 |
| | 72.5 | 66.2 | 10.0 | 67.3 | 1.02 | 10 | 13 | 23 | 195 |
| | 71.8 | 66.1 | 11.0 | 66.6 | 1.01 | 9 | 15 | 24 | 240 |
| | 69.2 | 63.1 | 9.5 | 66.0 | 1.05 | 10 | 18 | 28 | 285 |
| | 69.3 | 63.8 | 9.5 | 70.2 | 1.10 | 7 | 16 | 23 | 255 |
| | 74.2 | 67.5 | 11.2 | 66.2 | 0.98 | 11 | 22 | 33 | 335 |
| Average 2014-T6 | 71.6 | 65.7 | 10.4 | 67.4 | 1.03 | 10 | 16 | 26 | 250 |
| Alclad 2014-T6 | 72.4 | 66.2 | 11.0 | 59.4 | 0.90 | 8 | 11 | 19 | 170 |
| | 68.4 | 62.7 | 9.8 | 70.8 | 1.13 | 9 | 19 | 28 | 290 |
| | 68.3 | 61.9 | 11.0 | 74.5 | 1.20 | 14 | 21 | 35 | 340 |
| | 68.0 | 61.9 | 10.8 | 74.3 | 1.20 | 12 | 21 | 33 | 335 |
| Average Alclad 2014-T6 | 69.3 | 63.2 | 10.6 | 69.8 | 1.11 | 11 | 18 | 29 | 285 |
| 2020-O(a) | 28.6 | 9.9 | 20.5 | 33.0 | 3.33 | 26 | 44 | 70 | 695 |
| Alclad 2020-O(a) | 28.0 | 10.7 | 21.0 | 34.6 | 3.23 | 33 | 63 | 96 | 1000 |
| 2020-T4(a) | 50.0 | 34.2 | 16.5 | 64.1 | 1.87 | 28 | 70 | 98 | 1110 |
| 2020-T6(a) | 82.7 | 77.8 | 6.0 | 48.3 | 0.63 | 5 | 1 | 6 | 15 |
| | 94.8 | 80.4 | 7.0 | 42.2 | 0.52 | 4 | 0 | 4 | 0 |
| | 81.4 | 77.2 | 8.0 | 52.1 | 0.67 | 3 | 5 | 8 | 80 |
| | 80.2 | 75.9 | 7.8 | 50.2 | 0.66 | 4 | 4 | 8 | 65 |
| | 81.1 | 76.1 | 8.2 | 36.0 | 0.47 | 3 | 0 | 3 | 0 |
| Average 2020-T6 | 82.0 | 77.5 | 7.4 | 45.8 | 0.59 | 4 | 2 | 6 | 30 |
| Alclad 2020-T6(a) | 73.5 | 68.0 | 7.2 | 45.1 | 0.66 | 3 | 2 | 5 | 30 |
| 2024-T4 | 69.7 | 54.6 | 18.8 | 77.5 | 1.42 | 22 | 42 | 64 | 655 |
| | 69.6 | 45.8 | 21.5 | 77.8 | 1.70 | 20 | 43 | 63 | 665 |
| | 70.4 | 48.1 | 20.0 | 79.0 | 1.64 | 18 | 38 | 56 | 595 |
| | 69.0 | 44.2 | 20.8 | 77.2 | 1.75 | 26 | 57 | 83 | 900 |
| Average 2024-T4 | 69.7 | 48.2 | 20.3 | 77.9 | 1.63 | 22 | 45 | 67 | 705 |
| Alclad 2024-T4 | 62.4 | 42.4 | 21.5 | 67.7 | 1.60 | 21 | 50 | 71 | 770 |
| | 66.5 | 42.0 | 20.5 | 74.5 | 1.77 | 26 | 54 | 80 | 830 |
| | 66.5 | 45.4 | 20.5 | 71.1 | 1.57 | 19 | 46 | 65 | 725 |
| Average Alclad 2024-T4 | 65.1 | 43.3 | 20.8 | 71.1 | 1.65 | 22 | 50 | 72 | 775 |
| 2024-T3 | 68.2 | 50.0 | 19.2 | 75.7 | 1.51 | 24 | 46 | 70 | 725 |
| | 68.1 | 49.9 | 20.2 | 75.7 | 1.52 | 19 | 42 | 61 | 650 |
| | 70.4 | 55.8 | 18.2 | 74.2 | 1.33 | 17 | 35 | 52 | 565 |
| | 68.4 | 48.1 | 20.5 | 76.5 | 1.59 | 17 | 44 | 61 | 700 |
| | 69.7 | 53.0 | 20.0 | 76.9 | 1.45 | 16 | 44 | 60 | 705 |
| | 69.6 | 53.8 | 18.2 | 78.0 | 1.45 | 18 | 42 | 60 | 690 |
| | 71.0 | 53.5 | 20.5 | 81.1 | 1.52 | 23 | 63 | 86 | 950 |
| | 71.5 | 55.5 | 19.2 | 74.8 | 1.35 | 16 | 42 | 58 | 685 |
| Average 2024-T3 | 69.6 | 52.4 | 19.5 | 76.6 | 1.46 | 19 | 45 | 64 | 710 |
| Alclad 2024-T3 | 69.2 | 53.2 | 19.2 | 74.9 | 1.41 | 20 | 38 | 58 | 580 |
| | 69.4 | 54.4 | 17.2 | 78.4 | 1.44 | 18 | 34 | 52 | 540 |
| | 71.0 | 52.3 | 20.8 | 71.9 | 1.37 | 17 | 33 | 50 | 515 |
| | 67.8 | 52.0 | 20.0 | 71.0 | 1.37 | 16 | 32 | 48 | 510 |
| | 71.1 | 53.8 | 20.8 | 75.3 | 1.40 | 17 | 36 | 53 | 560 |
| | 69.8 | 51.1 | 20.0 | 73.0 | 1.43 | 19 | 35 | 54 | 555 |
| | 67.3 | 52.1 | 19.0 | 73.0 | 1.41 | 21 | 36 | 57 | 580 |
| | 66.2 | 47.8 | 20.2 | 73.8 | 1.54 | 10 | 33 | 43 | 530 |
| | 64.6 | 47.0 | 18.8 | 70.9 | 1.51 | 12 | 31 | 43 | 500 |
| | 65.2 | 45.4 | 19.0 | 74.6 | 1.64 | 14 | 32 | 46 | 515 |
| | 67.7 | 48.8 | 19.2 | 74.3 | 1.52 | 15 | 39 | 54 | 630 |
| | 66.0 | 51.5 | 18.8 | 76.4 | 1.48 | 20 | 39 | 59 | 610 |
| Average 2024-T3 | 67.9 | 50.8 | 19.4 | 74.0 | 1.46 | 17 | 35 | 51 | 550 |
| 2024-T36 | 78.0 | 67.0 | 15.8 | 69.8 | 1.04 | 11 | 17 | 28 | 265 |
| | 76.4 | 64.6 | 15.0 | 80.0 | 1.24 | 18 | 27 | 45 | 410 |
| | 76.3 | 65.1 | 16.5 | 78.8 | 1.21 | 17 | 27 | 44 | 410 |

(continued)

Each line represents the average of duplicate tear tests (Fig. A1.8) of an individual lot of sheet. For tensile yield strength, offset is 0.2%. (a) Obsolete alloy

**Table 6.2(a)** (continued)

| Alloy and temper | Tensile tests | | | Tear tests | | | | | |
|---|---|---|---|---|---|---|---|---|---|
| | Ultimate tensile strength (UTS), ksi | Tensile yield strength (TYS), ksi | Elongation in 2 in., % | Tear strength, ksi | Ratio tear strength to yield strength (TYR) | Initiate a crack, in.-lb | Propagate a crack, in.-lb | Total energy, in.-lb | Unit propagation energy, in.-lb/in.$^2$ |
| | 73.4 | 63.2 | 15.0 | 77.2 | 1.22 | 13 | 29 | 42 | 445 |
| | 74.0 | 61.6 | 14.5 | 75.8 | 1.23 | 14 | 27 | 41 | 430 |
| | 74.6 | 62.0 | 14.0 | 77.5 | 1.25 | 15 | 29 | 44 | 460 |
| | 72.8 | 61.8 | 14.8 | 81.9 | 1.33 | 15 | 34 | 49 | 540 |
| Average 2024-T36 | 75.1 | 63.6 | 15.1 | 77.3 | 1.22 | 15 | 27 | 42 | 425 |
| Alclad 2024-T36 | 74.8 | 63.6 | 15.8 | 70.2 | 1.04 | 12 | 20 | 32 | 310 |
| | 72.0 | 63.1 | 14.8 | 73.1 | 1.16 | 15 | 25 | 40 | 390 |
| | 71.2 | 60.7 | 16.5 | 75.6 | 1.25 | 16 | 35 | 51 | 510 |
| | 69.7 | 57.4 | 16.0 | 75.0 | 1.30 | 16 | 36 | 52 | 535 |
| | 70.4 | 58.9 | 16.8 | 73.2 | 1.24 | 12 | 27 | 39 | 405 |
| Average 2024-T36 | 71.6 | 60.7 | 16.0 | 73.4 | 1.20 | 14 | 29 | 43 | 430 |
| 2024-T6 | 67.2 | 53.2 | 9.5 | 64.2 | 1.21 | 10 | 17 | 27 | 275 |
| 2024-T81 | 72.7 | 68.0 | 6.5 | 63.4 | 0.93 | 8 | 11 | 19 | 175 |
| | 77.3 | 73.4 | 7.0 | 58.2 | 0.79 | 6 | 10 | 16 | 160 |
| | 72.6 | 68.0 | 6.2 | 61.0 | 0.90 | 9 | 11 | 20 | 180 |
| Average 2024-T81 | 74.2 | 69.8 | 6.6 | 60.9 | 0.87 | 8 | 11 | 18 | 170 |
| Alclad 2024-T81 | 66.8 | 60.4 | 6.5 | 69.4 | 1.15 | 11 | 17 | 28 | 275 |
| | 68.1 | 62.2 | 7.0 | 66.8 | 1.07 | 9 | 15 | 24 | 240 |
| | 70.3 | 65.8 | 7.0 | 64.8 | 0.98 | 12 | 11 | 23 | 170 |
| Average Alclad 2024-T81 | 68.4 | 62.8 | 6.8 | 67.0 | 1.07 | 11 | 14 | 25 | 230 |
| 2024-T86 | 77.2 | 72.4 | 6.5 | 58.3 | 0.81 | 6 | 8 | 14 | 120 |
| | 77.0 | 72.5 | 6.2 | 57.7 | 0.80 | 8 | 9 | 17 | 140 |
| Average 2024-T86 | 77.1 | 72.4 | 6.4 | 58.0 | 0.80 | 7 | 8 | 16 | 130 |
| Alclad 2024-T86 | 73.6 | 67.8 | 6.8 | 64.5 | 0.95 | 6 | 12 | 18 | 180 |
| | 75.6 | 71.0 | 6.8 | 60.9 | 0.86 | 7 | 11 | 18 | 170 |
| | 73.3 | 67.4 | 6.5 | 63.5 | 0.94 | 8 | 11 | 19 | 170 |
| | 69.8 | 65.0 | 6.5 | 59.4 | 0.91 | 7 | 12 | 19 | 185 |
| | 70.4 | 65.8 | 5.8 | 59.5 | 0.90 | 9 | 9 | 18 | 145 |
| | 74.4 | 71.6 | 6.0 | 55.0 | 0.77 | 5 | 7 | 12 | 105 |
| Average Alclad 2024-T86 | 72.8 | 68.1 | 6.4 | 60.5 | 0.89 | 7 | 10 | 17 | 160 |
| 2219-T4 | 55.4 | 37.0 | 21.0 | 72.4 | 1.96 | 36 | 92 | 128 | 1460 |
| 2219-T31 | 55.6 | 44.0 | 17.2 | 74.5 | 1.69 | 24 | 67 | 91 | 1065 |
| 2219-T37 | 61.9 | 54.2 | 9.0 | 83.0 | 1.53 | 19 | 42 | 61 | 690 |
| 2219-T62 | 54.6 | 35.8 | 9.0 | 63.3 | 1.77 | 18 | 37 | 55 | 565 |
| | 60.8 | 42.5 | 10.0 | 68.0 | 1.60 | 15 | 36 | 51 | 580 |
| Average 2219-T62 | 57.7 | 39.2 | 9.5 | 65.6 | 1.68 | 16 | 36 | 53 | 570 |
| 2219-T81 | 67.9 | 53.0 | 9.2 | 70.4 | 1.33 | 13 | 23 | 36 | 370 |
| | 64.9 | 51.8 | 10.0 | 72.4 | 1.40 | 14 | 26 | 40 | 410 |
| Average 2219-T81 | 66.4 | 52.4 | 9.6 | 71.4 | 1.36 | 14 | 24 | 38 | 390 |
| 2219-T87 | 71.5 | 59.2 | 9.2 | 65.7 | 1.11 | 10 | 15 | 25 | 235 |
| | 67.9 | 56.2 | 9.8 | 75.2 | 1.34 | 17 | ... | ... | ... |
| Average 2219-T87 | 69.7 | 57.7 | 9.5 | 70.4 | 1.22 | 14 | 15 | 25 | 235 |
| 2618-T6 | 61.3 | 56.2 | 6.2 | 66.1 | 1.18 | 12 | 17 | 29 | 270 |
| 6061-T4 | 38.4 | 26.9 | 21.0 | 53.4 | 1.99 | 24 | 66 | 90 | 1080 |
| | 38.8 | 27.6 | 20.0 | 52.8 | 1.91 | 24 | 62 | 86 | 1000 |
| | 37.1 | 26.2 | 20.5 | 53.5 | 2.04 | 27 | 71 | 98 | 1145 |
| | 35.9 | 26.2 | 19.2 | 57.7 | 2.20 | 24 | 71 | 95 | 1220 |
| | 37.1 | 29.6 | 17.5 | 54.8 | 1.85 | 25 | 66 | 91 | 1100 |
| | 36.8 | 26.2 | 20.0 | 51.6 | 1.97 | 30 | 79 | 109 | 1275 |
| Average 6061-T4 | 37.4 | 27.1 | 19.7 | 54.0 | 1.99 | 26 | 69 | 95 | 1135 |
| 6061-T6 | 46.6 | 43.4 | 11.5 | 65.5 | 1.51 | 16 | 56 | 72 | 910 |
| | 45.2 | 41.9 | 11.5 | 64.1 | 1.53 | 19 | 52 | 71 | 825 |
| | 47.0 | 44.2 | 12.0 | 68.3 | 1.55 | 25 | 46 | 71 | 725 |
| | 43.5 | 40.9 | 11.2 | 66.0 | 1.61 | 21 | 56 | 77 | 915 |
| | 46.6 | 42.4 | 12.5 | 66.9 | 1.58 | 20 | 54 | 74 | 845 |
| | 45.0 | 41.8 | 11.0 | 67.9 | 1.62 | 22 | 58 | 80 | 920 |
| | 46.8 | 44.0 | 10.8 | 68.7 | 1.56 | 19 | 53 | 72 | 855 |
| | 44.1 | 37.6 | 15.5 | 65.9 | 1.75 | 23 | 0 | 89 | 1090 |
| | 44.6 | 37.0 | 15.8 | 65.2 | 1.76 | 23 | 66 | 89 | 1040 |
| Average 6061-T6 | 45.5 | 41.5 | 12.4 | 66.5 | 1.61 | 21 | 56 | 77 | 900 |
| Alclad 6061-T6 | 43.2 | 40.4 | 11.0 | 63.7 | 1.58 | 22 | 54 | 76 | 840 |
| | 43.5 | 40.2 | 11.0 | 60.0 | 1.49 | 21 | 51 | 72 | 800 |
| | 43.4 | 41.0 | 11.8 | 63.9 | 1.56 | 20 | 52 | 72 | 855 |
| | 44.4 | 39.5 | 15.0 | 64.4 | 1.63 | 25 | 54 | 79 | 855 |

(continued)

Each line represents the average of duplicate tear tests (Fig. A1.8) of an individual lot of sheet. For tensile yield strength, offset is 0.2%. (a) Obsolete alloy

**Table 6.2(a)   (continued)**

| Alloy and temper | Tensile tests | | | Tear tests | | | | | |
|---|---|---|---|---|---|---|---|---|---|
| | Ultimate tensile strength (UTS), ksi | Tensile yield strength (TYS), ksi | Elongation in 2 in., % | Tear strength, ksi | Ratio tear strength to yield strength, (TYR) | Initiate a crack, in.-lb | Propagate a crack, in.-lb | Total energy, in.-lb | Unit propagation energy, in.-lb/in.$^2$ |
| | 42.0 | 36.6 | 15.5 | 61.1 | 1.67 | 22 | 59 | 81 | 930 |
| | 42.4 | 37.6 | 13.5 | 63.9 | 1.70 | 24 | 61 | 85 | 950 |
| Average Alclad 6061-T6 | 43.2 | 39.2 | 13.0 | 62.8 | 1.60 | 22 | 55 | 78 | 870 |
| 6066-T6 | 59.4 | 51.4 | 12.0 | 69.5 | 1.35 | 14 | 24 | 38 | 370 |
| 6071-T4 | 49.2 | 34.8 | 22.5 | 61.9 | 1.78 | 24 | 55 | 79 | 875 |
| 6071-T6(a) | 57.4 | 54.7 | 10.2 | 60.6 | 1.11 | 8 | 16 | 24 | 250 |
| X7002-T6(a) | 69.8 | 61.8 | 11.5 | 91.2 | 1.48 | 32 | 44 | 81 | 790 |
| X7005-T6 | 52.0 | 45.7 | 13.8 | 78.1 | 1.71 | 29 | 83 | 112 | 1290 |
| 7039-T6 | 63.0 | 54.8 | 11.8 | 94.9 | 1.55 | 26 | 64 | 90 | 1010 |
| 7075-T6 | 83.0 | 76.0 | 11.2 | 73.8 | 0.97 | 10 | 14 | 24 | 220 |
| | 82.3 | 76.1 | 11.5 | 74.2 | 0.98 | 12 | 23 | 35 | 360 |
| | 81.0 | 72.8 | 11.0 | 79.2 | 1.09 | 16 | 26 | 42 | 395 |
| | 83.3 | 75.6 | 11.0 | 78.1 | 1.03 | 12 | 17 | 29 | 270 |
| | 85.5 | 77.0 | 12.5 | 72.1 | 0.94 | 8 | 11 | 19 | 175 |
| | 82.3 | 75.6 | 10.8 | 78.1 | 1.03 | 11 | 20 | 31 | 310 |
| | 82.0 | 74.6 | 11.0 | 70.5 | 0.94 | 13 | 16 | 29 | 265 |
| | 81.6 | 73.6 | 11.0 | 72.3 | 0.98 | 12 | 16 | 28 | 255 |
| | 83.4 | 76.1 | 11.0 | 69.2 | 0.91 | 14 | 14 | 28 | 215 |
| | 81.6 | 73.9 | 11.5 | 79.7 | 1.08 | 10 | 18 | 28 | 285 |
| | 81.8 | 73.4 | 11.5 | 75.8 | 1.03 | 13 | 19 | 32 | 290 |
| | 81.7 | 73.4 | 11.2 | 79.1 | 1.08 | 16 | 17 | 33 | 260 |
| | 81.5 | 73.0 | 10.2 | 84.2 | 1.15 | 12 | 22 | 34 | 360 |
| | 81.0 | 72.8 | 11.0 | 79.0 | 1.09 | 16 | 26 | 42 | 395 |
| Average 7075-T6 | 82.3 | 74.9 | 11.2 | 76.1 | 1.02 | 12 | 18 | 31 | 290 |
| Alclad 7075-T6 | 77.0 | 69.8 | 11.2 | 71.4 | 1.02 | 11 | 17 | 28 | 275 |
| | 72.9 | 65.7 | 11.5 | 74.5 | 1.13 | 13 | 21 | 34 | 345 |
| | 76.2 | 68.4 | 11.5 | 71.6 | 1.05 | 11 | 20 | 31 | 320 |
| | 78.0 | 70.4 | 10.5 | 75.4 | 1.07 | 12 | 21 | 33 | 325 |
| | 78.0 | 71.3 | 11.0 | 74.8 | 1.05 | 12 | 18 | 30 | 290 |
| | 78.2 | 69.2 | 11.0 | 79.0 | 1.14 | 10 | 16 | 26 | 260 |
| | 74.6 | 67.4 | 10.5 | 74.8 | 1.11 | 12 | 22 | 34 | 360 |
| | 80.4 | 72.8 | 11.0 | 73.6 | 1.01 | 10 | 16 | 26 | 250 |
| | 77.3 | 69.3 | 10.5 | 74.9 | 1.08 | 15 | 18 | 33 | 285 |
| | 73.8 | 65.4 | 10.2 | 70.3 | 1.07 | 11 | 18 | 29 | 285 |
| Average Alclad 7075-T6 | 76.6 | 69.0 | 11.0 | 74.0 | 1.07 | 12 | 19 | 30 | 300 |
| 7075-T73 | 72.0 | 60.0 | 10.0 | 84.8 | 1.41 | 18 | 35 | 5.30 | 565 |
| | 72.0 | 61.0 | 10.2 | 75.0 | 1.23 | 16 | 27 | 4.30 | 420 |
| | 69.5 | 57.8 | 11.0 | 79.0 | 1.37 | 17 | 36 | 5.30 | 545 |
| | 73.0 | 62.5 | 11.0 | 77.5 | 1.24 | 16 | 30 | 4.60 | 515 |
| Average 7075-T73 | 71.6 | 60.3 | 10.6 | 79.1 | 1.00 | 17 | 32 | 4.9 | 510 |
| 7079-T6(a) | 77.0 | 70.2 | 10.8 | 77.5 | 1.10 | 14 | 25 | 39 | 390 |
| | 72.8 | 64.0 | 11.0 | 86.3 | 1.35 | 23 | 34 | 57 | 565 |
| | 78.2 | 71.6 | 11.0 | 80.8 | 1.13 | 16 | 36 | 52 | 565 |
| Average 7079-T6 | 76.0 | 68.6 | 10.9 | 81.5 | 1.19 | 18 | 32 | 49 | 510 |
| Alclad 7079-T6(a) | 68.9 | 61.2 | 11.4 | 68.9 | 1.13 | 14 | 25 | 39 | 395 |
| | 73.8 | 66.5 | 11.0 | 68.1 | 1.02 | 11 | 23 | 34 | 370 |
| | 78.0 | 71.2 | 10.8 | 73.9 | 1.04 | 13 | 29 | 42 | 420 |
| Average Alclad 7079-T6 | 73.6 | 66.3 | 11.1 | 70.3 | 1.06 | 13 | 26 | 49 | 395 |
| X7106-T6(a) | 61.2 | 53.9 | 11.0 | 87.3 | 1.62 | 28 | 64 | 92 | 1050 |
| | 67.0 | 60.7 | 9.8 | 81.0 | 1.34 | 16 | 39 | 55 | 635 |
| Average X7106-T6 | 64.1 | 57.3 | 10.4 | 84.2 | 1.48 | 22 | 52 | 74 | 840 |
| X7139-T6(a) | 65.2 | 56.1 | 11.0 | 81.2 | 1.45 | 16 | 50 | 66 | 795 |
| 7178-T6 | 86.1 | 79.8 | 11.5 | 66.8 | 0.94 | 8 | 12 | 20 | 185 |
| | 88.8 | 80.5 | 12.5 | 61.8 | 0.77 | 6 | 6 | 1.20 | 90 |
| | 90.2 | 81.0 | 13.0 | 64.6 | 0.80 | 6 | 11 | 1.70 | 175 |
| | 89.0 | 80.5 | 12.5 | 66.6 | 0.83 | 11 | 7 | 1.80 | 110 |
| | 89.0 | 81.3 | 12.5 | 69.2 | 0.85 | 11 | 8 | 1.90 | 125 |
| | 89.4 | 82.4 | 11.5 | 62.5 | 0.76 | 6 | 10 | 1.60 | 155 |
| Average 7178-T6 | 88.8 | 80.9 | 12.2 | 65.2 | 0.81 | 8 | 9 | 1.6 | 140 |
| Alclad 7178-T6 | 81.0 | 74.2 | 12.0 | 59.7 | 1.80 | 8 | 11 | 19 | 175 |
| | 81.0 | 73.8 | 12.0 | 60.6 | 0.82 | 12 | 7 | 19 | 110 |
| | 80.6 | 73.4 | 12.5 | 61.9 | 0.84 | 6 | 9 | 15 | 140 |
| Average Alclad 7178-T6 | 80.9 | 73.8 | 12.2 | 60.7 | 0.82 | 9 | 9 | 18 | 140 |

Each line represents the average of duplicate tear tests (Fig. A1.8) of an individual lot of sheet. For tensile yield strength, offset is 0.2%. (a) Obsolete alloy

**Table 6.2(b)  Results of tensile and tear tests of 0.063 in. thick heat-treated aluminum alloy sheet, transverse**

| Alloy and temper | Tensile tests | | | Tear tests | | | | | |
|---|---|---|---|---|---|---|---|---|---|
| | Ultimate tensile strength (UTS), ksi | Tensile yield strength (TYS), ksi | Elongation in 2 in., % | Tear strength, ksi | Ratio tear strength to yield strength, (TYR) | Energy required to: | | Total energy, in.-lb | Unit propagation energy, in.-lb/in.² |
| | | | | | | Initiate a crack, in.-lb | Propagate a crack, in.-lb | | |
| Alclad 2014-T3 | 61.6 | 37.6 | 21.5 | 67.1 | 1.78 | 21 | 47 | 68 | 725 |
| | 63.9 | 42.1 | 21.5 | 69.3 | 1.65 | 17 | 39 | 56 | 615 |
| | 62.7 | 37.6 | 20.2 | 67.8 | 1.80 | 21 | 43 | 64 | 690 |
| Average Alclad 2014-T3 | 62.7 | 39.1 | 21.1 | 68.1 | 1.74 | 20 | 43 | 63 | 675 |
| 2014-T6 | 72.6 | 65.6 | 9.8 | 63.9 | 0.97 | 8 | 11 | 18 | 175 |
| | 72.6 | 64.6 | 11.0 | 63.0 | 0.98 | 8 | 10 | 18 | 150 |
| | 71.1 | 64.1 | 9.5 | 64.0 | 1.00 | 8 | 12 | 20 | 190 |
| | 69.1 | 61.6 | 9.2 | 60.6 | 0.98 | 5 | 13 | 18 | 205 |
| | 68.5 | 61.6 | 9.0 | 66.3 | 1.08 | 7 | 11 | 18 | 175 |
| | 72.3 | 64.3 | 11.2 | 62.2 | 0.97 | 9 | 12 | 21 | 180 |
| Average 2014-T6 | 71.0 | 63.6 | 10.0 | 63.3 | 1.00 | 8 | 12 | 19 | 180 |
| Alclad 2014-T6 | 71.8 | 63.1 | 11.0 | 60.9 | 0.97 | 8 | 8 | 16 | 125 |
| | 67.8 | 59.8 | 10.5 | 65.5 | 1.10 | 10 | 14 | 24 | 215 |
| | 67.5 | 59.1 | 11.2 | 70.6 | 1.19 | 13 | 16 | 29 | 260 |
| | 67.0 | 58.3 | 11.0 | 70.5 | 1.21 | 10 | 17 | 27 | 270 |
| Average Alclad 2014-T6 | 68.5 | 60.1 | 10.9 | 66.9 | 1.12 | 10 | 14 | 24 | 220 |
| 2020-O(a) | 28.2 | 10.2 | 20.2 | 31.7 | 3.11 | 26 | 44 | 70 | 695 |
| Alclad 2020-O(a) | 38.1 | 11.3 | 22.0 | 34.0 | 3.01 | 32 | 51 | 83 | 810 |
| 2020-T4(a) | 49.4 | 31.6 | 16.5 | 62.5 | 1.98 | 28 | 67 | 95 | 1060 |
| 2020-T6(a) | 81.1 | 73.8 | 6.8 | 48.3 | 0.65 | 3 | 0 | 3 | 0 |
| | 83.2 | 76.2 | 6.8 | 40.9 | 0.54 | 4 | 0 | 4 | 0 |
| | 81.8 | 75.8 | 7.0 | 49.2 | 0.65 | 4 | 3 | 7 | 50 |
| | 81.1 | 75.8 | 7.5 | 41.0 | 0.54 | 3 | 2 | 5 | 30 |
| | 81.6 | 75.4 | 7.0 | 34.0 | 0.45 | 2 | 0 | 2 | 0 |
| Average 2020-T6 | 81.8 | 75.4 | 7.0 | 42.7 | 0.57 | 3 | 2 | 4 | 15 |
| Alclad 2020-T6(a) | 73.6 | 67.2 | 7.5 | 42.9 | 0.64 | 2 | 2 | 4 | 30 |
| 2024-T4 | 66.5 | 42.7 | 20.0 | 68.1 | 1.59 | 17 | 37 | 54 | 595 |
| | 67.8 | 48.2 | 17.5 | 73.8 | 1.53 | 21 | 34 | 58 | 580 |
| | 67.2 | 44.8 | 20.5 | 73.8 | 1.65 | 15 | 36 | 51 | 560 |
| | 69.3 | 47.6 | 19.2 | 74.6 | 1.57 | 15 | 33 | 48 | 515 |
| | 66.8 | 42.8 | 22.0 | 75.3 | 1.76 | 24 | 51 | 75 | 805 |
| Average 2024-T4 | 67.5 | 45.2 | 19.8 | 73.1 | 1.62 | 18 | 38 | 57 | 610 |
| Alclad 2024-T4 | 61.0 | 37.4 | 21.0 | 63.7 | 1.70 | 22 | 39 | 61 | 600 |
| | 64.7 | 42.0 | 22.5 | 72.0 | 1.71 | 24 | 49 | 73 | 750 |
| | 65.0 | 44.2 | 19.5 | 69.5 | 1.57 | 20 | 39 | 59 | 615 |
| Average Alclad 2024-T4 | 63.6 | 41.2 | 21.0 | 68.4 | 1.66 | 22 | 42 | 64 | 655 |
| 2024-T3 | 65.8 | 44.4 | 21.0 | 72.2 | 1.63 | 22 | 42 | 64 | 660 |
| | 65.9 | 45.4 | 19.0 | 71.3 | 1.57 | 18 | 36 | 54 | 570 |
| | 68.4 | 48.0 | 19.0 | 70.3 | 1.46 | 15 | 31 | 46 | 500 |
| | 66.8 | 45.8 | 20.0 | 73.0 | 1.59 | 19 | 36 | 55 | 570 |
| | 67.4 | 46.7 | 18.5 | 73.7 | 1.58 | 17 | 37 | 54 | 595 |
| | 67.0 | 46.5 | 20.0 | 80.9 | 1.74 | 14 | 42 | 56 | 690 |
| | 69.0 | 47.0 | 20.5 | 78.1 | 1.66 | 18 | 43 | 61 | 650 |
| | 68.9 | 47.8 | 19.5 | 72.1 | 1.51 | 18 | 35 | 53 | 565 |
| Average 2024-T3 | 67.4 | 46.4 | 19.7 | 74.0 | 1.59 | 18 | 38 | 55 | 600 |
| Alclad 2024-T3 | 67.0 | 46.2 | 19.2 | 70.4 | 1.52 | 18 | 34 | 52 | 525 |
| | 67.7 | 49.1 | 17.8 | 74.9 | 1.53 | 16 | 33 | 49 | 525 |
| | 69.2 | 45.5 | 20.0 | 67.5 | 1.48 | 16 | 30 | 46 | 470 |
| | 66.8 | 46.1 | 18.5 | 67.6 | 1.47 | 17 | 26 | 43 | 415 |
| | 67.9 | 44.9 | 19.5 | 69.5 | 1.55 | 16 | 32 | 48 | 495 |
| | 68.2 | 45.9 | 21.2 | 68.9 | 1.50 | 15 | 34 | 49 | 540 |
| | 65.6 | 44.8 | 18.8 | 72.6 | 1.62 | 14 | 35 | 49 | 565 |
| | 64.1 | 43.5 | 19.8 | 68.7 | 1.58 | 14 | 28 | 42 | 450 |
| | 62.0 | 42.0 | 18.0 | 65.0 | 1.55 | 12 | 24 | 36 | 390 |
| | 63.4 | 44.0 | 19.0 | 68.8 | 1.56 | 12 | 26 | 38 | 420 |
| | 65.9 | 44.7 | 20.0 | 72.3 | 1.62 | 15 | 34 | 49 | 550 |
| | 64.3 | 45.2 | 19.0 | 71.4 | 1.58 | 21 | 40 | 61 | 625 |
| Average Alclad 2024-T3 | 66.0 | 45.2 | 19.2 | 69.8 | 1.55 | 16 | 31 | 47 | 500 |
| 2024-T36 | 75.4 | 57.8 | 14.8 | 69.5 | 1.20 | 10 | 17 | 27 | 265 |
| | 74.8 | 56.8 | 15.8 | 74.5 | 1.31 | 16 | 27 | 43 | 410 |

(continued)

Each line represents the average of duplicate tear tests (Fig. A1.8) of an individual lot of sheet. For tensile yield strength, offset is 0.2%. (a) Obsolete alloy

**Table 6.2(b)   (continued)**

| Alloy and temper | Tensile tests | | | Tear tests | | | | | |
|---|---|---|---|---|---|---|---|---|---|
| | Ultimate tensile strength (UTS), ksi | Tensile yield strength (TYS), ksi | Elongation in 2 in., % | Tear strength, ksi | Ratio tear strength to yield strength, (TYR) | Energy required to: | | Total energy, in.-lb | Unit propagation energy, in.-lb/in.[2] |
| | | | | | | Initiate a crack, in.-lb | Propagate a crack, in.-lb | | |
| | 75.2 | 57.5 | 15.5 | 74.2 | 1.29 | 13 | 22 | 35 | 335 |
| | 71.8 | 55.2 | 15.0 | 75.7 | 1.37 | 11 | 24 | 35 | 370 |
| | 73.2 | 57.0 | 13.2 | 76.4 | 1.34 | 16 | 26 | 42 | 415 |
| | 72.8 | 56.5 | 15.0 | 76.3 | 1.35 | 17 | 26 | 43 | 410 |
| | 70.9 | 53.8 | 15.5 | 77.1 | 1.43 | 12 | 31 | 43 | 490 |
| Average 2024-T36 | 73.4 | 56.4 | 15.0 | 74.8 | 1.33 | 14 | 25 | 38 | 385 |
| Alclad 2024-T36 | 72.2 | 54.8 | 15.5 | 65.8 | 1.20 | 11 | 18 | 29 | 275 |
| | 70.7 | 56.0 | 14.0 | 72.5 | 1.29 | 14 | 24 | 38 | 375 |
| | 69.2 | 53.0 | 14.8 | 73.6 | 1.39 | 16 | 29 | 45 | 425 |
| | 67.8 | 51.8 | 14.2 | 75.3 | 1.45 | 18 | 33 | 51 | 490 |
| | 69.2 | 52.6 | 16.0 | 72.3 | 1.37 | 14 | 30 | 44 | 450 |
| Average Alclad 2024-T36 | 69.8 | 53.6 | 14.9 | 71.9 | 1.34 | 15 | 27 | 41 | 405 |
| 2024-T6 | 66.3 | 51.8 | 8.8 | 58.6 | 1.13 | 7 | 15 | 22 | 245 |
| 2024-T81 | 72.4 | 67.6 | 6.0 | 58.4 | 0.86 | 6 | 11 | 17 | 175 |
| | 76.8 | 72.6 | 6.0 | 52.1 | 0.72 | 5 | 8 | 13 | 130 |
| | 71.6 | 66.7 | 6.2 | 55.5 | 0.83 | 6 | 8 | 15 | 140 |
| Average 2024-T81 | 73.6 | 69.0 | 6.1 | 55.3 | 0.80 | 6 | 9 | 15 | 150 |
| Alclad 2024-T81 | 66.3 | 59.6 | 6.5 | 60.3 | 1.01 | 10 | 12 | 22 | 195 |
| | 66.6 | 60.3 | 6.8 | 65.2 | 1.08 | 11 | 11 | 22 | 175 |
| | 69.0 | 64.0 | 6.5 | 64.8 | 1.01 | 4 | 14 | 18 | 215 |
| Average Alclad 2024-T81 | 67.3 | 61.3 | 6.6 | 63.4 | 1.03 | 8 | 12 | 21 | 195 |
| 2024-T86 | 75.8 | 70.8 | 6.2 | 56.1 | 0.79 | 5 | 7 | 12 | 110 |
| | 76.4 | 71.6 | 6.0 | 58.2 | 0.81 | 7 | 7 | 14 | 120 |
| Average 2024-T86 | 76.1 | 71.2 | 6.1 | 57.2 | 0.80 | 6 | 7 | 13 | 115 |
| Alclad 2024-T86 | 73.1 | 67.6 | 6.0 | 58.8 | 0.87 | 6 | 9 | 15 | 135 |
| | 76.0 | 70.5 | 6.5 | 59.1 | 0.84 | 8 | 9 | 17 | 135 |
| | 72.4 | 65.8 | 6.2 | 65.0 | 0.99 | 7 | 9 | 16 | 140 |
| | 69.0 | 64.1 | 6.2 | 54.5 | 1.85 | 6 | 10 | 16 | 155 |
| | 69.8 | 65.1 | 5.2 | 52.5 | 0.81 | 6 | 7 | 13 | 110 |
| | 73.0 | 70.0 | 5.8 | 51.9 | 0.74 | 7 | 7 | 14 | 105 |
| Average Alclad 2024-T86 | 72.2 | 67.2 | 6.0 | 57.0 | 0.85 | 7 | 8 | 15 | 130 |
| 2219-T4 | 55.7 | 33.6 | 19.5 | 71.7 | 2.13 | 33 | 82 | 115 | 1300 |
| 2219-T31 | 56.2 | 39.0 | 17.0 | 73.6 | 1.89 | 21 | 50 | 71 | 795 |
| 2219-T37 | 63.2 | 50.8 | 11.2 | 76.1 | 1.50 | 14 | 31 | 45 | 510 |
| 2219-T62 | 56.1 | 38.6 | 9.5 | 60.9 | 1.58 | 12 | 29 | 41 | 445 |
| | 61.2 | 42.8 | 10.5 | 66.8 | 1.56 | 15 | 33 | 48 | 525 |
| Average 2219-T62 | 58.6 | 40.7 | 10.0 | 63.8 | 1.57 | 14 | 31 | 44 | 485 |
| 2219-T81 | 68.6 | 52.4 | 9.5 | 65.9 | 1.26 | 9 | 15 | 24 | 240 |
| | 65.5 | 52.4 | 10.2 | 70.4 | 1.34 | 14 | 27 | 41 | 430 |
| Average 2219-T81 | 67.0 | 52.4 | 9.8 | 68.2 | 1.30 | 12 | 21 | 32 | 335 |
| 2219-T87 | 71.8 | 59.0 | 9.2 | 64.2 | 1.09 | 10 | 16 | 26 | 250 |
| | 68.2 | 56.3 | 9.5 | 67.2 | 1.19 | 10 | 21 | 31 | 340 |
| Average 2219-T87 | 70.0 | 57.6 | 9.4 | 65.7 | 1.14 | 10 | 18 | 28 | 195 |
| 2618-T6 | 60.6 | 54.2 | 6.0 | 59.2 | 1.09 | 8 | 15 | 23 | 235 |
| 6061-T4 | 38.8 | 20.0 | 23.0 | 46.5 | 2.32 | 30 | 72 | 102 | 1105 |
| | 38.3 | 25.3 | 18.0 | 51.8 | 2.05 | 21 | 52 | 73 | 850 |
| | 38.0 | 25.2 | 19.0 | 52.2 | 2.07 | 21 | 58 | 79 | 935 |
| | 36.7 | 24.8 | 18.5 | 51.3 | 2.07 | 22 | 58 | 80 | 935 |
| | 34.8 | 23.7 | 19.0 | 56.2 | 2.37 | 22 | 66 | 88 | 1135 |
| | 36.0 | 26.6 | 17.0 | 53.5 | 2.01 | 19 | 60 | 79 | 1000 |
| | 36.4 | 23.4 | 21.5 | 50.3 | 2.15 | 27 | 70 | 97 | 1130 |
| Average 6061-T4 | 37.0 | 24.1 | 19.4 | 51.7 | 2.15 | 23 | 62 | 85 | 1015 |
| 6061-T6 | 46.1 | 41.8 | 11.5 | 62.4 | 1.49 | 14 | 40 | 54 | 650 |
| | 45.4 | 40.7 | 11.2 | 63.2 | 1.55 | 17 | 40 | 57 | 635 |
| | 47.6 | 42.3 | 12.2 | 64.3 | 1.52 | 18 | 32 | 50 | 505 |
| | 43.1 | 39.1 | 10.8 | 65.4 | 1.67 | 16 | 53 | 69 | 865 |
| | 46.4 | 40.6 | 12.5 | 65.0 | 1.60 | 17 | 48 | 65 | 750 |
| | 44.8 | 40.4 | 11.8 | 65.4 | 1.62 | 22 | 45 | 67 | 715 |
| | 46.6 | 42.4 | 11.5 | 66.5 | 1.57 | 17 | 42 | 59 | 680 |
| | 43.3 | 35.4 | 15.0 | 62.9 | 1.78 | 19 | 56 | 75 | 930 |
| | 43.8 | 35.2 | 15.8 | 63.0 | 1.79 | 24 | 58 | 82 | 915 |

(continued)

Each line represents the average of duplicate tear tests (Fig. A1.8) of an individual lot of sheet. For tensile yield strength, offset is 0.2%. (a) Obsolete alloy

**Table 6.2(b)  (continued)**

| Alloy and temper | Tensile tests | | | Tear tests | | | | | |
|---|---|---|---|---|---|---|---|---|---|
| | Ultimate tensile strength (UTS), ksi | Tensile yield strength (TYS), ksi | Elongation in 2 in., % | Tear strength, ksi | Ratio tear strength to yield strength (TYR) | Energy required to: | | Total energy, in.-lb | Unit propagation energy, in.-lb/in.$^2$ |
| | | | | | | Initiate a crack, in.-lb | Propagate a crack, in.-lb | | |
| Average 6061-T6 | 45.2 | 39.8 | 12.5 | 64.2 | 1.62 | 18 | 46 | 64 | 740 |
| Alclad 6061-T6 | 42.6 | 38.4 | 11.2 | 62.1 | 1.62 | 20 | 46 | 66 | 715 |
| | 43.2 | 38.8 | 11.2 | 60.3 | 1.55 | 17 | 40 | 57 | 625 |
| | 43.0 | 38.6 | 9.8 | 62.0 | 1.61 | 18 | 41 | 59 | 670 |
| | 42.4 | 35.6 | 18.8 | 61.6 | 1.73 | 22 | 50 | 72 | 795 |
| | 41.0 | 33.6 | 13.5 | 58.6 | 1.74 | 21 | 54 | 75 | 850 |
| | 41.9 | 35.7 | 14.0 | 60.7 | 1.70 | 23 | 50 | 73 | 780 |
| Average Alclad 6061-T6 | 42.4 | 36.8 | 13.1 | 60.9 | 1.66 | 20 | 47 | 67 | 740 |
| 6066-T6 | 58.2 | 51.0 | 11.2 | 62.6 | 1.23 | 11 | 16 | 27 | 250 |
| 6071-T4 | 47.4 | 30.5 | 22.5 | 59.2 | 1.94 | 19 | 44 | 63 | 700 |
| 6071-T6(a) | 56.2 | 52.5 | 10.0 | 60.5 | 1.15 | 9 | 13 | 22 | 200 |
| 6151-T6 | 49.2 | 44.5 | 11.2 | 64.6 | 1.45 | 16 | 30 | 46 | 460 |
| X7002-T6(a) | 69.3 | 59.6 | 10.8 | 83.4 | 1.40 | 19 | 44 | 63 | 710 |
| X7005-T6 | 52.2 | 44.8 | 11.8 | 80.1 | 1.79 | 36 | 82 | 118 | 1275 |
| 7039-T6 | 63.0 | 54.2 | 11.0 | 82.4 | 1.52 | 23 | 49 | 72 | 780 |
| 7075-T6 | 84.6 | 75.5 | 11.0 | 67.5 | 0.89 | 8 | 10 | 18 | 155 |
| | 82.6 | 73.5 | 11.0 | 67.7 | 0.92 | 10 | 14 | 24 | 220 |
| | 82.1 | 71.4 | 10.5 | 72.2 | 1.01 | 10 | 16 | 26 | 240 |
| | 82.4 | 73.1 | 10.5 | 72.4 | 0.99 | 9 | 12 | 21 | 190 |
| | 94.8 | 73.0 | 12.5 | 71.1 | 0.97 | 10 | 12 | 22 | 190 |
| | 84.3 | 75.1 | 10.5 | 70.7 | 0.94 | 11 | 10 | 21 | 155 |
| | 81.0 | 71.4 | 10.5 | 68.9 | 0.96 | 10 | 15 | 25 | 240 |
| | 81.2 | 71.1 | 10.5 | 66.2 | 0.93 | 10 | 12 | 22 | 190 |
| | 82.6 | 74.0 | 10.2 | 64.0 | 0.86 | 11 | 15 | 26 | 230 |
| | 80.8 | 71.3 | 11.5 | 74.0 | 1.04 | 13 | 18 | 31 | 285 |
| | 81.4 | 70.6 | 11.2 | 74.2 | 1.05 | 12 | 17 | 29 | 255 |
| | 81.6 | 71.6 | 10.8 | 74.2 | 1.04 | 12 | 15 | 27 | 230 |
| | 81.0 | 71.8 | 10.2 | 82.6 | 1.15 | 12 | 14 | 26 | 240 |
| | 82.1 | 71.4 | 10.5 | 72.2 | 1.01 | 10 | 16 | 26 | 240 |
| Average 7075-T6 | 82.3 | 72.5 | 10.8 | 71.3 | 0.98 | 11 | 14 | 25 | 220 |
| Alclad 7075-T6 | 76.7 | 67.8 | 10.2 | 66.9 | 0.99 | 9 | 12 | 21 | 195 |
| | 74.8 | 64.9 | 11.0 | 68.3 | 1.05 | 8 | 13 | 21 | 215 |
| | 75.2 | 66.2 | 10.5 | 67.7 | 1.02 | 11 | 12 | 23 | 190 |
| | 75.8 | 66.0 | 10.5 | 70.8 | 1.07 | 11 | 14 | 25 | 215 |
| | 75.8 | 70.2 | 10.5 | 72.3 | 1.03 | 10 | 10 | 20 | 160 |
| | 78.0 | 67.0 | 10.8 | 78.7 | 1.17 | 11 | 11 | 22 | 180 |
| | 75.8 | 66.8 | 10.5 | 69.5 | 1.04 | 10 | 17 | 27 | 280 |
| | 79.1 | 70.0 | 10.0 | 70.8 | 1.01 | 8 | 12 | 20 | 185 |
| | 78.1 | 69.3 | 10.5 | 72.4 | 1.04 | 10 | 14 | 24 | 225 |
| | 73.9 | 64.2 | 10.5 | 71.3 | 1.11 | 9 | 12 | 21 | 190 |
| Average Alclad 7075-T6 | 76.3 | 67.2 | 10.5 | 70.9 | 1.05 | 10 | 13 | 22 | 205 |
| 7075-T73 | 70.9 | 59.0 | 10.2 | 80.2 | 1.36 | 15 | 31 | 46 | 500 |
| | 74.0 | 62.9 | 10.5 | 70.7 | 1.12 | 9 | 20 | 29 | 310 |
| | 71.3 | 58.3 | 10.2 | 68.0 | 1.19 | 14 | 25 | 39 | 380 |
| | 75.5 | 63.8 | 10.2 | 76.8 | 1.20 | 16 | 24 | 40 | 415 |
| Average 7075-T73 | 72.9 | 61.0 | 10.3 | 73.9 | 1.22 | 14 | 25 | 38 | 400 |
| 7079-T6(a) | 76.1 | 67.3 | 10.5 | 75.0 | 1.11 | 13 | 18 | 31 | 280 |
| | 72.8 | 62.4 | 11.0 | 82.3 | 1.32 | 15 | 29 | 44 | 485 |
| | 78.8 | 70.0 | 10.8 | 75.6 | 1.08 | 14 | 22 | 36 | 345 |
| Average 7079-T6 | 75.9 | 66.6 | 10.8 | 77.6 | 1.17 | 14 | 23 | 37 | 370 |
| Alclad 7079-T6(a) | 67.7 | 58.7 | 11.0 | 64.1 | 1.09 | 9 | 19 | 28 | 300 |
| | 73.4 | 65.1 | 10.8 | 65.2 | 1.00 | 10 | 14 | 24 | 225 |
| | 75.9 | 66.6 | 10.8 | 66.3 | 1.00 | 9 | 15 | 24 | 220 |
| Average Alclad 7079-T6 | 72.3 | 63.5 | 10.9 | 65.2 | 1.03 | 9 | 16 | 25 | 250 |
| X7106-T6(a) | 62.4 | 54.0 | 11.0 | 83.7 | 1.55 | 23 | 49 | 72 | 805 |
| | 64.3 | 58.8 | 11.0 | 82.8 | 1.41 | 19 | 35 | 54 | 565 |
| Average X7106-T6 | 63.4 | 56.4 | 11.0 | 83.2 | 1.48 | 21 | 42 | 63 | 685 |
| X7139-T6(a) | 65.7 | 56.1 | 10.0 | 77.8 | 1.39 | 15 | 32 | 47 | 510 |
| 7178-T6 | 88.9 | 77.9 | 11.5 | 61.5 | 0.79 | 7 | 9 | 16 | 140 |
| | 88.4 | 78.0 | 11.5 | 57.8 | 0.74 | 6 | 6 | 12 | 90 |
| | 88.1 | 78.0 | 12.5 | 65.9 | 0.84 | 6 | 10 | 16 | 160 |
| | 87.4 | 77.0 | 12.5 | 63.0 | 0.82 | 7 | 9 | 16 | 145 |
| | 87.0 | 77.0 | 12.0 | 64.1 | 0.83 | 10 | 11 | 21 | 175 |
| | 88.0 | 77.8 | 11.2 | 56.3 | 0.73 | 6 | 6 | 12 | 85 |
| Average 7178-T6 | 88.0 | 77.6 | 11.9 | 61.4 | 0.79 | 7 | 8 | 16 | 130 |
| Alclad 7178-T6 | 80.9 | 71.8 | 11.8 | 54.9 | 0.76 | 6 | 6 | 12 | 95 |
| | 79.9 | 70.6 | 11.5 | 57.5 | 0.81 | 3 | 7 | 10 | 110 |
| | 80.5 | 71.5 | 11.4 | 56.1 | 0.78 | 4 | 7 | 11 | 105 |
| Average Alclad 7178-T6 | 80.4 | 71.2 | 11.6 | 56.2 | 0.78 | 4 | 7 | 11 | 105 |

Each line represents the average of duplicate tear tests (Fig. A1.8) of an individual lot of sheet. For tensile yield strength, offset is 0.2%. (a) Obsolete alloy

**Table 6.3(a)    Results of tensile and tear tests of aluminum alloy plate, longitudinal**

| Alloy and temper | Thickness, in. | Ultimate tensile strength (UTS), ksi | Tensile yield strength (TYS), ksi | Elongation in 2 in., % | Tear strength, ksi | Ratio tear strength to yield strength, (TYR) | Initiate a crack, in.-lb | Propagate a crack, in.-lb | Total energy, in.-lb | Unit propagation energy, in.-lb/in.$^2$ |
|---|---|---|---|---|---|---|---|---|---|---|
| 2014-T651 | 0.25 | 69.7 | 64.3 | 11.6 | 78.4 | 1.22 | 15 | 22 | 37 | 355 |
| | 0.25 | 67.7 | 62.1 | 11.0 | 71.5 | 1.15 | 10 | 23 | 33 | 350 |
| | 0.25 | 70.3 | 65.0 | 11.0 | 72.7 | 1.12 | 16 | 37 | 53 | 375 |
| | 1.00 | 69.0 | 63.5 | 10.2 | 67.4 | 1.06 | 17 | 27 | 44 | 270 |
| 2020-T651(a) | 0.25 | 81.6 | 77.4 | 8.5 | 59.0 | 0.76 | 5 | 6 | 11 | 90 |
| | 1.38 | 81.9 | 76.3 | 6.0 | 51.7 | 0.68 | 7 | 12 | 19 | 115 |
| 2024-T351 | 1.00 | 72.7 | 58.2 | 18.0 | 83.3 | 1.43 | 38 | 72 | 110 | 720 |
| | 1.50 | 69.4 | 53.0 | 19.5 | 80.3 | 1.52 | 38 | 74 | 111 | 740 |
| 2024-T851 | 0.25 | 71.8 | 65.2 | 10.0 | 67.0 | 1.03 | 7 | 12 | 19 | 195 |
| | 0.25 | 73.0 | 66.4 | 8.8 | 69.6 | 1.05 | 7 | 14 | 21 | 215 |
| | 0.88 | 71.9 | 67.9 | 8.0 | 67.2 | 0.99 | 16 | 23 | 39 | 230 |
| | 1.38 | 71.8 | 65.6 | 7.5 | 64.9 | 0.99 | 13 | 15 | 28 | 155 |
| 2024-T86 | 0.88 | 76.5 | 72.8 | 8.5 | 73.0 | 1.00 | 13 | 23 | 35 | 230 |
| 2219-T62 | 1.00 | 58.2 | 39.8 | 13.2 | 67.6 | 1.70 | 23 | 50 | 72 | 500 |
| 2219-T851 | 1.00 | 65.8 | 52.8 | 11.2 | 75.0 | 1.42 | 21 | 47 | 68 | 470 |
| | 1.25 | 65.8 | 50.8 | 11.0 | 69.2 | 1.36 | 20 | 36 | 55 | 355 |
| | 1.38 | 66.6 | 52.0 | 11.0 | 70.2 | 1.36 | 22 | 53 | 75 | 530 |
| 2219-T87 | 0.25 | 69.3 | 57.6 | 10.5 | 70.5 | 1.22 | 18 | 33 | 50 | 330 |
| | 0.25 | 69.4 | 56.0 | 10.8 | 73.0 | 1.30 | 13 | 26 | 38 | 410 |
| | 0.38 | 68.0 | 56.0 | 11.0 | 73.4 | 1.31 | 13 | 23 | 37 | 370 |
| | 1.00 | 68.4 | 57.0 | 11.8 | 70.6 | 1.24 | 16 | 52 | 68 | 525 |
| | 1.00 | 68.4 | 57.1 | 11.5 | 72.8 | 1.27 | 22 | ... | ... | ... |
| 2618-T651 | 1.38 | 62.5 | 58.0 | 9.3 | 65.1 | 1.12 | 13 | 28 | 41 | 280 |
| 5083-O | 0.38 | 47.6 | 23.0 | 23.0 | 54.0 | 2.35 | 55 | 113 | 167 | 1125 |
| | 0.75 | 45.4 | 21.8 | 23.0 | 53.4 | 2.45 | 57 | 108 | 164 | 1075 |
| | 1.00 | 43.2 | 19.0 | 24.5 | 50.0 | 2.63 | 55 | 121 | 176 | 1215 |
| | 7.00 | 45.0 | 19.7 | 24.2 | 51.3 | 2.61 | 50 | 108 | 158 | 1075 |
| | 7.70 | 39.6 | 17.6 | 22.2 | 44.0 | 2.38 | 44 | 83 | 127 | 830 |
| 5083-H321 | 0.38 | 48.8 | 35.0 | 18.5 | 66.0 | 1.89 | 45 | 113 | 158 | 1125 |
| | 0.38 | 51.0 | 34.4 | 17.2 | 63.4 | 1.84 | 42 | 85 | 127 | 855 |
| | 0.50 | 50.7 | 36.7 | 18.4 | 64.4 | 1.75 | 38 | 82 | 120 | 820 |
| | 0.75 | 49.8 | 40.8 | 15.5 | 67.0 | 1.64 | 32 | 82 | 114 | 820 |
| 5083-H131 | 1.50 | 49.4 | 42.6 | 10.5 | 63.9 | 1.50 | 46 | 63 | 108 | 626 |
| 5083-H115 | 1.38 | ... | ... | ... | 66.6 | ... | 29 | 60 | 89 | 600 |
| 5086-O | 0.38 | 37.5 | 17.3 | 31.6 | 43.6 | 2.52 | 73 | 139 | 212 | 1390 |
| | 0.50 | 38.0 | 18.4 | 30.0 | 44.6 | 2.42 | 69 | 134 | 203 | 1340 |
| | 0.75 | 36.6 | 16.0 | 32.0 | 44.6 | 2.79 | 73 | 127 | 200 | 1270 |
| | 0.75 | 41.7 | 20.5 | 25.0 | 46.5 | 2.27 | 66 | 114 | 179 | 1135 |
| 5086-H32 | 0.75 | 43.3 | 31.1 | 16.0 | 62.2 | 2.13 | 60 | 86 | 146 | 860 |
| 5086-H34 | 0.75 | 50.7 | 38.2 | 12.5 | 64.0 | 1.68 | 37 | 65 | 101 | 645 |
| 5154-O | 0.75 | 38.0 | 19.5 | 27.5 | 49.2 | 2.52 | 62 | 138 | 199 | 1375 |
| | 0.75 | 35.1 | 16.1 | 30.7 | 45.1 | 2.80 | 80 | 135 | 215 | 1350 |
| 5154-H34 | 0.75 | 42.0 | 30.3 | 16.2 | 57.5 | 1.90 | 62 | 92 | 154 | 920 |
| 5356-O | 0.75 | 42.3 | 19.5 | 18.5 | 49.8 | 2.55 | 66 | 148 | 213 | 1475 |
| | 0.75 | 43.5 | 18.9 | 28.8 | 50.8 | 1.76 | 65 | 141 | 206 | 1405 |
| 5356-H321 | 0.75 | 51.0 | 35.7 | 15.3 | 68.0 | 1.90 | 66 | 105 | 171 | 1050 |
| | 0.75 | 53.3 | 34.7 | 16.0 | 65.6 | 1.89 | 60 | 87 | 147 | 865 |
| 5454-O | 0.25 | 36.3 | 18.2 | 22.0 | 49.4 | 2.71 | 41 | 53 | 94 | 830 |
| | 0.50 | 39.2 | 23.6 | 20.8 | 53.9 | 2.28 | 78 | 120 | 199 | 1205 |
| | 0.50 | 37.5 | 19.6 | 21.6 | 48.0 | 2.45 | 39 | 67 | 106 | 1040 |
| | 0.38 | 38.8 | 22.7 | 21.2 | 51.9 | 2.29 | 69 | 107 | 176 | 1070 |
| | 0.75 | 41.1 | 23.2 | 22.0 | 53.2 | 2.29 | 67 | 114 | 181 | 1140 |
| | 0.75 | 35.9 | 16.6 | 25.0 | 46.7 | 2.81 | 75 | 129 | 204 | 1290 |
| | 1.00 | 38.6 | 19.8 | 21.5 | 47.5 | 2.40 | 38 | 76 | 114 | 1190 |
| 5454-H32 | 0.25 | 42.1 | 31.6 | 17.5 | 57.2 | 1.81 | 30 | 52 | 82 | 810 |
| 5454-H34 | 0.38 | 41.1 | 27.1 | 20.0 | 55.0 | 2.03 | 72 | 111 | 183 | 1110 |
| | 0.50 | 41.6 | 35.0 | 15.2 | 61.6 | 1.76 | 43 | 92 | 135 | 920 |
| | 0.75 | 40.9 | 28.9 | 15.7 | 59.1 | 2.04 | 61 | 101 | 162 | 1010 |
| 5456-O | 0.38 | 50.0 | 26.6 | 20.8 | 58.6 | 2.20 | 57 | 99 | 157 | 995 |
| | 0.75 | 49.9 | 23.4 | 22.5 | 51.6 | 2.21 | 44 | 99 | 142 | 985 |
| | 0.75 | 50.8 | 24.1 | 20.0 | 55.5 | 2.30 | 50 | 112 | 162 | 1120 |
| | 0.75 | 49.0 | 23.2 | 21.8 | 52.1 | 2.25 | 55 | 106 | 160 | 1055 |
| | 1.00 | 47.5 | 22.6 | 20.8 | 52.6 | 2.33 | 56 | 93 | 149 | 930 |

(continued)

Each line of data represents a separate lot of material; average of duplicate or triplicate tests. Specimens per Fig. A1.8, generally 0.100 in. thick; in a few cases, 0.063-in. thick specimens were used. For yield strengths, offset is 0.2%. (a) Obsolete alloy

**Table 6.3(a)   (continued)**

| Alloy and temper | Thickness, in. | Tensile tests | | | Tear tests | | | | | |
|---|---|---|---|---|---|---|---|---|---|---|
| | | Ultimate tensile strength (UTS), ksi | Tensile yield strength (TYS), ksi | Elongation in 2 in., % | Tear strength, ksi | Ratio tear strength to yield strength, (TYR) | Energy required to: | | Total energy, in.-lb | Unit propagation energy, in.-lb/in.$^2$ |
| | | | | | | | Initiate a crack, in.-lb | Propagate a crack, in.-lb | | |
| 5456-H321 | 0.38 | 51.4 | 32.3 | 19.6 | 60.6 | 1.88 | 42 | 92 | 114 | 920 |
| | 0.50 | 55.8 | 35.5 | 16.0 | 68.4 | 1.92 | 54 | 104 | 158 | 1040 |
| | 0.50 | ... | ... | ... | 68.1 | ... | 39 | 73 | 112 | 725 |
| | 0.75 | 57.5 | 35.6 | 14.8 | 68.8 | 1.93 | 47 | 75 | 122 | 750 |
| | 0.75 | 56.3 | 34.5 | 13.5 | 65.3 | 1.89 | 42 | 90 | 132 | 900 |
| | 0.75 | ... | ... | ... | 65.0 | ... | 42 | 83 | 126 | 835 |
| | 1.00 | 52.0 | 36.7 | 12.5 | 64.9 | 1.77 | 41 | 86 | 126 | 860 |
| | 1.25 | ... | ... | ... | 65.6 | ... | 32 | 62 | 95 | 630 |
| | 1.25 | 55.4 | 33.8 | 13.2 | 65.7 | 1.95 | 49 | 88 | 137 | 875 |
| 6061-T651 | 1.24 | 44.9 | 42.2 | 16.5 | 71.0 | 1.69 | 47 | 91 | 138 | 905 |
| 7001-T75(a) | 0.75 | 88.0 | 83.0 | 8.8 | 43.9 | 0.53 | 5 | 10 | 15 | 95 |
| | 1.00 | 94.2 | 89.5 | 9.5 | 55.4 | 0.62 | 7 | 12 | 19 | 130 |
| 7001-T7551(a) | 1.00 | 81.9 | 74.8 | 11.0 | 63.7 | 0.85 | 11 | 17 | 28 | 175 |
| | 1.00 | 80.6 | 73.0 | 11.0 | 58.9 | 0.81 | 9 | 12 | 21 | 120 |
| | 1.00 | 81.4 | 74.4 | 10.2 | 56.6 | 0.76 | 9 | 13 | 22 | 125 |
| | 1.38 | 81.8 | 72.2 | 9.5 | 69.9 | 0.97 | 14 | 19 | 32 | 185 |
| | 1.38 | 80.6 | 70.6 | 9.5 | 69.7 | 0.99 | 14 | 16 | 30 | 160 |
| | 1.38 | 80.6 | 70.6 | 9.5 | 72.6 | 1.03 | 13 | 18 | 31 | 180 |
| 7005-T6351 | 1.00 | 54.2 | 47.2 | 17.0 | 80.7 | 1.71 | 62 | 101 | 164 | 1015 |
| 7075-T651 | 0.25 | 82.3 | 75.8 | 12.2 | 75.0 | 0.99 | 18 | 31 | 49 | 315 |
| | 0.25 | 83.0 | 77.3 | 14.5 | 81.2 | 1.05 | 14 | 22 | 36 | 345 |
| | 0.25 | 85.0 | 78.8 | 13.5 | 80.2 | 1.02 | 11 | 23 | 34 | 360 |
| | 0.25 | 83.9 | 78.2 | 13.0 | 80.2 | 1.03 | 10 | 18 | 28 | 280 |
| | 0.25 | 84.3 | 78.0 | 14.0 | 78.6 | 1.01 | 17 | 23 | 40 | 225 |
| | 0.38 | 81.9 | 75.8 | 13.8 | 88.4 | 1.17 | 26 | 47 | 73 | 470 |
| | 0.50 | 83.0 | 72.6 | 12.0 | ... | ... | ... | ... | ... | ... |
| | 0.50 | 86.9 | 79.4 | 11.5 | 76.7 | 0.97 | 16 | 45 | 61 | 450 |
| | 1.00 | 91.5 | 83.2 | 10.5 | 75.8 | 0.91 | 13 | 28 | 41 | 285 |
| | 1.25 | 90.4 | 81.6 | 10.0 | 70.3 | 0.86 | 14 | 20 | 34 | 200 |
| | 1.38 | 88.8 | 80.4 | 9.8 | 77.1 | 0.96 | 20 | 21 | 41 | 210 |
| | 2.50 | 85.4 | 76.8 | 7.8 | 68.6 | 0.89 | 15 | 21 | 36 | 215 |
| | 2.75 | 82.6 | 69.6 | 10.0 | 70.0 | 1.01 | 16 | 25 | 41 | 250 |
| 7075-T7351 | 0.25 | 71.4 | 60.4 | 13.0 | 79.3 | 1.32 | 28 | 49 | 77 | 495 |
| | 0.25 | 71.8 | 61.4 | 12.5 | 80.8 | 0.13 | 25 | 52 | 77 | 525 |
| | 0.25 | 70.3 | 59.2 | 13.5 | 83.7 | 1.41 | 23 | 39 | 61 | 610 |
| | 0.50 | 73.2 | 62.5 | 12.8 | 79.0 | 1.26 | 27 | 46 | 73 | 470 |
| | 1.00 | 72.8 | 61.6 | 11.2 | 77.4 | 1.26 | 26 | 41 | 67 | 410 |
| | 1.00 | 74.5 | 62.8 | 11.8 | 80.4 | 1.28 | 24 | 39 | 63 | 395 |
| | 1.38 | 76.2 | 66.2 | 10.2 | 82.2 | 1.24 | 27 | 39 | 65 | 385 |
| | 2.50 | 70.0 | 58.3 | 9.8 | 74.9 | 1.28 | 22 | 42 | 64 | 425 |
| 7079-T651(a) | 0.25 | 80.2 | 74.7 | 11.0 | 75.8 | 1.01 | 17 | 25 | 41 | 245 |
| | 0.25 | 79.0 | 73.8 | 14.0 | 86.8 | 1.18 | 14 | 24 | 38 | 240 |
| | 0.25 | 79.4 | 74.0 | 12.8 | 84.2 | 1.14 | 23 | 34 | 56 | 335 |
| | 1.50 | 84.2 | 76.8 | 10.0 | 79.6 | 1.04 | 18 | 39 | 57 | 390 |
| | 3.00 | 83.5 | 76.3 | 10.0 | 76.8 | 1.01 | 18 | 37 | 55 | 370 |
| | 4.00 | 79.0 | 72.6 | 11.0 | 69.4 | 0.96 | 14 | 21 | 35 | 210 |
| 7106-T6351(a) | 0.25 | 61.0 | 54.5 | 14.5 | 83.8 | 1.54 | 22 | 56 | 78 | 895 |
| | 0.50 | 65.6 | 57.7 | 14.8 | 86.8 | 1.50 | 47 | 105 | 152 | 1040 |
| | 1.50 | 66.0 | 59.4 | 12.6 | 87.3 | 1.53 | 42 | 89 | 131 | 880 |
| | 3.00 | 62.2 | 57.0 | 15.0 | 86.0 | 1.51 | 39 | 91 | 130 | 915 |
| 7139-T6351(a) | 1.00 | 70.5 | 61.2 | 13.0 | 85.0 | 1.39 | 27 | 62 | 89 | 615 |
| 7178-T651 | 0.25 | 88.7 | 84.3 | 13.0 | 70.8 | 0.84 | 9 | 6 | 15 | 95 |
| | 0.31 | 89.4 | 84.0 | 12.0 | 70.8 | 0.84 | 12 | 24 | 36 | 240 |
| | 0.31 | 89.0 | 84.9 | 11.5 | 60.3 | 0.71 | 6 | 20 | 26 | 200 |
| | 0.34 | 85.8 | 79.2 | 11.5 | 64.2 | 0.81 | 13 | 14 | 28 | 145 |
| | 0.44 | 88.3 | 80.1 | 11.2 | 66.2 | 0.83 | 11 | 25 | 36 | 250 |
| | 0.50 | 90.6 | 83.2 | 12.4 | 70.2 | 0.84 | 13 | 25 | 38 | 250 |

(continued)

Each line of data represents a separate lot of material; average of duplicate or triplicate tests. Specimens per Fig. A1.8, generally 0.100 in. thick; in a few cases, 0.063-in. thick specimens were used. For yield strengths, offset is 0.2%. (a) Obsolete alloy

## Table 6.3(a)  (continued)

| Alloy and temper | Thickness, in. | Tensile tests | | | Tear tests | | | | | |
|---|---|---|---|---|---|---|---|---|---|---|
| | | Ultimate tensile strength (UTS), ksi | Tensile yield strength (TYS), ksi | Elongation in 2 in., % | Tear strength, ksi | Ratio tear strength to yield strength (TYR) | Energy required to: | | Total energy, in.-lb | Unit propagation energy, in.-lb/in.$^2$ |
| | | | | | | | Initiate a crack, in.-lb | Propagate a crack, in.-lb | | |
| | 0.50 | 90.3 | 83.4 | 12.0 | 72.6 | 0.87 | 14 | 24 | 38 | 235 |
| | 0.63 | 88.2 | 81.2 | 9.1 | 67.1 | 0.83 | 13 | 20 | 33 | 205 |
| | 1.00 | 93.8 | 87.2 | 9.5 | 61.5 | 0.71 | 9 | 18 | 27 | 180 |
| | 1.25 | 93.6 | 84.2 | 9.0 | 61.8 | 0.73 | 13 | 17 | 30 | 170 |
| 7178-T7651 | 0.25 | 77.4 | 68.4 | 11.0 | 80.3 | 1.18 | 28 | 29 | 49 | 290 |
| | 0.31 | 75.9 | 65.8 | 12.0 | 73.3 | 1.11 | 18 | 16 | 34 | 155 |
| | 1.00 | 80.6 | 71.7 | 11.0 | 81.1 | 1.13 | 27 | 18 | 45 | 175 |
| | 1.00 | 80.2 | 71.2 | 10.2 | 79.0 | 1.11 | 22 | 33 | 55 | 325 |

Each line of data represents a separate lot of material; average of duplicate or triplicate tests. Specimens per Fig. A1.8, generally 0.100 in. thick; in a few cases, 0.063-in. thick specimens were used. For yield strengths, offset is 0.2%. (a) Obsolete alloy

## Table 6.3(b)  Results of tensile and tear tests of aluminum alloy plate, transverse

| Alloy and temper | Thickness, in. | Tensile tests | | | Tear tests | | | | | |
|---|---|---|---|---|---|---|---|---|---|---|
| | | Ultimate tensile strength (UTS), ksi | Tensile yield strength (TYS), ksi | Elongation in 2 in., % | Tear strength, ksi | Ratio tear strength to yield strength (TYR) | Energy required to: | | Total energy, in.-lb | Unit propagation energy, in.-lb/in.$^2$ |
| | | | | | | | Initiate a crack, in.-lb | Propagate a crack, in.-lb | | |
| 2014-T651 | 0.25 | 70.0 | 62.2 | 10.2 | 67.8 | 1.09 | 9 | 12 | 21 | 185 |
| | 0.25 | 68.4 | 60.7 | 10.0 | 68.8 | 1.13 | 8 | 14 | 22 | 220 |
| | 0.25 | 69.6 | 62.8 | 10.5 | 63.2 | 1.00 | 10 | 16 | 25 | 155 |
| | 1.00 | 69.5 | 62.7 | 8.8 | 61.0 | 0.97 | 13 | 18 | 31 | 180 |
| 2020-T651(a) | 0.25 | 83.1 | 78.0 | 6.0 | 35.0 | 0.45 | 2 | 5 | 7 | 50 |
| | 1.38 | 82.2 | 77.4 | 2.4 | 36.7 | 0.47 | 3 | 5 | 8 | 50 |
| 2024-T42 | 0.50 | 67.4 | 43.2 | 20.0 | 71.4 | 1.65 | 25 | 53 | 78 | 525 |
| 2024-T351 | 0.50 | 67.4 | 50.0 | 18.8 | 71.6 | 1.43 | 24 | 44 | 68 | 440 |
| | 1.00 | 72.4 | 52.0 | 16.5 | 80.2 | 1.54 | 37 | 52 | 88 | 515 |
| | 1.50 | 68.4 | 47.4 | 17.2 | 71.1 | 1.50 | 25 | 49 | 74 | 490 |
| 2024-T36 | 0.50 | 72.0 | 64.8 | 12.6 | 67.2 | 1.04 | 14 | 19 | 34 | 195 |
| 2024-T62 | 0.50 | 70.2 | 57.4 | 10.6 | 62.8 | 1.09 | 13 | 13 | 26 | 135 |
| 2024-T851 | 0.25 | 72.0 | 66.2 | 8.0 | 64.1 | 0.97 | 6 | 10 | 16 | 160 |
| | 0.25 | 72.4 | 65.8 | 7.5 | 68.6 | 1.04 | 8 | 10 | 18 | 160 |
| | 0.50 | 70.2 | 65.5 | 7.2 | 60.8 | 0.93 | 12 | 11 | 23 | 109 |
| | 0.88 | 71.5 | 66.9 | 7.0 | 56.8 | 0.85 | 10 | 17 | 27 | 170 |
| | 1.38 | 71.2 | 64.8 | 6.0 | 56.7 | 0.88 | 10 | 7 | 18 | 70 |
| 2024-T86 | 0.50 | 75.1 | 71.6 | 4.8 | 49.6 | 0.69 | 6 | 7 | 13 | 70 |
| | 0.88 | 73.9 | 69.9 | 6.0 | 56.0 | 0.80 | 14 | 18 | 32 | 180 |
| 2219-T62 | 1.00 | 58.8 | 39.8 | 11.0 | 65.7 | 1.65 | 18 | 34 | 51 | 335 |
| 2219-T851 | 1.00 | 66.5 | 50.4 | 10.0 | 66.9 | 1.33 | 12 | 30 | 43 | 305 |
| | 1.25 | 65.8 | 49.1 | 11.0 | 65.6 | 1.34 | 21 | 46 | 68 | 460 |
| | 1.38 | 65.8 | 49.3 | 9.3 | 65.2 | 1.32 | 14 | 24 | 35 | 245 |
| 2219-T87 | 0.25 | 70.2 | 57.2 | 10.5 | 71.6 | 1.25 | 19 | 22 | 41 | 220 |
| | 0.25 | 69.5 | 55.9 | 10.0 | 69.4 | 1.24 | 10 | 21 | 31 | 330 |
| | 0.38 | 69.0 | 55.5 | ... | 68.6 | 1.24 | 11 | 18 | 29 | 280 |
| | 1.00 | 68.8 | 56.7 | 9.9 | 63.5 | 1.12 | 12 | 27 | 39 | 270 |
| | 1.00 | 69.1 | 57.1 | 9.0 | 61.5 | 1.08 | 12 | 22 | 34 | 225 |
| 2618-T651 | 1.38 | 63.2 | 56.4 | 10.0 | 66.3 | 1.17 | 13 | 26 | 39 | 260 |
| 5083-O | 0.38 | 47.4 | 23.0 | 25.2 | 54.0 | 2.35 | 57 | 97 | 153 | 965 |
| | 0.75 | 43.3 | 23.6 | 20.8 | 51.0 | 2.16 | 47 | 99 | 146 | 990 |
| | 1.00 | 43.2 | 19.7 | 24.2 | 49.2 | 2.50 | 57 | 109 | 166 | 1090 |
| | 7.00 | 45.0 | 20.8 | 18.8 | 49.1 | 2.35 | 37 | 83 | 110 | 830 |
| | 7.70 | 39.4 | 18.1 | 21.0 | 41.4 | 2.27 | 31 | 62 | 93 | 625 |
| 5083-H321 | 0.38 | 49.6 | 31.3 | 21.8 | 63.2 | 2.02 | 44 | 84 | 128 | 840 |
| | 0.38 | 49.9 | 31.6 | 20.0 | 61.7 | 1.95 | 43 | 81 | 123 | 805 |
| | 0.50 | 50.3 | 32.9 | 19.2 | 63.0 | 1.91 | 38 | 74 | 112 | 745 |

(continued)

Each line of data represents a separate lot of material; average of duplicate or triplicate tests. Specimens per Fig. A1.8. generally 0.100 in. thick; in a few cases, 0.063-in. thick specimens were tested. (a) Obsolete alloy

## Table 6.3(b)  (continued)

| Alloy and temper | Thickness, in. | Ultimate tensile strength (UTS), ksi | Tensile yield strength (TYS), ksi | Elongation in 2 in., % | Tear strength, ksi | Ratio tear strength to yield strength (TYR) | Initiate a crack, in.-lb | Propagate a crack, in.-lb | Total energy, in.-lb | Unit propagation energy, in.-lb/in.² |
|---|---|---|---|---|---|---|---|---|---|---|
| | 0.75 | 51.2 | 36.6 | 14.5 | 64.8 | 1.77 | 33 | 56 | 89 | 560 |
| 5083-H131 | 1.50 | 50.1 | 37.5 | 13.0 | 63.5 | 1.69 | 35 | 52 | 87 | 520 |
| 5083-H115 | 1.38 | 48.7 | 34.8 | 17.0 | 63.0 | 1.81 | 26 | 50 | 75 | 495 |
| 5086-O | 0.38 | 38.1 | 18.1 | 31.2 | 43.3 | 2.39 | 68 | 129 | 196 | 1287 |
| | 0.50 | 38.2 | 18.9 | 28.8 | 44.6 | 2.36 | 73 | 117 | 190 | 1170 |
| | 0.75 | 37.6 | 17.3 | 30.5 | 45.2 | 2.61 | 67 | 124 | 191 | 1240 |
| | 0.75 | 41.1 | 20.6 | 27.8 | 46.0 | 2.23 | 64 | 92 | 156 | 920 |
| 5086-H32 | 0.75 | 42.4 | 29.0 | 22.0 | 60.4 | 2.08 | 51 | 80 | 131 | 800 |
| 5086-H34 | 0.75 | 52.4 | 37.4 | 15.7 | 61.7 | 1.65 | 33 | 40 | 73 | 395 |
| 5154-O | 0.75 | 38.8 | 19.2 | 27.8 | 49.6 | 2.58 | 65 | 123 | 188 | 1230 |
| | 0.75 | 36.1 | 16.2 | 29.6 | 44.1 | 2.72 | 77 | 115 | 192 | 1145 |
| 5154-H34 | 0.75 | 42.2 | 30.5 | 20.0 | 58.4 | 1.91 | 66 | 87 | 152 | 865 |
| 5356-O | 0.75 | 43.4 | 19.8 | 29.3 | 48.8 | 2.46 | 63 | 120 | 182 | 1195 |
| | 0.75 | 44.7 | 21.7 | 27.7 | 49.2 | 2.27 | 58 | 114 | 172 | 1140 |
| 5356-H321 | 0.75 | 48.1 | 33.7 | 22.2 | 64.2 | 1.90 | 46 | 66 | 112 | 660 |
| | 0.75 | 51.8 | 33.2 | 21.0 | 62.1 | 1.87 | 46 | 70 | 116 | 700 |
| 5454-O | 0.25 | 36.5 | 18.7 | 21.5 | 50.0 | 2.67 | 36 | 61 | 97 | 950 |
| | 0.50 | 38.6 | 23.8 | 22.8 | 55.0 | 2.31 | 79 | 120 | 198 | 1200 |
| | 0.50 | 38.1 | 20.6 | 24.0 | 48.6 | 2.36 | 50 | 73 | 123 | 1140 |
| | 0.38 | 38.4 | 23.2 | 24.0 | 52.8 | 2.28 | 76 | 107 | 183 | 1075 |
| | 0.75 | 40.1 | 23.2 | 22.0 | 50.8 | 2.19 | 55 | 86 | 141 | 860 |
| | 0.75 | 35.6 | 16.8 | 24.0 | 45.6 | 2.71 | 75 | 111 | 186 | 1105 |
| | 1.00 | 38.8 | 20.4 | 22.0 | 47.0 | 2.30 | 39 | 61 | 100 | 960 |
| 5454-H32 | 0.25 | 41.0 | 29.8 | 18.0 | 59.9 | 2.01 | 39 | 58 | 97 | 905 |
| 5454-H34 | 0.38 | 40.7 | 26.4 | 24.4 | 57.5 | 2.18 | 72 | 120 | 192 | 1200 |
| | 0.50 | 42.6 | 33.9 | 16.8 | 62.3 | 1.84 | 48 | 65 | 113 | 650 |
| | 0.75 | 41.4 | 29.4 | 18.8 | 60.5 | 2.06 | 68 | 91 | 160 | 915 |
| 5456-O | 0.38 | 50.4 | 27.4 | 21.2 | 58.1 | 2.12 | 50 | 91 | 142 | 910 |
| | 0.75 | 50.4 | 23.7 | 22.3 | 49.2 | 2.08 | 42 | 82 | 124 | 820 |
| | 0.75 | 48.5 | 24.0 | 22.0 | 54.0 | 2.25 | 44 | 92 | 136 | 925 |
| | 0.75 | 48.8 | 24.0 | 21.2 | 49.9 | 2.08 | 47 | 81 | 128 | 810 |
| | 1.00 | 48.3 | 22.8 | 22.2 | 52.4 | 2.30 | 53 | 93 | 1247 | 935 |
| 5456-H321 | 0.38 | 51.5 | 30.6 | 22.0 | 61.0 | 1.99 | 46 | 79 | 125 | 785 |
| | 0.50 | 54.4 | 34.4 | 21.2 | 66.0 | 1.92 | 48 | 81 | 129 | 810 |
| | 0.50 | ... | ... | ... | 65.5 | ... | 31 | 58 | 89 | 575 |
| | 0.75 | 55.8 | 34.4 | 19.3 | 62.0 | 1.80 | 35 | 62 | 97 | 620 |
| | 0.75 | 55.9 | 33.6 | 19.0 | 61.8 | 1.84 | 38 | 58 | 96 | 580 |
| | 0.75 | ... | ... | ... | 62.2 | ... | 32 | 66 | 98 | 660 |
| | 1.00 | 50.1 | 33.0 | 17.0 | 63.8 | 1.93 | 39 | 71 | 110 | 710 |
| | 1.25 | ... | ... | ... | 60.9 | ... | 32 | 67 | 99 | 675 |
| | 1.25 | 55.4 | 33.3 | 16.5 | 62.8 | 1.89 | 41 | 80 | 121 | 800 |
| 6061-T651 | 0.50 | 45.0 | 40.2 | 16.8 | 66.1 | 1.64 | 31 | 49 | 79 | 485 |
| | 1.24 | 44.9 | 40.4 | 15.2 | 68.9 | 1.71 | 36 | 77 | 114 | 775 |
| 7001-T75(a) | 0.75 | 89.6 | 80.6 | 10.2 | 38.3 | 0.48 | 4 | <5 | <9 | <50 |
| | 1.00 | 92.8 | 85.8 | 7.0 | 39.0 | 0.45 | 3 | <5 | <9 | <50 |
| 7001-T7551(a) | 1.00 | 81.8 | 73.7 | 8.5 | 45.2 | 0.61 | 5 | 14 | 18 | 130 |
| | 1.00 | 80.0 | 71.2 | 9.0 | 47.6 | 0.67 | 6 | 11 | 17 | 110 |
| | 1.00 | 81.8 | 73.4 | 9.2 | 42.6 | 0.58 | 5 | 12 | 17 | 120 |
| | 1.38 | 80.8 | 71.3 | 8.8 | 51.6 | 0.72 | 7 | 16 | 23 | 100 |
| | 1.38 | 79.9 | 69.6 | 9.0 | 57.8 | 0.83 | 8 | 7 | 16 | 75 |
| | 1.38 | 80.5 | 70.6 | 8.8 | 57.6 | 0.82 | 8 | 18 | 26 | 180 |
| 7005-T6351 | 1.00 | 53.3 | 46.5 | 16.2 | 79.8 | 1.72 | 53 | 86 | 139 | 855 |
| 7075-T651 | 0.25 | 84.8 | 74.8 | 11.2 | 65.7 | 0.88 | 16 | 23 | 39 | 235 |
| | 0.25 | 84.8 | 74.2 | 13.0 | 77.2 | 1.04 | 12 | 14 | 26 | 215 |
| | 0.25 | 75.8 | 75.3 | 13.2 | 79.8 | 1.06 | 12 | 9 | 20 | 140 |
| | 0.25 | 84.0 | 72.0 | 13.0 | 79.3 | 1.10 | 10 | 12 | 23 | 190 |
| | 0.25 | 86.1 | 74.8 | 12.5 | 67.0 | 0.90 | 12 | 12 | 24 | 120 |
| | 0.38 | 83.8 | 74.6 | 12.0 | 77.8 | 1.04 | 20 | 15 | 35 | 150 |
| | 0.50 | 82.4 | 71.6 | 11.2 | 73.4 | 1.03 | 17 | 14 | 31 | 140 |
| | 0.50 | 86.6 | 77.9 | 11.5 | 72.6 | 0.93 | 14 | 17 | 31 | 170 |

(continued)

Each line of data represents a separate lot of material; average of duplicate or triplicate tests. Specimens per Fig. A1.8. generally 0.100 in. thick; in a few cases, 0.063-in. thick specimens were tested. (a) Obsolete alloy

## Table 6.3(b)  (continued)

| Alloy and temper | Thickness, in. | Ultimate tensile strength (UTS), ksi | Tensile yield strength (TYS), ksi | Elongation in 2 in., % | Tear strength, ksi | Ratio tear strength to yield strength, (TYR) | Initiate a crack, in.-lb | Propagate a crack, in.-lb | Total energy, in.-lb | Unit propagation energy, in.-lb/in.² |
|---|---|---|---|---|---|---|---|---|---|---|
| | 1.00 | 88.0 | 79.2 | 9.8 | 63.3 | 0.80 | 11 | 17 | 28 | 175 |
| | 1.25 | 88.0 | 78.8 | 10.0 | 50.5 | 0.64 | 7 | 12 | 19 | 120 |
| | 1.38 | 86.6 | 77.4 | 10.0 | 67.9 | 0.88 | 15 | 13 | 28 | 135 |
| | 2.50 | 83.4 | 71.8 | 7.5 | 57.9 | 0.81 | 12 | 12 | 24 | 125 |
| | 2.75 | 79.0 | 66.6 | 9.0 | 54.6 | 0.82 | 9 | 12 | 21 | 120 |
| 7075-T7351 | 0.25 | 72.4 | 60.4 | 11.0 | 76.5 | 1.27 | 22 | 32 | 54 | 325 |
| | 0.25 | 73.2 | 61.6 | 11.0 | 75.8 | 1.23 | 22 | 32 | 54 | 325 |
| | 0.25 | 71.8 | 59.4 | 12.0 | 76.8 | 1.29 | 13 | 26 | 40 | 410 |
| | 0.50 | 73.0 | 62.1 | 11.2 | 73.6 | 1.19 | 18 | 27 | 45 | 275 |
| | 1.00 | 72.4 | 61.8 | 10.0 | 74.2 | 1.20 | 20 | 30 | 50 | 300 |
| | 1.00 | 73.6 | 62.4 | 9.8 | 72.4 | 1.16 | 20 | 30 | 50 | 305 |
| | 1.38 | 75.6 | 65.2 | 10.0 | 72.4 | 1.11 | 19 | 14 | 33 | 145 |
| | 2.50 | 69.5 | 57.1 | 9.2 | 67.0 | 1.17 | 16 | 25 | 41 | 255 |
| 7079-T651(a) | 0.25 | 81.0 | 72.6 | 11.5 | 67.7 | 0.93 | 14 | 21 | 35 | 210 |
| | 0.25 | 79.6 | 71.0 | 12.5 | 81.0 | 1.14 | 21 | 20 | 40 | 200 |
| | 0.25 | 80.2 | 71.2 | 12.0 | 77.2 | 1.08 | 17 | 18 | 35 | 175 |
| | 1.50 | 83.4 | 74.4 | 11.2 | 73.9 | 0.99 | 15 | 23 | 38 | 230 |
| | 3.00 | 75.6 | 66.5 | 8.0 | 57.4 | 0.86 | 10 | 13 | 23 | 130 |
| | 4.00 | 79.4 | 70.5 | 7.5 | 52.8 | 0.75 | 10 | 13 | 23 | 130 |
| 7106-T6351(a) | 0.25 | 62.2 | 54.4 | 13.4 | 84.4 | 1.55 | 21 | 44 | 65 | 695 |
| | 0.50 | 65.4 | 57.7 | 14.4 | 84.4 | 1.46 | 39 | 93 | 131 | 935 |
| | 1.50 | 65.2 | 58.5 | 12.2 | 87.3 | 1.53 | 39 | 68 | 107 | 685 |
| | 3.00 | 58.9 | 53.2 | 11.0 | 80.4 | 1.51 | 30 | 53 | 83 | 535 |
| 7139-T6351(a) | 1.00 | 68.6 | 59.0 | 12.0 | 76.6 | 1.30 | 21 | 22 | 44 | 220 |
| 7178-T651 | 0.25 | 90.4 | 80.4 | 12.5 | 65.5 | 0.82 | 8 | 5 | 13 | 50 |
| | 0.31 | 90.2 | 80.8 | 10.8 | 62.2 | 0.77 | 9 | 5 | 14 | 50 |
| | 0.31 | 90.3 | 82.3 | 11.0 | 52.3 | 0.64 | 4 | 16 | 20 | 165 |
| | 0.34 | 88.3 | 77.0 | 12.0 | 55.4 | 0.72 | 8 | 7 | 15 | 75 |
| | 0.44 | 87.2 | 76.7 | 12.0 | 54.6 | 0.71 | 6 | 12 | 18 | 115 |
| | 0.50 | 92.4 | 82.0 | 11.2 | 55.4 | 0.68 | 9 | 11 | 20 | 110 |
| | 0.50 | 90.2 | 80.8 | 11.2 | 60.7 | 0.75 | 9 | 13 | 22 | 130 |
| | 0.63 | 89.8 | 81.5 | 9.7 | 51.4 | 0.63 | 8 | 10 | 18 | 100 |
| | 1.00 | 89.6 | 80.9 | 8.2 | 42.3 | 0.52 | 2 | 5 | 7 | 50 |
| | 1.25 | 94.2 | 83.0 | 9.0 | 53.1 | 0.64 | 7 | 12 | 19 | 125 |
| 7178-T7651 | 0.25 | 77.5 | 67.0 | 11.0 | 73.1 | 1.09 | 20 | 14 | 34 | 145 |
| | 0.31 | 78.5 | 68.0 | 11.0 | 69.2 | 1.02 | 14 | 15 | 29 | 150 |
| | 1.00 | 78.3 | 67.7 | 11.0 | 69.0 | 1.02 | 16 | 15 | 31 | 150 |
| | 1.00 | 79.8 | 70.5 | 10.3 | 70.8 | 1.00 | 16 | 16 | 32 | 155 |

Each line of data represents a separate lot of material; average of duplicate or triplicate tests. Specimens per Fig. A1.8. generally 0.100 in. thick; in a few cases, 0.063-in. thick specimens were tested. (a) Obsolete alloy

**Table 6.4(a)  Results of tensile and tear tests of aluminum alloy extruded shapes. longitudinal**

| | | Tensile tests | | | Tear tests | | | | | |
|---|---|---|---|---|---|---|---|---|---|---|
| | | | | | | Ratio tear strength to yield strength, (TYR) | Energy required to: | | | |
| Alloy and temper | Thickness, in. | Ultimate tensile strength (UTS), ksi | Tensile yield strength (TYS), ksi | Elongation in 2 in., % | Tear strength, ksi | | Initiate a crack, in.-lb | Propagate a crack, in.-lb | Total energy, in.-lb | Unit propagation energy, in.-lb/in.² |
| 1350-H12 | 0.25 | 15.0 | 11.1 | 32.2 | 23.8 | 2.14 | 62 | 96 | 158 | 960 |
| 2014-T6 | 0.25 | 67.3 | 63.9 | 12.2 | 74.8 | 1.17 | 53 | 100 | 153 | 395 |
| | 0.50 | 76.1 | 69.9 | 10.8 | 71.4 | 1.02 | 13 | 36 | 49 | 270 |
| 2020-T6510(a) | 0.25 | 77.4 | 72.3 | 10.2 | 78.2 | 1.08 | 28 | 38 | 66 | 380 |
| | 0.25 | 76.0 | 70.8 | 9.8 | 84.6 | 1.20 | 25 | 37 | 61 | 365 |
| | 0.25 | 85.8 | 81.7 | 7.0 | 54.6 | 0.67 | 13 | 15 | 27 | 60 |
| | 0.69 | 77.2 | 75.4 | 7.0 | 75.9 | 1.01 | 18 | 40 | 58 | 405 |
| | 2.00 | 80.6 | 74.7 | 7.5 | 63.4 | 0.85 | 10 | 12 | 22 | 120 |
| | 2.00 | 89.3 | 83.8 | 7.8 | 87.2 | 1.04 | 25 | 55 | 91 | 660 |
| | 2.00 | 92.9 | 88.2 | 6.8 | 59.1 | 0.67 | 9 | 12 | 21 | 125 |
| 2024-T4 | 0.25 | 69.8 | 50.4 | 21.0 | 78.0 | 1.55 | 74 | 153 | 227 | 620 |
| 2024-T8511 | 2.00 | 71.4 | 64.0 | 10.2 | 68.3 | 1.07 | 16 | 33 | 49 | 330 |
| 2219-T8511 | 0.13 | 64.7 | 48.4 | 10.0 | 76.8 | 1.59 | 30 | 72 | 102 | 725 |
| 5456-H311 | 0.19 | 52.1 | 31.3 | 17.0 | 47.2 | 1.51 | 46 | 130 | 176 | 1300 |
| | 0.19 | 53.6 | 31.1 | 16.0 | 60.1 | 1.93 | 48 | 139 | 188 | 1880 |
| | 0.25 | 54.1 | 32.9 | 15.0 | 62.9 | 1.91 | 59 | 114 | 173 | 1140 |
| | 0.25 | 53.4 | 33.5 | 16.0 | 65.2 | 1.95 | 49 | 211 | 260 | 2110 |
| | 0.46 | 55.8 | 33.5 | 15.4 | 34.8 | 1.04 | 48 | 144 | 192 | 1440 |
| | 0.44 | 55.2 | 32.0 | 18.0 | 59.6 | 1.86 | 43 | 134 | 177 | 1340 |
| | 0.69 | 53.8 | 27.8 | 23.0 | 60.1 | 2.16 | 44 | 142 | 186 | 1420 |
| 6005-T51 | 0.13 | 41.8 | 36.4 | 10.5 | 51.5 | 1.40 | 18 | 27 | 45 | 225 |
| | 0.19 | 42.0 | 36.7 | 13.0 | 55.2 | 1.50 | 15 | 23 | 38 | 230 |
| | 0.19 | 42.4 | 36.7 | 11.0 | 55.0 | 1.50 | 19 | 47 | 66 | 470 |
| 6005-T6 | 0.13 | 45.0 | 41.4 | 11.0 | 65.6 | 1.59 | 40 | 70 | 110 | 560 |
| | 0.13 | 45.0 | 41.4 | 11.0 | 63.8 | 1.54 | 35 | 63 | 98 | 505 |
| | 0.13 | 45.2 | 41.5 | 11.8 | 63.8 | 1.53 | 29 | 62 | 91 | 510 |
| 6051-T6(a) | 0.25 | 46.6 | 43.8 | 8.5 | 60.9 | 1.39 | 12 | 20 | 32 | 310 |
| 6061-T51 | 0.19 | 41.5 | 32.1 | 13.5 | 58.7 | 1.83 | 27 | 78 | 105 | 780 |
| | 0.19 | 43.8 | 37.0 | 11.5 | 61.7 | 1.67 | 28 | 59 | 87 | 590 |
| | 0.19 | 42.7 | 38.0 | 12.0 | 56.8 | 1.50 | 18 | 41 | 59 | 410 |
| | 0.19 | 34.9 | 27.9 | 12.0 | 53.2 | 1.91 | 18 | 49 | 67 | 495 |
| | 0.19 | 40.9 | 36.9 | 11.5 | 56.7 | 1.53 | 19 | 39 | 58 | 390 |
| | 0.19 | 41.8 | 37.2 | 12.5 | 57.4 | 1.54 | 25 | 62 | 87 | 615 |
| | 0.19 | 43.6 | 41.3 | 12.0 | 68.0 | 1.65 | 42 | 113 | 155 | 1125 |
| | 0.19 | 36.4 | 31.8 | 12.0 | 53.8 | 1.69 | 26 | 87 | 113 | 860 |
| 6061-T6 | 0.13 | 46.5 | 44.0 | 12.0 | 73.0 | 1.66 | 48 | 135 | 183 | 1350 |
| | 0.13 | 44.8 | 41.5 | 11.5 | 66.0 | 1.58 | 38 | 74 | 122 | 590 |
| | 0.13 | 50.0 | 44.2 | 16.0 | 65.0 | 1.47 | 42 | 98 | 140 | 975 |
| | 0.13 | 50.4 | 43.4 | 15.9 | 64.1 | 1.48 | 37 | 95 | 132 | 940 |
| 6062-T6 | 0.25 | 47.3 | 43.9 | 16.0 | 73.6 | 1.68 | 105 | 328 | 433 | 1320 |
| | 0.25 | 47.6 | 45.2 | 12.2 | 65.9 | 1.46 | 14 | 32 | 46 | 500 |
| 6063-T5 | 0.25 | 24.2 | 19.6 | 13.5 | 36.9 | 1.88 | 19 | 62 | 81 | 960 |
| 6063-T6 | 0.25 | ... | ... | ... | 64.0 | ... | 31 | 110 | 141 | 1100 |
| 6066-T6 | 0.13 | ... | ... | ... | 73.8 | ... | 26 | 45 | 71 | 705 |
| | 0.25 | 55.1 | 36.8 | 20.0 | 75.9 | 2.06 | 22 | 25 | 47 | 390 |
| | 0.50 | 52.4 | 46.4 | 16.0 | 68.1 | 1.47 | 15 | 35 | 50 | 545 |
| 6066-T6511 | 0.50 | 58.0 | 51.8 | 12.2 | 70.1 | 1.35 | 20 | 41 | 61 | 410 |
| 6070-T6 | 0.25 | 55.1 | 55.9 | 13.0 | 76.6 | 1.37 | 91 | 220 | 312 | 870 |
| | 0.25 | 52.8 | 52.4 | 8.4 | 78.6 | 1.47 | 83 | 229 | 312 | 870 |
| | 0.25 | 57.6 | 54.6 | 10.0 | 78.6 | 1.44 | 29 | 69 | 98 | 690 |
| | 0.25 | 55.3 | 52.5 | 12.0 | 66.4 | 1.26 | 12 | 36 | 49 | 560 |
| 6101-T6 | 0.25 | 29.8 | 25.8 | 16.5 | 47.9 | 1.86 | 41 | 143 | 184 | 1430 |
| 6151-T6 | 0.06 | 48.8 | 44.0 | 14.0 | 72.1 | 1.64 | 28 | 79 | 108 | 990 |
| | 0.75 | 53.4 | 50.1 | 16.5 | 75.9 | 1.52 | 36 | 93 | 130 | 930 |
| | 0.75 | 50.0 | 46.0 | 16.0 | 70.8 | 1.54 | 35 | 71 | 106 | 705 |
| | 0.75 | 51.4 | 46.8 | 18.0 | 72.6 | 1.55 | 43 | 101 | 144 | 1005 |
| | 0.75 | 54.8 | 50.8 | 17.0 | 75.6 | 1.49 | 38 | 80 | 117 | 800 |
| 6351-T51 | 0.19 | 44.1 | 38.0 | 13.5 | 67.6 | 1.78 | 37 | 79 | 116 | 785 |
| | 0.19 | 40.8 | 34.5 | 13.0 | 59.6 | 1.73 | 36 | 91 | 127 | 910 |

(continued)

Specimens per Fig. A1.8. Each line of data represents average of duplicate or triplicate tests for an individual lot of material. Specimens were generally about 0.100 in. thick; for shapes less than 0.2 in. in thickness, full-thickness specimens were sometimes used. For yield strengths, off set is 0.2%. (a) Obsolete alloy. (b) Crack path was diagonal; propagation values may be unrealistically high.

## Table 6.4(a)   (continued)

| Alloy and temper | Thickness, in. | Tensile tests | | | Tear tests | | | | | |
|---|---|---|---|---|---|---|---|---|---|---|
| | | Ultimate tensile strength (UTS), ksi | Tensile yield strength (TYS), ksi | Elongation in 2 in., % | Tear strength, ksi | Ratio tear strength to yield strength, (TYR) | Energy required to: | | Total energy, in.-lb | Unit propagation energy, in.-lb/in.$^2$ |
| | | | | | | | Initiate a crack, in.-lb | Propagate a crack, in.-lb | | |
| | 0.19 | 42.8 | 37.5 | 12.0 | 62.2 | 1.66 | 34 | 88 | 122 | 880 |
| | 0.19 | 43.2 | 38.0 | 13.0 | 66.4 | 1.75 | 42 | 84 | 126 | 845 |
| | 0.19 | 40.3 | 34.5 | 11.5 | 61.6 | 1.79 | 42 | 92 | 134 | 920 |
| | 0.19 | 38.3 | 32.7 | 12.5 | 59.6 | 1.82 | 40 | 102 | 142 | 1025 |
| | 0.19 | 42.5 | 37.5 | 12.5 | 63.6 | 1.72 | 35 | 86 | 121 | 860 |
| | 0.19 | 41.7 | 37.5 | 12.5 | 62.0 | 1.65 | 34 | 89 | 123 | 880 |
| | 0.19 | 42.5 | 38.9 | 11.5 | 65.8 | 1.69 | 38 | 79 | 135 | 970 |
| | 0.19 | 38.3 | 34.0 | 12.5 | 58.0 | 1.70 | 40 | 114 | 154 | 1150 |
| | 0.19 | 36.6 | 31.6 | 13.0 | 54.5 | 1.73 | 40 | 111 | 151 | 1120 |
| | 0.19 | 41.1 | 38.7 | 12.5 | 67.5 | 1.70 | 45 | 121 | 166 | 1205 |
| 6351-T6 | 0.13 | 48.7 | 45.8 | 10.5 | 73.4 | 1.60 | 54 | 132 | 186 | 1060 |
| | 0.06 | 53.4 | 48.7 | 12.5 | 76.6 | 1.57 | 30 | 88 | 118 | 1100 |
| | 0.25 | 50.4 | 47.4 | 12.0 | 67.9 | 1.43 | 18 | 43 | 61 | 675 |
| 7001-T6(a) | 2.00 | 101.8 | 95.6 | 10.0 | 94.0 | 0.98 | 36 | <10 | <46 | <100 |
| 7001-T73(a) | 2.00 | 84.0 | 75.6 | 10.0 | 88.4 | 1.17 | 37 | 40 | 77 | 410 |
| 7001-T75(a) | 0.19 | 84.9 | 76.4 | 9.2 | 61.1 | 0.80 | 11 | 19 | 30 | 185 |
| 7005-T53 | 0.25 | 59.8 | 51.6 | 14.3 | 86.4 | 1.67 | 67 | 104 | 171 | 1040 |
| | 0.75 | 55.8 | 49.4 | 18.5 | 82.7 | 1.67 | 73 | 113 | 186 | 1130 |
| 7005-T6 | 1.50 | 58.0 | 50.7 | 14.0 | 85.2 | 1.68 | 71 | 97 | 168 | 970 |
| | 1.50 | 57.8 | 49.6 | 17.0 | 87.1 | 1.76 | 83 | 114 | 198 | 1140 |
| 7005-T63 | 0.25 | 62.0 | 54.5 | 14.2 | 87.8 | 1.61 | 62 | 114 | 176 | 1140 |
| | 1.50 | 59.5 | 52.2 | 17.0 | 84.2 | 1.61 | 57 | 110 | 168 | 1105 |
| X7006-T53 | 0.25 | 60.8 | 55.6 | 13.5 | 84.6 | 1.52 | 48 | 122 | 170 | 1225 |
| X7006-T63 | 0.25 | 63.0 | 57.2 | 14.8 | 90.8 | 1.59 | 66 | 184 | 250 | 1860 |
| 7039-T53 | 0.25 | 63.8 | 55.9 | 13.3 | 84.4 | 1.51 | 52 | 112 | 164 | 1130 |
| 7039-T63 | 0.25 | 62.7 | 56.6 | 14.3 | 88.6 | 1.57 | 46 | 125 | 125 | 1260 |
| 7075-T651X | 0.13 | 90.0 | 81.6 | 11.2 | 78.0 | 0.95 | 19 | 36 | 55 | 360 |
| | 0.13 | 89.2 | 85.2 | 10.5 | 74.0 | 0.87 | 14 | 40 | 54 | 400 |
| | 0.19 | 93.9 | 86.4 | 10.8 | 86.5 | 1.00 | 21 | 35 | 56 | 350 |
| | 0.30 | 85.1 | 77.0 | 11.5 | 73.8 | 1.03 | 14 | 42 | 56 | 415 |
| | 0.30 | 79.2 | 71.4 | 5.2 | 80.2 | 1.12 | 16 | 55 | 72 | 550 |
| | 0.31 | 86.7 | 78.6 | 12.0 | 77.7 | 0.99 | 14 | 28(b) | 42 | 435(b) |
| | 0.46 | 93.0 | 84.2 | 12.0 | 78.4 | 0.93 | 27 | 47 | 73 | 470 |
| | 0.50 | 85.5 | 76.5 | 11.8 | 80.4 | 1.05 | 20 | 46(b) | 65 | 460(b) |
| | 0.50 | 87.2 | 77.9 | 10.5 | 75.8 | 0.91 | 16 | 41(b) | 57 | 410(b) |
| | 0.69 | 90.4 | 83.8 | 10.7 | 75.4 | 0.90 | 16 | 46 | 62 | 460 |
| | 0.69 | 88.2 | 80.2 | 11.8 | 75.9 | 0.95 | 18 | 28(b) | 46 | 280(b) |
| | 0.75 | 92.2 | 85.4 | 10.5 | 81.8 | 0.96 | 31 | <10 | <41 | <100 |
| | 0.75 | 88.4 | 80.8 | 10.5 | 77.2 | 0.96 | 17 | 37(b) | 53 | 365(b) |
| | 0.75 | 88.6 | 81.6 | 10.8 | 82.0 | 1.00 | 18 | 33(b) | 51 | 330(b) |
| | 0.75 | 88.4 | 81.0 | 11.0 | 76.6 | 0.95 | 16 | 25 | 41 | 245 |
| | 0.75 | 84.9 | 77.6 | 12.2 | 84.0 | 1.08 | 21 | 32 | 53 | 280 |
| | 0.75 | 84.8 | 77.6 | 12.0 | 87.8 | 1.13 | 25 | 29 | 54 | 285 |
| | 0.94 | 89.6 | 83.1 | 12.5 | 78.2 | 0.94 | 19 | 38(b) | 56 | 375(b) |
| | 0.94 | 88.8 | 82.8 | 12.0 | 82.1 | 0.99 | 20 | 47(b) | 66 | 465(b) |
| | 0.94 | 87.5 | 81.0 | 11.2 | 80.0 | 0.99 | 19 | 37 | 56 | 370 |
| | 0.94 | 89.2 | 83.6 | 12.0 | 78.4 | 0.94 | 18 | 36 | 53 | 360 |
| | 1.13 | 90.0 | 82.8 | 11.0 | 75.1 | 0.91 | 14 | 23(b) | 37 | 230(b) |
| | 1.25 | 87.0 | 79.4 | 12.8 | 78.0 | 0.98 | 16 | 28 | 44 | 275 |
| | 1.25 | 86.4 | 79.4 | 12.0 | 73.6 | 0.93 | 14 | 24 | 38 | 240 |
| | 3.00 | 88.2 | 80.8 | 11.4 | 79.8 | 0.99 | 16 | 27 | 43 | 270 |
| 7075-T7351X | 0.06 | 78.6 | 69.1 | 9.0 | 81.2 | 1.18 | 16 | 35 | 51 | 505 |
| | 0.25 | 77.4 | 68.0 | 12.0 | 86.1 | 1.27 | 21 | 34 | 55 | 535 |
| | 0.50 | 78.9 | 69.6 | 13.0 | 81.6 | 1.17 | 24 | 50 | 74 | 495 |
| | 0.50 | 79.8 | 70.4 | 11.0 | 79.4 | 1.13 | 21 | 41 | 62 | 415 |
| | 0.50 | 80.8 | 71.8 | 14.0 | 91.2 | 1.54 | 33 | 64 | 96 | 635 |
| | 0.69 | 73.2 | 61.5 | 12.9 | 79.7 | 1.30 | 25 | 62 | 87 | 615 |
| | 0.75 | 79.3 | 70.0 | 11.5 | 82.6 | 1.18 | 23 | 50 | 73 | 500 |
| | 0.75 | 81.0 | 72.8 | 11.0 | 83.6 | 1.15 | 23 | 39 | 61 | 385 |
| | 0.75 | 81.8 | 73.4 | 11.0 | 82.8 | 1.13 | 22 | 36 | 57 | 355 |

(continued)

Specimens per Fig. A1.8. Each line of data represents average of duplicate or triplicate tests for an individual lot of material. Specimens were generally about 0.100 in. thick; for shapes less than 0.2 in. in thickness, full-thickness specimens were sometimes used. For yield strengths, offset is 0.2%. (a) Obsolete alloy. (b) Crack path was diagonal; propagation values may be unrealistically high.

**Table 6.4(a)** (continued)

| Alloy and temper | Thickness, in. | Tensile tests | | | Tear tests | | | | | |
|---|---|---|---|---|---|---|---|---|---|---|
| | | Ultimate tensile strength (UTS), ksi | Tensile yield strength (TYS), ksi | Elongation in 2 in., % | Tear strength, ksi | Ratio tear strength to yield strength (TYR) | Energy required to: | | Total energy, in.-lb | Unit propagation energy, in.-lb/in.² |
| | | | | | | | Initiate a crack, in.-lb | Propagate a crack, in.-lb | | |
| | 0.75 | 80.4 | 72.0 | 12.5 | 85.5 | 1.39 | 23 | 44 | 67 | 440 |
| | 0.75 | 81.4 | 73.5 | 12.5 | 82.8 | 1.13 | 23 | 38 | 62 | 380 |
| | 1.13 | 79.5 | 70.7 | 12.0 | 85.4 | 1.21 | 25 | 34 | 59 | 335 |
| | 1.25 | 77.7 | 67.6 | 13.5 | 86.7 | 1.28 | 31 | 75(b) | 106 | 755(b) |
| | 1.25 | 74.0 | 63.6 | 12.8 | 82.6 | 1.30 | 25 | 52 | 78 | 520 |
| | 3.00 | 76.2 | 68.0 | 12.0 | 82.0 | 1.21 | 17 | 31 | 48 | 490 |
| 7075-T7651X | 0.30 | 82.4 | 73.5 | 11.5 | 77.9 | 1.06 | 18 | 24 | 42 | 240 |
| | 0.50 | 80.1 | 71.1 | 13.0 | 82.5 | 1.16 | 26 | 55(b) | 81 | 550(b) |
| | 0.50 | 79.0 | 69.2 | 12.0 | 77.7 | 1.12 | 20 | 40(b) | 60 | 400(b) |
| | 0.69 | 79.0 | 69.6 | 11.4 | 78.7 | 1.13 | 22 | 32 | 54 | 315 |
| | 0.75 | 79.7 | 70.6 | 11.5 | 83.7 | 1.19 | 22 | 34 | 56 | 340 |
| | 0.75 | 80.3 | 71.7 | 12.0 | 84.2 | 1.17 | 21 | 37 | 58 | 370 |
| | 0.75 | 80.8 | 72.4 | 11.5 | 84.4 | 1.17 | 23 | 31 | 53 | 300 |
| | 0.75 | 80.2 | 71.7 | 13.0 | 86.1 | 1.20 | 25 | 40 | 65 | 400 |
| | 0.75 | 81.0 | 72.7 | 12.5 | 84.8 | 1.17 | 22 | 31 | 54 | 310 |
| | 1.13 | 80.7 | 72.3 | 12.0 | 85.6 | 1.19 | 24 | 42 | 66 | 420 |
| | 1.25 | 77.1 | 67.8 | 13.5 | 84.5 | 1.25 | 24 | 60 | 84 | 605 |
| | 1.25 | 76.0 | 66.5 | 12.0 | 84.5 | 1.27 | 27 | 38 | 65 | 385 |
| 7079-T6(a) | 0.13 | 82.5 | 74.1 | 10.0 | 74.6 | 1.01 | 18 | 54 | 72 | 540 |
| | 0.13 | 81.2 | 73.9 | 9.4 | 81.0 | 1.10 | 22 | 43 | 65 | 430 |
| | 0.19 | 84.8 | 77.6 | 9.9 | 76.4 | 0.98 | 17 | 33 | 50 | 330 |
| | 0.70 | 86.7 | 80.2 | 10.0 | 71.2 | 0.89 | 14 | 36 | 50 | 360 |
| X7106-T53(a) | 0.19 | 61.0 | 54.4 | 12.2 | 86.2 | 1.58 | 49 | 120 | 169 | 1185 |
| | 0.25 | 65.1 | 56.6 | 13.5 | 91.6 | 1.62 | 49 | 114 | 162 | 1130 |
| X7139-T53(a) | 0.25 | 66.7 | 55.7 | 12.2 | 85.8 | 1.54 | 40 | 82 | 122 | 815 |
| X7139-T63(a) | 0.25 | 70.3 | 62.3 | 12.0 | 90.7 | 1.46 | 43 | 88 | 131 | 890 |
| 7178-T651X | 0.25 | 96.0 | 89.0 | 10.0 | 49.0 | 0.55 | 12 | <10 | <22 | <100 |
| | 0.25 | 93.0 | 86.1 | 1.5 | 56.8 | 0.66 | 7 | 10 | 17 | 95 |
| | 1.25 | 91.8 | 86.4 | 1.5 | 62.2 | 0.72 | 10 | 13 | 23 | 135 |
| 7178-T7651X | 0.07 | 83.5 | 73.2 | ... | 81.2 | 1.11 | 20 | 39(b) | 58 | 560(b) |
| | 0.19 | 82.2 | 72.5 | 9.5 | 72.7 | 1.00 | 13 | 28 | 41 | 290 |
| | 0.19 | 80.3 | 70.7 | 10.0 | 77.2 | 1.09 | 19 | 43(b) | 61 | 430(b) |
| | 0.19 | 79.3 | 68.7 | 8.5 | 76.5 | 1.11 | 20 | 46 | 65 | 460 |
| | 0.19 | 79.5 | 70.1 | 9.1 | 74.0 | 1.06 | 16 | 31(b) | 47 | 315(b) |
| | 0.38 | 81.8 | 72.5 | ... | 74.0 | 1.02 | 19 | 40 | 54 | 355 |
| | 0.50 | 83.4 | 74.4 | ... | 77.5 | 1.04 | 18 | 35 | 53 | 345 |
| | 0.69 | 83.2 | 73.0 | 11.6 | 76.5 | 1.05 | 19 | 23 | 42 | 235 |
| | 1.25 | 77.8 | 68.5 | 11.0 | 79.4 | 1.16 | 21 | 38 | 58 | 375 |
| | 1.25 | 82.0 | 74.4 | ... | 77.5 | 1.04 | 19 | 31 | 50 | 310 |
| | 1.25 | 83.2 | 74.0 | 8.8 | 72.8 | 0.98 | 10 | 19(b) | 39 | 290(b) |
| | 1.25 | 91.9 | 73.4 | 10.2 | 72.4 | 0.99 | 15 | 28 | 43 | 280 |

Specimens per Fig. A1.8. Each line of data represents average of duplicate or triplicate tests for an individual lot of material. Specimens were generally about 0.100 in. thick; for shapes less than 0.2 in. in thickness, full-thickness specimens were sometimes used. For yield strengths, off set is 0.2%. (a) Obsolete alloy. (b) Crack path was diagonal; propagation values may be unrealistically high.

**Table 6.4(b)   Results of tensile and tear tests of aluminum alloy extruded shapes, transverse**

| | | Tensile tests | | | Tear tests | | | | | |
|---|---|---|---|---|---|---|---|---|---|---|
| | | | | | | Ratio tear strength to yield strength, (TYR) | Energy required to: | | | Unit propagation energy, in.-lb/in.$^2$ |
| Alloy and temper | Thickness, in. | Ultimate tensile strength (UTS), ksi | Tensile yield strength (TYS), ksi | Elongation in 2 in., % | Tear strength, ksi | | Initiate a crack, in.-lb | Propagate a crack, in.-lb | Total energy, in.-lb | |
| 1350-H12 | 0.25 | 13.8 | 11.2 | 11.0 | 24.3 | 2.17 | 51 | 95 | 146 | 945 |
| 2014-T6 | 0.25 | ... | ... | ... | 56.4 | ... | 26 | 41 | 67 | 160 |
| | 0.50 | 74.8 | 69.0 | 10.0 | 53 | 0.77 | 6 | 9 | 15 | 85 |
| 2020-T6510(a) | 0.25 | 79.5 | 75.5 | 2.0 | 44.4 | 0.59 | 7 | 9 | 16 | 90 |
| | 0.25 | ... | ... | ... | 47.2 | ... | 10 | <5 | <15 | <50 |
| | 0.69 | 82.2 | 80.6 | 0.8 | 53 | 0.66 | 7 | 9 | 16 | 85 |
| | 2.00 | 81.7 | 76.2 | 4.7 | 45.4 | 0.60 | 5 | 7 | 12 | 65 |
| 2024-T4 | 0.25 | ... | ... | ... | 77.8 | ... | 71 | 116 | 187 | 460 |
| 2024-T8511 | 2.00 | 71.4 | 65.7 | 6.4 | 44.6 | 0.68 | 7 | 13 | 20 | 125 |
| 2219-T8511 | 0.13 | 64.7 | 48.4 | 10.0 | 73.1 | 1.51 | 28 | 35 | 63 | 350 |
| | 0.25 | 53.4 | 29.0 | 20.0 | 60.8 | 2.10 | 53 | 99 | 152 | 990 |
| | 0.25 | 54.8 | 30.2 | 16.8 | 62.6 | 2.07 | 59 | 106 | 165 | 1060 |
| | 0.46 | 53.2 | 29.8 | 24.0 | 61.4 | 2.06 | 43 | 87 | 130 | 870 |
| | 0.44 | 51.7 | 28.6 | 24.5 | 59.4 | 2.08 | 45 | 97 | 142 | 970 |
| | 0.69 | 51.4 | 28.6 | 23.0 | 57.7 | 2.02 | 41 | 93 | 134 | 930 |
| 6005-T51 | 0.13 | 43.8 | 36.0 | 9.5 | 50.8 | 1.39 | 13 | 19 | 32 | 155 |
| | 0.19 | 43.1 | 37.4 | 11.0 | 55.0 | 1.47 | 19 | 35 | 54 | 355 |
| | 0.19 | 40.3 | 33.7 | 14.5 | 55.4 | 1.64 | 18 | 33 | 51 | 330 |
| 6005-T6 | 0.13 | 46.3 | 40.1 | 11.5 | 65.2 | 1.62 | 32 | 51 | 83 | 405 |
| | 0.13 | 46.2 | 40.2 | 12.0 | 63.2 | 1.57 | 28 | 42 | 70 | 330 |
| | 0.13 | 47.0 | 39.4 | 10.0 | 65.0 | 1.58 | 25 | 46 | 71 | 380 |
| 6061-T51 | 0.19 | 43.7 | 33.9 | 14.0 | 63.0 | 1.86 | 34 | 56 | 90 | 560 |
| | 0.19 | 46.2 | 38.4 | 12.0 | 64.2 | 1.67 | 27 | 53 | 80 | 535 |
| | 0.19 | 42.7 | 38.0 | 13.0 | 56.2 | 1.48 | 16 | 23 | 39 | 225 |
| | 0.19 | 41.0 | 34.0 | 10.0 | 53.2 | 1.57 | 15 | 27 | 42 | 265 |
| | 0.19 | 43.6 | 38.6 | 10.0 | 54.7 | 1.48 | 14 | 25 | 38 | 235 |
| | 0.19 | 40.6 | 35.1 | 15.5 | 57.2 | 1.63 | 17 | 28 | 45 | 275 |
| | 0.19 | 44.7 | 40.2 | 15.0 | 70.8 | 1.76 | 34 | 57 | 91 | 565 |
| | 0.19 | 36.0 | 30.6 | 18.0 | 55.6 | 1.86 | 23 | 47 | 70 | 470 |
| 6061-T6 | 0.13 | 45.4 | 42.2 | 17.0 | 74.3 | 1.76 | 48 | 78 | 126 | 780 |
| | 0.13 | 47.2 | 40.8 | 10.0 | 59.8 | 1.48 | 19 | 37 | 56 | 300 |
| 6062-T6 | 0.25 | ... | ... | ... | 76.6 | ... | 108 | 179 | 287 | 720 |
| 6063-T6 | 0.25 | ... | ... | ... | 56.6 | ... | 31 | 68 | 99 | 680 |
| 6066-T6 | 0.25 | 53.2 | 34.7 | 15.0 | 61.8 | 1.78 | 11 | 11 | 22 | 175 |
| | 0.50 | 58.8 | 53.4 | 12.0 | 59.5 | 1.11 | 10 | 10 | 20 | 150 |
| 6066-T6511 | 0.50 | 59.3 | 52.7 | 10.0 | 61.6 | 1.17 | 12 | 14 | 26 | 135 |
| 6070-T6 | 0.25 | ... | ... | ... | 67.0 | ... | 64 | 77 | 140 | 305 |
| | 0.25 | ... | ... | ... | 65.8 | ... | 36 | 49 | 85 | 190 |
| | 0.25 | 58.7 | 55.6 | 14.0 | 79.7 | 1.43 | 25 | 38 | 63 | 370 |
| 6101-T6 | 0.25 | 30.2 | 26.1 | 14.0 | 49.6 | 1.90 | 50 | 141 | 192 | 1410 |
| 6151-T6 | 0.06 | ... | ... | ... | 76.8 | ... | 29 | 49 | 79 | 615 |
| | 0.75 | 51.0 | 48.0 | 16.0 | 74.2 | 1.54 | 36 | 60 | 97 | 600 |
| | 0.75 | 47.4 | 43.3 | 17.0 | 69.4 | 1.60 | 34 | 76 | 109 | 755 |
| | 0.75 | 49.2 | 44.1 | 18.0 | 71.8 | 1.63 | 38 | 64 | 102 | 630 |
| | 0.75 | 52.4 | 48.6 | 18.0 | 76.0 | 1.56 | 41 | 55 | 96 | 550 |
| 6351-T51 | 0.19 | 46.0 | 39.8 | 14.5 | 70.4 | 1.77 | 41 | 64 | 105 | 640 |
| | 0.19 | 41.4 | 34.3 | 15.5 | 62.8 | 1.83 | 40 | 87 | 127 | 860 |
| | 0.19 | 42.7 | 36.8 | 13.0 | 63.6 | 1.73 | 32 | 80 | 112 | 800 |
| | 0.19 | 43.4 | 38.0 | 16.0 | 69.1 | 1.82 | 47 | 76 | 123 | 765 |
| | 0.19 | 40.5 | 34.5 | 12.5 | 63.2 | 1.83 | 36 | 70 | 106 | 690 |
| | 0.19 | 40.0 | 33.8 | 13.5 | 61.2 | 1.81 | 38 | 58 | 96 | 575 |
| | 0.19 | 44.4 | 38.2 | 13.0 | 67.2 | 1.76 | 42 | 95 | 137 | 950 |
| | 0.19 | 42.6 | 36.4 | 13.5 | 64.9 | 1.78 | 36 | 71 | 107 | 705 |
| | 0.19 | 43.8 | 38.2 | 15.0 | 69.6 | 1.82 | 44 | 88 | 132 | 875 |
| | 0.19 | 38.0 | 32.3 | 17.5 | 61.0 | 1.88 | 39 | 76 | 115 | 755 |
| | 0.19 | 35.3 | 29.0 | 18.0 | 58.2 | 2.01 | 36 | 78 | 114 | 780 |
| | 0.19 | 42.4 | 38.5 | 15.5 | 69.8 | 1.81 | 43 | 83 | 126 | 835 |
| 6351-T6 | 0.13 | 50.0 | 45.6 | 9.5 | 76.3 | 1.67 | 57 | 95 | 152 | 760 |
| | 0.06 | ... | ... | ... | 78.8 | ... | 30 | 62 | 92 | 770 |
| 7001-T6(a) | 2.00 | 89.4 | 82.3 | 4.0 | 40.2 | 0.49 | 5 | <5 | <10 | <50 |

(continued)

Specimens per Fig. A1.8. Each line of data represents average of duplicate or triplicate tests for an individual lot of material. Specimens were generally about 0.100 in. thick; for shapes less than 0.2 in. in thickness, full-thickness specimens were sometimes used. For yield strengths, offset is 0.2%. (a) Obsolete alloy

**Table 6.4(b)** (continued)

| | | Tensile tests | | | Tear tests | | | | | |
|---|---|---|---|---|---|---|---|---|---|---|
| | | Ultimate tensile strength (UTS), ksi | Tensile yield strength (TYS), ksi | Elongation in 2 in., % | Tear strength, ksi | Ratio tear strength to yield strength (TYR) | Energy required to: | | Total energy, in.-lb | Unit propagation energy, in.-lb/in.$^2$ |
| Alloy and temper | Thickness, in. | | | | | | Initiate a crack, in.-lb | Propagate a crack, in.-lb | | |
| 7001-T73(a) | 2.00 | 78.4 | 70.8 | 6.0 | 43.4 | 0.61 | 6 | <5 | <10 | <50 |
| 7001-T75(a) | 0.19 | 85.1 | 76.5 | 8.0 | 52.0 | 0.68 | 7 | <5 | <10 | <50 |
| 7005-T53 | 0.25 | 58.0 | 49.9 | 16.0 | 86.8 | 1.74 | 64 | 84 | 148 | 840 |
| | 0.75 | 54.2 | 47.2 | 18.5 | 82.7 | 1.75 | 66 | 90 | 156 | 895 |
| 7005-T6 | 1.50 | 55.7 | 48.4 | 17.2 | 84.4 | 1.74 | 63 | 84 | 146 | 840 |
| | 1.50 | 55.4 | 47.6 | 20.5 | 86.4 | 1.82 | 79 | 102 | 181 | 1020 |
| 7005-T63 | 0.25 | 59.1 | 51.1 | 17.5 | 87.3 | 1.71 | 54 | 82 | 136 | 820 |
| | 1.50 | 56.3 | 48.6 | 19.0 | 82.8 | 1.70 | 53 | 67 | 121 | 675 |
| X7006-T53 | 0.25 | 60.0 | 53.1 | 18.0 | 86.8 | 1.63 | 49 | 100 | 149 | 1015 |
| X7006-T63 | 0.25 | 62.2 | 54.4 | 20.0 | 94.0 | 1.73 | 68 | 149 | 217 | 1490 |
| 7039-T53 | 0.25 | 63.6 | 53.8 | 17.5 | 83.6 | 1.55 | 35 | 68 | 103 | 685 |
| 7039-T63 | 0.25 | 64.4 | 55.4 | 19.0 | 90.9 | 1.64 | 46 | 93 | 139 | 930 |
| 7075-T651X | 0.13 | 86.5 | 77.6 | 13.9 | 70.4 | 0.91 | 13 | 20 | 33 | 200 |
| | 0.13 | 86.5 | 77.3 | 9.5 | 70.6 | 0.91 | 12 | 28 | 40 | 280 |
| | 0.19 | 86.4 | 76.4 | 13.0 | 83.4 | 1.09 | 18 | 25 | 43 | 250 |
| | 0.30 | 84.3 | 73.8 | 12.5 | 58.8 | 0.80 | 8 | 19 | 27 | 195 |
| | 0.30 | 86.8 | 75.6 | 12.0 | 67.0 | 0.89 | 22 | 14 | 35 | 140 |
| | 0.31 | 82.6 | 74.0 | 12.0 | 65.9 | 0.89 | 7 | 11 | 19 | 180 |
| | 0.46 | 85.6 | 76.4 | 12.0 | 54.8 | 0.72 | 4 | 13 | 17 | 130 |
| | 0.50 | 83.4 | 73.8 | 15.2 | 76.6 | 0.77 | 17 | 16 | 33 | 165 |
| | 0.50 | 84.5 | 74.2 | 14.0 | 68.8 | 0.93 | 13 | 13 | 26 | 130 |
| | 0.69 | 82.6 | 74.2 | 8.6 | 56.7 | 0.77 | 7 | 15 | 22 | 150 |
| | 0.69 | 85.7 | 75.8 | 13.6 | 65.6 | 0.87 | 14 | 10 | 24 | 100 |
| | 0.75 | 84.3 | 74.5 | 11.0 | 53.6 | 0.72 | 9 | 13 | 22 | 130 |
| | 0.75 | 85.8 | 77.6 | 10.5 | 68.6 | 0.89 | 15 | 14 | 29 | 140 |
| | 0.75 | 87.0 | 78.8 | 11.0 | 66.0 | 0.84 | 12 | 13 | 24 | 125 |
| | 0.75 | 86.8 | 79.0 | 10.0 | 66.0 | 0.83 | 11 | 11 | 22 | 110 |
| | 0.75 | 82.8 | 74.0 | 13.5 | 74.9 | 1.01 | 16 | 17 | 33 | 175 |
| | 0.75 | 82.4 | 73.6 | 13.0 | 77.2 | 1.05 | 15 | 23 | 38 | 230 |
| | 0.94 | 84.6 | 77.7 | 11.4 | 54.8 | 0.71 | 9 | 10 | 19 | 95 |
| | 0.94 | 83.4 | 76.9 | 11.4 | 59.0 | 0.77 | 11 | 13 | 23 | 125 |
| | 0.94 | 83.2 | 75.9 | 11.2 | 53.4 | 0.71 | 8 | 10 | 18 | 100 |
| | 0.94 | 83.8 | 75.8 | 10.9 | 55.4 | 1.17 | 9 | 10 | 19 | 95 |
| | 1.13 | 86.0 | 77.8 | 13.0 | 69.0 | 0.89 | 10 | 12 | 22 | 160 |
| | 1.25 | 85.2 | 75.8 | 14.0 | 68.6 | 0.91 | 13 | 16 | 29 | 160 |
| | 1.25 | 84.0 | 75.4 | 13.0 | 63.4 | 0.84 | 11 | 10 | 21 | 100 |
| | 3.00 | 83.6 | 75.4 | 11.4 | 58.2 | 0.77 | 11 | 9 | 19 | 95 |
| 7075-T7351X | 0.06 | 80.4 | 72.4 | 8.2 | 81.5 | 1.13 | 17 | 25 | 42 | 355 |
| | 0.25 | 78.3 | 68.7 | 15.0 | 78.4 | 1.14 | 13 | 19 | 32 | 295 |
| | 0.50 | 78.5 | 70.0 | 12.5 | 75.8 | 1.08 | 18 | 20 | 38 | 200 |
| | 0.50 | 77.5 | 67.6 | 12.5 | 72.8 | 1.08 | 17 | 20 | 37 | 200 |
| | 0.50 | 76.4 | 66.0 | 14.0 | 85.6 | 1.30 | ... | ... | ... | ... |
| | 0.69 | 72.6 | 60.9 | 12.9 | 75.2 | 1.23 | 22 | 27 | 49 | 270 |
| | 0.75 | 78.2 | 69.4 | 14.0 | 77.6 | 1.12 | 19 | 20 | 38 | 195 |
| | 0.75 | 80.2 | 72.0 | 12.0 | 78.2 | 1.09 | 20 | 18 | 38 | 185 |
| | 0.75 | 81.2 | 72.8 | 11.5 | 75.3 | 1.04 | 18 | 18 | 36 | 180 |
| | 0.75 | 79.8 | 71.4 | 13.0 | 79.7 | 1.12 | 20 | 17 | 37 | 165 |
| | 0.75 | 80.6 | 72.3 | 13.5 | 77.2 | 0.75 | 18 | 23 | 41 | 230 |
| | 1.13 | 78.7 | 70.2 | 12.0 | 79.4 | 1.13 | 24 | 19 | 43 | 190 |
| | 1.25 | 76.0 | 66.3 | 14.0 | 82.4 | 1.24 | 20 | 32 | 52 | 325 |
| | 1.25 | 72.6 | 62.2 | 11.2 | 75.4 | 1.20 | 18 | 22 | 39 | 215 |
| | 3.00 | 72.2 | 62.0 | 8.5 | 62.0 | 1.00 | 7 | 11 | 18 | 170 |
| 7075-T7651X | 0.30 | 82.6 | 73.8 | 11.5 | 61.2 | 0.83 | 8 | 15 | 23 | 155 |
| | 0.50 | 78.6 | 69.4 | 13.0 | 75.5 | 1.09 | 17 | 21 | 38 | 205 |
| | 0.50 | 78.6 | 69.0 | 12.0 | 71.8 | 1.04 | 16 | 19 | 35 | 190 |
| | 0.69 | 77.5 | 67.6 | 11.4 | 76.5 | 1.13 | 20 | 19 | 40 | 190 |
| | 0.75 | 79.0 | 70.3 | 11.5 | 77.6 | 1.10 | 20 | 18 | 39 | 180 |
| | 0.75 | 79.8 | 71.2 | 12.5 | 75.5 | 1.06 | 18 | 19 | 36 | 190 |
| | 0.75 | 79.6 | 70.9 | 12.0 | 78.0 | 1.10 | 18 | 19 | 38 | 190 |
| | 0.75 | 79.1 | 70.4 | 13.5 | 77.4 | 1.10 | 16 | 24 | 40 | 235 |
| | 0.75 | 80.4 | 72.0 | 13.5 | 76.4 | 1.06 | 17 | 20 | 37 | 200 |
| | 1.13 | 78.9 | 70.4 | 12.0 | 75.0 | 1.07 | 17 | 19 | 35 | 190 |

(continued)

Specimens per Fig. A1.8. Each line of data represents average of duplicate or triplicate tests for an individual lot of material. Specimens were generally about 0.100 in. thick; for shapes less than 0.2 in. in thickness, full-thickness specimens were sometimes used. For yield strengths, offset is 0.2%. (a) Obsolete alloy

**Table 6.4(b)   (continued)**

| Alloy and temper | Thickness, in. | Ultimate tensile strength (UTS), ksi | Tensile yield strength (TYS), ksi | Elongation in 2 in., % | Tear strength, ksi | Ratio tear strength to yield strength, (TYR) | Initiate a crack, in.-lb | Propagate a crack, in.-lb | Total energy, in.-lb | Unit propagation energy, in.-lb/in.$^2$ |
|---|---|---|---|---|---|---|---|---|---|---|
| | | Tensile tests | | | Tear tests | | Energy required to: | | | |
| | 1.25 | 76.2 | 66.6 | 12.0 | 82.0 | 1.23 | 23 | 29 | 52 | 290 |
| | 1.25 | 74.5 | 64.4 | 12.2 | 73.2 | 1.14 | 17 | 17 | 34 | 175 |
| 7079-T6(a) | 0.13 | 78.8 | 68.9 | 7.5 | 70.0 | 1.02 | 14 | 23 | 37 | 230 |
| | 0.13 | 79.2 | 71.6 | 13.9 | 72.8 | 1.02 | 16 | 23 | 39 | 230 |
| | 0.19 | 81.7 | 72.2 | 12.0 | 70.0 | 0.97 | 11 | 22 | 33 | 220 |
| | 0.70 | 79.6 | 71.0 | 8.0 | 60.2 | 0.85 | 7 | 22 | 29 | 220 |
| X7106-T53(a) | 0.19 | 65.6 | 57.2 | 17.0 | 94.2 | 1.65 | 54 | 67 | 122 | 670 |
| X7139-T53(a) | 0.25 | 66.6 | 56.8 | 15.0 | 87.8 | 1.55 | 41 | 43 | 84 | 435 |
| X7139-T63(a) | 0.25 | 68.2 | 59.0 | 16.5 | 92.5 | 1.57 | 42 | 44 | 86 | 455 |
| 7178-T651X | 0.25 | ... | ... | ... | 44.0 | ... | 9 | <5 | <14 | <50 |
| | 0.25 | 92.2 | 85.8 | 3.5 | 55.6 | 0.65 | 7 | <5 | <12 | <50 |
| | 1.25 | 89.5 | 82.0 | 10.2 | 45.5 | 0.55 | 5 | 7 | 12 | 70 |
| 7178-T7651X | 0.07 | 96.0 | 90.2 | 8.5 | 50.0 | 0.55 | 6 | 7 | 12 | 90 |
| | 0.07 | ... | ... | ... | 76.7 | ... | 14 | 11 | 25 | 165 |
| | 0.19 | 81.0 | 71.2 | 9.5 | 64.7 | 0.91 | 10 | 16 | 26 | 165 |
| | 0.19 | 79.0 | 69.1 | 9.8 | 68.9 | 1.00 | 13 | 19 | 32 | 190 |
| | 0.19 | 78.5 | 68.4 | 10.0 | 68.9 | 1.01 | 13 | 18 | 31 | 135 |
| | 0.19 | 79.4 | 71.2 | 8.8 | 68.0 | 0.96 | 14 | 13 | 28 | 135 |
| | 0.38 | 80.4 | 69.9 | ... | 62.9 | 0.90 | 11 | 13 | 24 | 125 |
| | 0.50 | 78.8 | 70.0 | ... | 65.8 | 0.94 | 14 | 17 | 31 | 170 |
| | 0.69 | 81.9 | 71.6 | 10.4 | 65.9 | 0.92 | 14 | 12 | 26 | 120 |
| | 1.25 | 77.0 | 67.6 | 9.0 | 60.9 | 0.90 | 12 | 13 | 24 | 125 |
| | 1.25 | 84.0 | 75.8 | 5.8 | 64.7 | 0.85 | 6 | 20 | 26 | 200 |
| | 1.25 | 80.2 | 72.0 | 7.2 | 60.3 | 0.84 | 10 | 14 | 24 | 135 |

Specimens per Fig. A1.8. Each line of data represents average of duplicate or triplicate tests for an individual lot of material. Specimens were generally about 0.100 in. thick; for shapes less than 0.2 in. in thickness, full-thickness specimens were sometimes used. For yield strengths, offset is 0.2%. (a) Obsolete alloy

**Table 6.5(a)    Results of tensile and tear tests of aluminum alloy forgings, longitudinal**

| | | Tensile tests | | | Tear tests | | | | | |
|---|---|---|---|---|---|---|---|---|---|---|
| Alloy and temper | Thickness, in. | Ultimate tensile strength (UTS), ksi | Tensile yield strength (TYS), ksi | Elongation in 2 in., % | Tear strength, ksi | Ratio tear strength to yield strength (TYR) | Energy required to: Initiate a crack, in.-lb | Propagate a crack, in.-lb | Total energy, in.-lb | Unit propagation energy, in.-lb/in.$^2$ |
| 2014-T652 Hand | 2.00 | 66.5 | 59.7 | 10.8 | 70.6 | 1.18 | 18 | 24 | 42 | 245 |
| forgings | 5.00 | 66.2 | 58.8 | 11.0 | 66.2 | 1.13 | 17 | 26 | 43 | 265 |
| 2024-T6 Hand | 2.00 | 66.1 | 53.9 | 11.0 | 72.2 | 1.34 | 19 | 36 | 55 | 355 |
| forgings | 3.00 | 63.0 | 49.6 | 10.8 | 68.4 | 1.38 | 18 | 29 | 47 | 295 |
| | 3.00 | 65.8 | 52.0 | 10.2 | 71.8 | 1.38 | 22 | 33 | 55 | 335 |
| | 3.75 | 64.6 | 50.2 | 11.0 | 75.4 | 1.50 | 26 | 40 | 66 | 405 |
| | 4.75 | 62.5 | 47.1 | 10.5 | 66.7 | 1.42 | 21 | 38 | 59 | 385 |
| | 5.50 | 65.4 | 51.8 | 10.8 | 81.8 | 1.58 | 31 | 48 | 79 | 475 |
| 2024-T852 Hand | 2.00 | 69.9 | 63.0 | 7.5 | 67.9 | 1.20 | 16 | 18 | 34 | 175 |
| forgings | 3.00 | 65.6 | 56.5 | 6.8 | 62.9 | 1.11 | 13 | 14 | 27 | 140 |
| | 3.00 | 66.8 | 57.8 | 9.0 | 63.8 | 1.10 | 12 | 12 | 24 | 120 |
| | 3.75 | 69.8 | 61.7 | 7.8 | 63.6 | 1.03 | 14 | 12 | 26 | 125 |
| | 4.75 | 64.2 | 54.2 | 8.8 | 64.0 | 1.18 | 16 | 25 | 41 | 250 |
| | 5.00 | 67.1 | 58.2 | 9.2 | 67.2 | 1.15 | 9 | 13 | 22 | 210 |
| | 5.50 | 67.3 | 57.4 | 10.0 | 71.4 | 1.24 | 20 | 22 | 42 | 220 |
| | 6.25 | 63.5 | 50.1 | 9.5 | 62.7 | 1.25 | 13 | 26 | 39 | 260 |
| 2024-T852 Die | 3.00 | 64.9 | 59.6 | 7.0 | 64.2 | 1.04 | 15 | 17 | 32 | 170 |
| forgings | 3.00 | 66.1 | 59.9 | 9.0 | 61.5 | 1.15 | 11 | 24 | 35 | 245 |
| | 3.00 | 71.5 | 64.2 | ... | 71.1 | 1.09 | 18 | 27 | 45 | 270 |
| | 3.00 | 71.5 | 65.1 | ... | 70.3 | 1.08 | 16 | 34 | 50 | 345 |
| 2025-T6 Die | 0.38 | 55.6 | 36.2 | 20.0 | 71.6 | 2.04 | 46 | 104 | 150 | 1035 |
| forgings | 1.125 | 55.6 | 36.2 | 20.0 | 69.8 | 1.92 | 38 | 102 | 140 | 1025 |
| 2219-T6-Hand | 7.00 | ... | 38.6 | ... | 72.1 | 1.82 | 38 | 76 | 114 | 770 |
| 7001-T75 Die | 0.25 | 80.2 | 71.7 | 10.0 | 68.8 | 0.96 | 13 | 23 | 36 | 230 |
| forgings(a) | 0.50 | 80.2 | 71.7 | 10.0 | 70.8 | 0.99 | 17 | 23 | 40 | 270 |
| 7001-T6-Die(a) | 0.50 | 91.9 | 83.2 | 12.0 | 70.6 | 0.85 | 10 | 28 | 38 | 280 |
| 7075-T6-Hand | 2.00 | 78.6 | 78.5 | 12.5 | 69.2 | 1.03 | 16 | 21 | 37 | 210 |
| 7075-T652 Hand | 2.00 | 83.0 | 7.22 | 10.0 | 74.6 | 1.02 | 10 | 48 | 59 | 480(b) |
| forgings | 3.00 | 80.7 | 71.4 | 10.0 | 72.4 | 1.01 | 24 | 27 | 54 | 260(b) |
| 7075-T73-Hand | 2.00 | 70.8 | 61.6 | 11.2 | 71.8 | 1.17 | 19 | 32 | 51 | 320 |
| 7075-T7352 Hand | 2.00 | 63.6 | 51.2 | 12.8 | 78.8 | 1.54 | 29 | 58 | 88 | 585 |
| forgings | 4.00 | 71.2 | 59.6 | 13.0 | 83.7 | 1.41 | 36 | 61 | 96 | 600 |
| 7076-T61 Die | 1.00 | 77.8 | 73.3 | 14.0 | 83.2 | 1.20 | 27 | 38 | 65 | 375 |
| forgings | 4.00 | 77.8 | 73.3 | 14.0 | 86.0 | 1.17 | 28 | 62 | 90 | 630 |
| 7079-T6-Die(a) | 0.38 | 78.0 | 68.3 | 14.0 | 79.7 | 1.17 | 26 | 40 | 66 | 400 |
| 7079-T6-Hand(a) | 2.00 | 79.4 | 68.7 | 12.5 | 74.6 | 1.09 | 17 | 23 | 40 | 230 |
| 7079-T652 Hand | 2.00 | 76.2 | 69.1 | 9.2 | 81.0 | 1.17 | 25 | 39 | 64 | 390 |
| forgings(a) | 5.00 | 75.6 | 66.5 | 11.0 | 88.9 | 1.34 | 34 | (c) | (c) | (c) |
| 7080-T7 Hand | 4.00 | 70.2 | 63.4 | 14.0 | 81.4 | 1.30 | 27 | 27 | 54 | 275 |
| forgings(a) | 7.00 | 68.0 | 60.5 | 14.0 | 92.0 | 1.52 | 36 | (c) | (c) | (c) |

Each line of data represents a separate lot of material; average of duplicate or triplicate tests. Specimens per Fig.A1.8, 0.100 in. thick. For yield strengths. offset is 0.2%. (a) Obsolete alloy. (b) Crack path was diagonal; propagation values may be unrealistically high. (c) Crack path erratic; energy values not meaningful

**Table 6.5(b)   Results of tensile and tear tests of aluminum alloy forgings, long transverse**

| Alloy and temper | Thickness, in. | Tensile tests | | | Tear tests | | | | | |
|---|---|---|---|---|---|---|---|---|---|---|
| | | Ultimate tensile strength (UTS), ksi | Tensile yield strength (TYS), ksi | Elongation in 2 in., % | Tear strength, ksi | Ratio tear strength to yield strength, (TYR) | Energy required to: | | Total energy, in.-lb | Unit propagation energy, in.-lb/in.$^2$ |
| | | | | | | | Initiate a crack, in.-lb | Propagate a crack, in.-lb | | |
| 2014-T652 Hand | 2.00 | 68.7 | 61.0 | 10.0 | 58.8 | 0.96 | 11 | 13 | 24 | 130 |
| forgings | 5.00 | 66.1 | 59.7 | 4.0 | 48.5 | 0.87 | 7 | 11 | 18 | 1100 |
| 2024-T6 Hand | 2.00 | 64.3 | 54.9 | 5.0 | 59.7 | 1.09 | 13 | 12 | 25 | 115 |
| forgings | 3.00 | 61.8 | 49.6 | 5.5 | 51.6 | 1.04 | 9 | 14 | 23 | 135 |
| | 3.00 | 64.2 | 52.5 | 7.0 | 60.6 | 1.16 | 12 | 11 | 23 | 110 |
| | 3.75 | 60.4 | 49.7 | 5.2 | 50.0 | 1.01 | 8 | 12 | 20 | 120 |
| | 4.75 | 57.8 | 46.0 | 5.5 | 43.2 | 0.94 | 6 | 9 | 15 | 95 |
| | 5.50 | 59.4 | 48.3 | 4.8 | 50.5 | 1.05 | 7 | 10 | 17 | 100 |
| 2024-T852 Hand | 2.00 | 70.8 | 63.2 | 6.5 | 61.1 | 0.97 | 12 | 12 | 24 | 125 |
| forgings | 3.00 | 68.9 | 60.9 | 7.0 | 53.5 | 0.88 | 8 | 10 | 18 | 105 |
| | 3.00 | 68.4 | 61.0 | 5.5 | 48.2 | 0.79 | 6 | 8 | 14 | 80 |
| | 3.75 | 68.6 | 64.1 | 2.8 | 32.8 | 0.51 | 3 | 7 | 10 | 65 |
| | 4.75 | 63.8 | 56.6 | 3.8 | 39.6 | 0.70 | 5 | 8 | 14 | 80 |
| | 5.00 | 66.1 | 57.3 | 7.0 | 56.3 | 0.97 | 7 | 8 | 16 | 120 |
| | 5.50 | 67.0 | 58.5 | 4.8 | 39.8 | 0.68 | 5 | 10 | 15 | 95 |
| | 6.25 | 63.8 | 54.0 | 7.0 | 56.8 | 1.05 | 9 | 10 | 19 | 100 |
| 2025-T6 Die | 0.38 | 56.4 | 37.3 | 1.6 | 69.1 | ... | 26 | 50 | 76 | 500 |
| forgings | 1.125 | 56.4 | 37.3 | 1.6 | 67.4 | 1.81 | 28 | 50 | 78 | 500 |
| 2219-T6-Hand | 7.00 | ... | 41.2 | ... | 59.8 | 1.45 | 16 | 26 | 42 | 260 |
| 7001-T75 Die | 0.25 | 74.7 | 65.8 | 5.5 | 55.8 | 0.85 | 8 | 12 | 20 | 120 |
| forgings (a) | 0.50 | 74.7 | 65.8 | 5.5 | 41.2 | 0.63 | 4 | 12 | 16 | 120 |
| 7001-T75 Hand forgings (a) | 2.00 | 67.2 | 63.4 | 1.8 | 43.4 | 0.63 | 6 | <5 | <11 | <50 |
| 7001-T6-Die (a) | 0.50 | 47.0 | 39.4 | 10.0 | ... | ... | ... | ... | ... | ... |
| 7075-T6-Hand | 2.00 | 80.3 | 69.7 | 10.0 | 59.6 | 0.85 | 10 | 14 | 24 | 140 |
| 7075-T652 | 2.00 | 78.2 | 67.6 | 10.2 | 62.4 | 0.92 | 6 | 18 | 28 | 185 |
| Hand forgings | 3.00 | 78.5 | 66.3 | 12.0 | 46.8 | 0.73 | 4 | 14 | 18 | 140 |
| 7075-T73-Hand | 2.00 | 71.7 | 62.5 | 11.0 | 71.8 | 1.15 | 20 | 36 | 56 | 365 |
| 7075-T7352 Hand | 2.00 | 63.4 | 50.9 | 9.0 | 68.1 | 1.34 | 21 | 35 | 56 | 355 |
| forgings | 4.00 | 67.7 | 56.9 | 7.0 | 49.4 | 0.87 | 7 | 12 | 20 | 125 |
| 7076-T61 Die | 1.00 | 75.4 | 69.4 | 8.8 | 78.9 | 1.10 | 23 | 18 | 41 | 18 |
| forgings | 4.00 | 75.4 | 69.4 | 8.8 | 83.4 | 1.20 | 25 | 23 | 48 | 235 |
| 7079-T6-Die(a) | 0.38 | 78.1 | 67.8 | 15.5 | 72.2 | 1.06 | 17 | 30 | 47 | 300 |
| 7079-T6-Hand(a) | 2.00 | 76.8 | 67.0 | 14.5 | 75.4 | 1.13 | 17 | 21 | 38 | 210 |
| 7079-T652 Hand | 2.00 | 78.0 | 68.9 | 9.8 | 66.4 | 0.96 | 15 | 22 | 37 | 220 |
| forgings(a) | 5.00 | 78.2 | 67.6 | 9.0 | 55.8 | 0.83 | 4 | 19 | 23 | 195 |
| 7080-T7 Hand | 4.00 | 67.5 | 59.3 | 9.0 | 57.3 | 0.97 | 11 | 16 | 27 | 160 |
| forgings(a) | 7.00 | 67.5 | 59.5 | 10.0 | 68.8 | 1.16 | 14 | 17 | 31 | 165 |
| X7106-T6352(a) Hand forgings | 12.00 | 59.8 | 51.7 | 7.5 | 67 | 1.28 | 18 | 28 | 46 | 275 |

Each line of data represents a separate lot of material; averages of duplicate or triplicate tests. Specimens per Fig.A1.8, 0.100 in. thick. For yield strengths, offset is 0.2%. (a) Obsolete alloy

**Table 6.5(c)  Results of tensile and tear tests of aluminum alloy forgings, short transverse**

| Alloy and temper | Thickness, in. | Tensile tests | | | Tear tests | | | | | |
|---|---|---|---|---|---|---|---|---|---|---|
| | | Ultimate tensile strength (UTS), ksi | Tensile yield strength (TYS), ksi | Elongation in 2 in., % | Tear strength, ksi | Ratio tear strength to yield strength (TYR) | Energy required to: | | Total energy, in.-lb | Unit propagation energy, in.-lb/in.$^2$ |
| | | | | | | | Initiate a crack, in.-lb | Propagate a crack, in.-lb | | |
| 2014-T652 Hand | 2.00 | 66.4 | 59.4 | 8.0 | 49.4 | 0.83 | 9 | <5 | <14 | <50 |
| forgings | 5.00 | 66.1 | 59.7 | 4.0 | 34.6 | 0.58 | 4 | 7 | 11 | 75 |
| 2024-T6 Hand | 2.00 | 64.1 | 54.9 | 6.0 | 52.4 | 0.95 | 9 | 11 | 20 | 105 |
| forgings | 3.00 | 62.0 | 54.5 | 4.5 | 50.6 | 0.93 | 9 | 8 | 17 | 85 |
| | 3.00 | 61.7 | 52.9 | 4.5 | 45.4 | 0.86 | 6 | 9 | 15 | 85 |
| | 3.75 | 61.0 | 50.0 | 5.7 | 48.1 | 0.95 | 7 | 8 | 15 | 80 |
| | 4.75 | 58.0 | 46.8 | 5.5 | 49.6 | 1.06 | 8 | 10 | 18 | 95 |
| | 5.50 | 63.2 | 52.1 | 5.8 | 46.2 | 0.89 | 7 | 8 | 15 | 85 |
| 2024-T852 Hand | 2.00 | 70.6 | 60.6 | 4.0 | 39.6 | 0.65 | 5 | 7 | 12 | 70 |
| forgings | 3.00 | 66.4 | 59.0 | 4.0 | 35.7 | 0.61 | 4 | 7 | 11 | 65 |
| | 3.00 | 63.8 | 57.5 | 3.0 | 33.8 | 0.59 | 3 | 7 | 10 | 70 |
| | 3.75 | 66.2 | 59.0 | 2.9 | 30.5 | 0.52 | 3 | 6 | 9 | 60 |
| | 4.75 | 62.5 | 52.2 | 4.8 | 39.0 | 0.75 | 4 | 8 | 12 | 75 |
| | 5.00 | 64.1 | 54.0 | 2.8 | 45.8 | 0.85 | 3 | 6 | 9 | 90 |
| | 5.50 | 66.2 | 55.4 | 5.8 | 38.1 | 0.69 | 4 | 7 | 11 | 70 |
| | 6.25 | 62.0 | 50.1 | 3.0 | 37.4 | 0.75 | 4 | 5 | 9 | 50 |
| 2024-T852 Die | 3.00 | ... | ... | ... | 45.1 | ... | <5 | <5 | <10 | <50 |
| forgings | 3.00 | ... | ... | ... | 32.8 | ... | 3 | 3 | 6 | 35 |
| | 3.00 | ... | ... | ... | 47.0 | ... | 7 | 7 | 14 | 75 |
| | 3.00 | ... | ... | ... | 46.2 | ... | 5 | 10 | 15 | 105 |
| 2219-T6 Hand | 7.00 | ... | 43.0 | ... | 58.9 | 1.37 | 14 | 17 | 31 | 170 |
| 7001-T75(a) Hand | 2.00 | 67.2 | 63.4 | 1.8 | 38.9 | 0.60 | 3 | <5 | <8 | <50 |
| forgings | 2.00 | 68.6 | 63.6 | 1.6 | 38.0 | 0.60 | 5 | 5 | 10 | 45 |
| 7001-T6(a)-Die | 0.50 | 90.5 | 80.4 | 3.1 | 35.6 | 0.44 | 3 | <5 | <8 | <50 |
| 7075-T6-Hand | 2.00 | 47.0 | 39.4 | 10.0 | 42.6 | 0.62 | 4 | 4 | 8 | 40 |
| 7075-T652 Hand | 2.00 | 75.6 | 62.0 | 8.0 | 42.4 | 0.68 | 2 | <5 | <7 | <50 |
| forgings | 3.00 | 75.0 | 64.2 | 6.0 | 43.4 | 0.65 | 2 | <5 | <7 | <50 |
| 7075-T73-Hand | 2.00 | 68.8 | 59.1 | 8.0 | 49.0 | 0.83 | 7 | 21 | 28 | 210 |
| 7075-T7352 Hand | 2.00 | 63.2 | 47.6 | 8.0 | 59.9 | 1.26 | 15 | 25 | 40 | 250 |
| forgings | 4.00 | 67.8 | 55.1 | 6.3 | 45.6 | 0.83 | 6 | 11 | 17 | 110 |
| 7076-T61 Die | 1.00 | 74.2 | 67.5 | 6.5 | ... | ... | ... | ... | ... | ... |
| forgings | 4.00 | 74.2 | 67.5 | 6.5 | 42.3 | 0.63 | 6 | 6 | 12 | 55 |
| 7079-T6(a)-Hand | 2.00 | 75.6 | 64.9 | 9.0 | 48.0 | 0.74 | 5 | 5 | 10 | 50 |
| 7079-T652(a) Hand | 2.00 | 77.2 | 64.6 | 9.0 | 46.4 | 0.72 | 5 | <5 | <10 | <50 |
| forgings | 5.00 | 71.5 | 57.8 | 7.0 | 47.2 | 0.82 | 3 | 12 | 15 | 115 |
| 7080-T7(a) Hand | 4.00 | 67.4 | 60.0 | 5.4 | 44.8 | 0.75 | 5 | 10 | 15 | 105 |
| forgings | 7.00 | 68.5 | 61.0 | 9.6 | 43.7 | 0.72 | 4 | 10 | 14 | 100 |

Each line of data represents a separate lot of material; average of duplicate or triplicate tests. Specimens per Fig.A1. 8, 0.100 in. thick. For yield strengths, offset is 0.2%. (a) Obsolete alloy

**Table 6.6  Results of tensile and tear tests of aluminum alloy castings**

| Alloy and temper | Tensile tests | | | Tear tests | | | | | |
|---|---|---|---|---|---|---|---|---|---|
| | Ultimate tensile strength (UTS), ksi | Tensile yield strength (TYS), ksi | Elongation in 2 in., % | Tear strength, ksi | Ratio tear strength to yield strength, (TYR) | Energy required to: | | Total energy, in.-lb | Unit propagation energy, in.-lb/in.$^2$ |
| | | | | | | Initiate a crack, in.-lb | Propagate a crack, in.-lb | | |
| **Sand casting** | | | | | | | | | |
| X335.0-T6 | 37.3 | 23.4 | 8.6 | 38.4 | 1.64 | 8 | 19 | 27 | 190 |
| | 35.3 | 22.6 | 6.8 | 39.3 | 1.74 | 11 | 19 | 30 | 195 |
| Average X335.0-T6 | 35.4 | 22.9 | 7.7 | 38.8 | 1.69 | 10 | 19 | 29 | 192 |
| 356.0-T4 | 31.1 | 19.8 | 4.4 | 32.8 | 1.66 | 7 | 13 | 20 | 130 |
| | 29.4 | 17.6 | 2.3 | 30.6 | 1.74 | 6 | 12 | 18 | 125 |
| Average 356.0-T4 | 30.2 | 18.7 | 3.4 | 31.7 | 1.70 | 6 | 12 | 19 | 128 |
| 356.0-T6 | 38.6 | 32.6 | 2.2 | 32.7 | 1.00 | 4 | 8 | 12 | 75 |
| 356.0-T7 | 37.8 | 33.7 | 1.6 | 39.2 | 0.87 | 3 | 7 | 10 | 70 |
| 356.0-T71 | 31.9 | 24.2 | 3.8 | 34.8 | 1.44 | 7 | 14 | 21 | 140 |
| | 29.4 | 20.7 | 4.1 | 32.2 | 1.55 | 7 | 11 | 18 | 110 |
| | 28.8 | 20.2 | 2.0 | 30.6 | 1.51 | 6 | 11 | 17 | 110 |
| Average 356.0-T71 | 30.0 | 21.7 | 3.3 | 52.6 | 1.50 | 7 | 12 | 19 | 120 |
| A356.0-T7 | 37.1 | 30.5 | 4.4 | 34.2 | 1.12 | 5 | 8 | 13 | 75 |
| | 37.6 | 33.2 | 2.1 | 33.6 | 1.01 | 4 | 10 | 14 | 100 |
| Average A356.0-T7 | 37.4 | 31.8 | 3.2 | 33.9 | 1.06 | 4 | 9 | 13 | 88 |
| B535.0-F (218-F) | 41.2 | 21.2 | 12.9 | 47.2 | 2.23 | 35 | 108 | 143 | 1075 |
| | 42.6 | 21.0 | 12.6 | 47.8 | 2.28 | 30 | 74 | 104 | 740 |
| Average B535.0-F | 41.9 | 21.1 | 12.8 | 47.5 | 2.26 | 32 | 91 | 124 | 908 |
| **Permanent-mold casting** | | | | | | | | | |
| X335.0-T61 | 40.8 | 28.4 | 8.5 | 43.2 | 1.52 | 9 | 22 | 31 | 220 |
| | 35.6 | 25.6 | 3.5 | 43.0 | 1.68 | 9 | 23 | 32 | 235 |
| Average X335.0-T61 | 38.2 | 23.8 | 6.0 | 43.1 | 1.60 | 9 | 22 | 31 | 228 |
| 354.0-T62 | 50.1 | 45.5 | 1.1 | 46.5 | 1.02 | 6 | 13 | 19 | 130 |
| | 47.8 | 44.3 | 0.9 | 45.9 | 1.04 | 6 | 13 | 19 | 125 |
| Average 354.0-T62 | 4.9 | 44.9 | 1.0 | 46.2 | 1.03 | 6 | 13 | 19 | 128 |
| C355.0-T7 | 37.0 | 31.0 | 2.1 | 36.9 | 1.19 | 6 | 8 | 14 | 85 |
| | 41.0 | 30.4 | 2.5 | 33.2 | 1.04 | 4 | 8 | 12 | 75 |
| Average C355.0-T7 | 39.0 | 30.7 | 2.3 | 35 | 1.12 | 5 | 8 | 13 | 80 |
| 356.0-T6 | 35.8 | 31.1 | 1.4 | 40.6 | 1.30 | 8 | 10 | 18 | 105 |
| | 41.4 | 32.2 | 4.2 | 34.2 | 1.06 | 4 | 6 | 10 | 55 |
| Average 356.0-T6 | 38.6 | 31.6 | 2.8 | 37.4 | 1.18 | 6 | 8 | 14 | 80 |
| 356.0-T7 | 28.4 | 21.4 | 4.3 | 37.0 | 1.73 | 10 | 16 | 26 | 165 |
| | 31.7 | 22.6 | 7.5 | 38.3 | 1.70 | 10 | 17 | 27 | 170 |
| | 29.6 | 22.0 | 3.2 | 35.2 | 1.60 | 10 | 17 | 27 | 170 |
| Average 356.0-T7 | 29.8 | 22.0 | 5.0 | 36.8 | 1.68 | 10 | 17 | 27 | 168 |
| A356.0-T61 | 39.4 | 30.8 | 4.3 | 43.2 | 1.40 | 9 | 14 | 23 | 140 |
| | 41.7 | 30.4 | 7.5 | 44.4 | 1.46 | 9 | 12 | 21 | 120 |
| Average A356.0-T61 | 40.6 | 30.6 | 5.9 | 43.8 | 1.43 | 9 | 13 | 22 | 130 |
| A356.0-T62 | 40.9 | 36.7 | 2.1 | 46.2 | 1.26 | 9 | 14 | 23 | 145 |
| | 43.6 | 36.3 | 3.9 | 46.4 | 1.28 | 9 | 13 | 22 | 130 |
| Average A356.0-T62 | 42.2 | 36.5 | 3.0 | 46.3 | 1.27 | 9 | 14 | 23 | 138 |
| A356.0-T7 | 28.2 | 21.4 | 5.3 | 29.8 | 1.86 | 14 | 30 | 44 | 295 |
| | 30.7 | 22.2 | 8.5 | 39.2 | 1.77 | 14 | 25 | 39 | 250 |
| Average A356.0-T7 | 29.4 | 21.8 | 6.9 | 34.5 | 1.82 | 14 | 28 | 42 | 272 |
| 359.0-T62 | 46.2 | 43.2 | 1.2 | 44.7 | 1.04 | 7 | 16 | 23 | 155 |
| | 47.4 | 43.1 | 1.6 | 41.0 | 0.95 | 4 | 12 | 16 | 115 |
| Average 359.0-T62 | 46.7 | 43.2 | 1.4 | 42.8 | 1.00 | 6 | 14 | 20 | 135 |
| A444.0-F | 23.2 | 9.7 | 22.2 | 27.6 | 2.85 | 19 | 39 | 58 | 390 |
| | 22.5 | 9.6 | 15.7 | 28.2 | 2.94 | 28 | 50 | 78 | 495 |
| Average A444.0-F | 22.8 | 9.6 | 19.0 | 27.9 | 2.90 | 24 | 44 | 68 | 442 |
| **Premium strength casting** | | | | | | | | | |
| A444.0-T4 | 23.0 | 8.0 | 24.4 | 27.1 | 3.39 | 30 | 57 | 107 | 580 |
| C355.0-T61 | 43.6 | 30.3 | 6.4 | 51.8 | 1.71 | 14 | 28 | 40 | 275 |
| A356.0-T6 | 41.6 | 30.2 | 8.8 | 51.6 | 1.71 | 18 | 34 | 52 | 345 |
| A357.0-T61 | 51.2 | 40.0 | 11.4 | 54.2 | 1.35 | 10 | 19 | 29 | 190 |
| A357.0-T62 | 53.9 | 46.4 | 5.3 | 53.8 | 1.16 | 11 | 12 | 23 | 125 |

Specimens per Fig. A1.8. Each line represents average results of tests of duplicate specimens of one individual lot of material. For yield strength, offset is 0.2%.

**Table 6.7(a)    Tensile tests of groove welds in wrought aluminum alloy sheet, plate, and extrusions**

| Alloy and temper combination | Sheet, plate thickness, in. | Specimen orientation(a) | Filler alloy | Post-weld thermal treatment | Ultimate tensile strength (UTS), ksi | Tensile yield strength (TYS), ksi | Elongation in 2 in., % |
|---|---|---|---|---|---|---|---|
| 1100-H112 As welded | 1.00 | Cross weld | 1100 | None | 11.6 | 6.1 | 26.5 |
| 3303-H112 As welded | 1.00 | Cross weld | 1100 | None | 16.1 | 7.6 | 24.0 |
| 2219-T62 Parent alloy | 0.063 | L | ... | ... | 60.8 | 42.5 | 10.0 |
| | 0.063 | T | ... | ... | 61.2 | 42.8 | 10.5 |
| 2219-T62 Post-weld heat treated | 0.063 | Cross weld | 2319 | HTA | 61.4 | 42.8 | 8.8 |
| 2219-T81 Parent alloy | 0.063 | L | ... | ... | 65.0 | 51.8 | 10.0 |
| | 0.063 | T | ... | ... | 65.5 | 52.3 | 10.2 |
| 2219-T81 As welded | 0.063 | Cross weld | 2319 | None | 46.6 | 33.2 | 1.8 |
| 2219-T81 Post-weld aged | 0.063 | Cross weld | 2319 | Aged | 48.4 | 40.2 | 1.5 |
| 2219-T87 Parent alloy | 0.063 | L | ... | ... | 67.9 | 56.2 | 9.8 |
| | 0.063 | T | ... | ... | 68.2 | 56.3 | 9.5 |
| 2219-T81 As welded | 0.063 | Cross weld | 2319 | None | 46.2 | 31.8 | 2.2 |
| 2219-T81 Post-weld Aged | 0.063 | Cross weld | 2319 | Aged | 52.6 | 40.4 | 2.0 |
| 5052-H112 As welded | 1.00 | Cross weld | 5052 | None | 29.1 | 13.9 | 18.0 |
| 5154-H112 As welded | 1.00 | Cross weld | 5154 | None | 32.6 | 14.5 | 17.0 |
| 5083-O Plate as welded | 0.38 | Cross weld | 5183 | None | 42.4 | (b) | ... |
| | 1.00 | Cross weld(c) | 5183 | None | 40.6 | 20.8 | 10.5 |
| | | Cross weld(d) | 5183 | None | 44.2 | 20.5 | 16.5 |
| | | Cross weld(e) | 5183 | None | 44.6 | 20.7 | 16.5 |
| | | Cross weld(f) | 5183 | None | 43.6 | 19.8 | 16.5 |
| 5083-O Extrusion as welded | 0.75 | Cross weld(c) | 5183 | None | 43.2 | 21.4 | 16.2 |
| | | Along HAZ | 5183 | None | 41.8 | 20.5 | 19.6 |
| 5083-H113 As welded | 1.00 | Cross weld | 5183 | None | 43.1 | (b) | ... |
| | 0.88 | Cross weld | 5556 | None | 41.2 | 21.2 | 12.5 |
| 5456-O As welded | 0.38 | Cross weld | 5556 | None | 46.8 | (b) | ... |
| 5456-H321 As welded | 1.00 | Cross weld | 5456 | None | 46.8 | 30.4 | 6.8 |
| 6061-T6 As welded | 1.00 | Cross weld | 4043 | None | 25.0 | 14.0 | 13.0 |
| 6061-T6 Post-weld heat treated | 1.00 | Cross weld | 4043 | HTA | 35.6 | 27.3 | 5.5 |
| 7005-T63 As welded | 1.25 | Cross weld | 5039 | None | 48.4 | 32.3 | 11.5 |
| 7005-T6351 As welded | 1.25 | Cross weld | 5356 | None | 42.1 | 28.2 | 6.8 |

Sheet or plate unless noted otherwise. Specimens per Fig. A1.8. Each line represents average results of tests of duplicate or triplicate specimens of each type. Joint yield strength not determined; ratio of tear strength to yield strength not available. L, longitudinal; T, transverse; HTA, heat treated and artificially aged after welding; HAZ, heat-affected zone. A and B designations in column for tear specimen type are as defined in Fig. A1.8. (a) Commercial gas metal arc welding or gas tungsten arc welding procedures unless otherwise noted. (b) Joint yield strength not determined; ratio of tear strength to yield strength not available. (c) Semiautomatic, horizontal position. (d) Semiautomatic, vertical position. (e) Semiautomatic, horizontal position. (f) Automatic, flat position. Matching tear test results are in Table 6.7(b).

**Table 6.7(b)  Tear tests of groove welds in wrought aluminum alloy sheet, plate, and extrusions**

| Alloy and temper combination | Sheet, plate thickness, in. | Specimen orientation (a) | Filler alloy | Post-weld thermal treatment | Tear specimen type (Fig. A1.8) | Tear Strength, ksi | Ratio tear strength to yield strength (TYR) | Energy required to: Initiate a crack, in.-lb | Propagate a crack, in.-lb | Total energy, in.-lb | Unit propagation energy, in.-lb/in.² |
|---|---|---|---|---|---|---|---|---|---|---|---|
| 1100-H112 As welded | 1.00 | Cross weld | 1100 | None | A | 19.5 | 3.20 | 48 | 76 | 124 | 755 |
| 3303-H112 As welded | 1.00 | Cross weld | 1100 | None | A | 24.0 | 3.16 | 40 | 78 | 118 | 785 |
| 2219-T62 Parent alloy | 0.063 | L | ... | ... | ... | 68.0 | 1.60 | 15 | 36 | 51 | 528 |
|  | 0.063 | T | ... | ... | ... | 66.8 | 1.56 | 15 | 33 | 48 | 527 |
| 2219-T62 Post-weld heat treated | 0.063 | Cross weld | 2319 | HTA | A | 87.2 | 2.04 | 33 | 44 | 77 | 705 |
| 2219-T81 Parent alloy | 0.063 | L | ... | ... | ... | 72.4 | 1.40 | 14 | 26 | 40 | 410 |
|  | 0.063 | T | ... | ... | ... | 70.4 | 1.34 | 14 | 27 | 41 | 429 |
| 2219-T81 As welded | 0.063 | Cross weld | 2319 | None | A | 70.8 | 2.13 | 31 | 20 | 51 | 324 |
| 2219-T81 Post-weld aged | 0.063 | Cross weld | 2319 | Aged | A | 74.1 | 1.84 | 20 | 23 | 43 | 363 |
| 2219-T87 Parent alloy | 0.063 | L | ... | ... | ... | ... | ... | ... | ... | ... | ... |
|  | 0.063 | T | ... | ... | ... | 67.2 | 1.19 | 10 | 21 | 31 | 338 |
| 2219-T81 As welded | 0.063 | Cross weld | 2319 | None | A | 67.0 | 2.10 | 24 | 22 | 46 | 352 |
| 2219-T81 Post-weld aged | 0.063 | Cross weld | 2319 | Aged | A | 72.4 | 1.79 | 17 | 26 | 43 | 419 |
| 5052-H112 As welded | 1.00 | Cross weld | 5052 | None | A | 37.0 | 2.66 | 45 | 108 | 153 | 1085 |
| 5154-H112 As welded | 1.00 | Cross weld | 5154 | None | A | 36.2 | 2.50 | 50 | 104 | 154 | 1040 |
| 5083-O Plate As welded | 0.38 | Cross weld | 5183 | None | A | 50.2 | (b) | 38 | 97 | 135 | 970 |
|  | 1.00 | Cross weld(c) | 5183 | None | A | 50.4 | 2.42 | 51 | 89 | 140 | 890 |
|  |  | Cross weld(d) | 5183 | None | A | 48.2 | 2.35 | 38 | 89 | 127 | 895 |
|  |  | Cross weld(e) | 5183 | None | A | 51.0 | 2.46 | 45 | 98 | 143 | 985 |
|  |  | Cross weld(f) | 5183 | None | A | 50.8 | 2.56 | 53 | 89 | 141 | 890 |
| 5083-O Extrusion As welded | 0.75 | Cross weld(c) | 5183 | None | A | 49.6 | 2.32 | 38 | 96 | 134 | 975 |
|  |  | Along HAZ | 5183 | None | B | 48.9 | 2.38 | 38 | 91 | 130 | 910 |
| 5083-H113 As welded | 1.00 | Cross weld | 5183 | None | A | 51.6 | (b) | 33 | 99 | 132 | 990 |
|  | 0.88 | Cross weld | 5556 | None | A | 48.2 | 2.27 | 36 | 85 | 121 | 850 |
| 5456-O As welded | 0.38 | Cross weld | 5556 | None | A | 51.7 | (b) | 38 | 91 | 129 | 910 |
| 5456-H321 As welded | 1.00 | Cross weld | 5456 | None | A | 51.7 | 1.70 | 46 | 92 | 138 | 920 |
| 6061-T6 As welded | 1.00 | Cross weld | 4043 | None | A | 35.5 | 2.34 | 19 | 38 | 57 | 380 |
| 6061-T6 Post-weld heat treated | 1.00 | Cross weld | 4043 | HTA | A | 36.1 | 0.93 | 5 | 8 | 13 | 80 |
| 7005-T63 As welded | 1.25 | Cross weld | 5039 | None | A | 60.0 | 1.86 | 30 | 95 | 125 | 950 |
| 7005-T6351 As welded | 1.25 | Cross weld | 5356 | None | A | 51.4 | 1.82 | 28 | 94 | 122 | 945 |

Sheet or plate unless noted otherwise. Specimens per Fig. A1.8. Each line represents average results of tests of duplicate or triplicate specimens of each type. Joint yield strength not determined; ratio of tear strength to yield strength not available. L, longitudinal; T, transverse; HTA, heat treated and artificially aged after welding; HAZ, heat-affected zone. A and B designations in column for tear specimen type are as defined in Fig. A1.8. (a) Commercial gas metal arc welding or gas tungsten arc welding procedures unless otherwise noted. (b) Joint yield strength not determined; ratio of tear strength to yield strength not available. (c) Semiautomatic, flat position. (d) Semiautomatic, vertical position. (e) Semiautomatic, horizontal position. (f) Automatic, flat position. Matching tensile test results are in Table 6.7(a).

**Table 6.8  Tear tests of groove welds in cast-to-cast and cast-to-wrought aluminum alloys**

| Alloy and temper combination | Filler alloy | Post-weld thermal treatment | Reduced section tensile strength, ksi | Joint yield strength (JYS), ksi | Free bend elongation, % | Tear specimen type (Fig. A1.8) | Tear strength, ksi | Ratio tear strength to yield strength (TYR) | Initiate a crack, in.-lb | Propagate a crack, in.-lb | Total energy, in.-lb | Unit propagated energy, in.-lb/in.$^2$ |
|---|---|---|---|---|---|---|---|---|---|---|---|---|
| **Sand casting** | | | | | | | | | | | | |
| A444.0-F to A444.0-F | 5556 | None | 37.8 | (a) | 18.8 | A | 51.0 | (a) | 60 | 103 | 163 | 1030 |
| | | | | | | B | 46.9 | | 38 | 82 | 120 | 820 |
| | | | | | | C | 50.6 | | 64 | 94 | 158 | 935 |
| A444.0-F to 6061-T6 | 5556 | None | 32.5 | (a) | 12.2 | A | 49.5 | (a) | 56 | 105 | 161 | 1050 |
| | | | | | | B | 45.0 | | 32 | 77 | 109 | 770 |
| | | | | | | C | 47.7 | | 46 | 92 | 138 | 920 |
| A444.0-F to 5456-H321 | 5556 | None | 42.6 | (a) | 12.2 | A | 53.0 | (a) | 66 | 115 | 181 | 1185 |
| | | | | | | B | 49.6 | | 38 | 91 | 129 | 910 |
| | | | | | | C | 50.1 | | 61 | 99 | 160 | 990 |
| 356.0-T4 to 6063-T4 | 4043 | None | 28.5 | (a) | 4.1 | A | 26.9 | (a) | 3 | 16 | 19 | 160 |
| | | | | | | B | 37.4 | | 11 | 24 | 34 | 240 |
| | | | | | | C | 34.2 | | 6 | 32 | 38 | 325 |
| 356.0-T6 to 6061-T6 | 5556 | None | 28.4 | (a) | 2.0 | A | 29.6 | (a) | 6 | 15 | 21 | 150 |
| | | | | | | B | 34.4 | | 9 | 21 | 30 | 210 |
| | | | | | | C | 31.8 | | 9 | 18 | 27 | 185 |
| | 4043 | None | 27 | (a) | 6.9 | A | 30.3 | (a) | 7 | 16 | 23 | 175 |
| | | | | | | B | 34.4 | | 10 | 31 | 41 | 310 |
| | | | | | | C | 25.7 | | 6 | 33 | 39 | 330 |
| 356.0-T7 to 6061-T6 | 5556 | None | 26.8 | (a) | 6.9 | A | 30.3 | (a) | 7 | 16 | 23 | 175 |
| | | | | | | B | 34.4 | | 10 | 31 | 41 | 310 |
| | | | | | | C | 25.7 | | 6 | 33 | 39 | 330 |
| | 4043 | None | 25.8 | (a) | 7.2 | A | 29.7 | (a) | 6 | 22 | 28 | 220 |
| | | | | | | B | 32.8 | | 11 | 30 | 41 | 295 |
| | | | | | | C | 26.6 | | 8 | 38 | 46 | 380 |
| 356.0-T71 to 356.0-T71 | 4043 | None | 26.5 | (a) | 6.1 | A | 28.8 | (a) | 8 | 18 | 26 | 175 |
| | | | | | | B | 32.4 | | 14 | 32 | 46 | 320 |
| | | | | | | C | 27.8 | | 8 | 24 | 32 | 245 |
| 356.0-T71 to 6061-T6 | 4043 | None | 26.7 | (a) | 9.4 | A | 29.4 | (a) | 8 | 20 | 28 | 205 |
| | | | | | | B | 33.4 | | 15 | 30 | 45 | 305 |
| | | | | | | C | 27.6 | | 9 | 44 | 53 | 435 |
| 356.0-T71 to 5456-H321 | 4043 | None | 25.7 | (a) | 5.7 | A | 30.8 | (a) | 8 | 14 | 22 | 140 |
| | | | | | | B | 32.8 | | 8 | 18 | 26 | 185 |
| | | | | | | C | 29.6 | | 11 | 16 | 27 | 165 |
| A357.0-T7 to 6061.T6 | 5556 | None | 25.9 | (a) | 8.2 | A | 51.2 | (a) | 61 | 112 | 173 | 1120 |
| | | | | | | B | 50.3 | | 42 | 100 | 142 | 995 |
| | | | | | | C | 50.2 | | 60 | 101 | 161 | 1010 |
| **Permanent mold castings** | | | | | | | | | | | | |
| C356.0-T7 to 6061-T6 | 4043 | None | 32.0 | (a) | 16.6 | A | 36.4 | (a) | 12 | 45 | 57 | 445 |
| | | | | | | B | 35.1 | | 10 | 28 | 38 | 275 |
| | | | | | | C | 32.7 | | 11 | 26 | 37 | 265 |
| 356.0-T6 to 356.0-T6 | 4043 | None | 28.1 | (a) | 11.4 | A | 34.0 | (a) | 14 | 34 | 48 | 340 |
| | | | | | | B | 32.8 | | 13 | 34 | 47 | 340 |
| | | | | | | C | 32.7 | | 11 | 26 | 37 | 265 |
| 356.0-T6 to 5456-H321 | 4043 | None | 29.5 | (a) | 14.3 | A | 39.4 | (a) | 28 | 41 | 71 | 410 |
| | | | | | | B | 32.7 | | 6 | 18 | 24 | 175 |
| | | | | | | C | 39.4 | | 17 | 42 | 58 | 420 |
| 356.0-T7 to 356.0-T7 | 4043 | None | 25.6 | (a) | 8.2 | A | 34.2 | (a) | 22 | 31 | 53 | 310 |
| | | | | | | B | 32.7 | | 6 | 18 | 71 | 410 |
| | | | | | | C | 39.4 | | 17 | 42 | 59 | 420 |
| 356.0-T7 to 6061-T6 | 4043 | None | 27.6 | (a) | 9.8 | A | 38.2 | (a) | 27 | 38 | 65 | 375 |
| | | | | | | B | 34.3 | | 15 | 36 | 51 | 355 |
| | | | | | | C | 43.6 | | 34 | 70 | 104 | 700 |
| 356.0-T7 to 5456-H321 | 4043 | None | 26.6 | (a) | 4.0 | A | 37.4 | (a) | 38 | 30 | 68 | 295 |
| | | | | | | B | 24.3 | | 2 | 10 | 12 | 105 |
| | | | | | | C | 33.0 | | 13 | 31 | 44 | 310 |
| A356.0-T61 to 6061-T6 | 4043 | None | 28.6 | (a) | 9.8 | A | 36.8 | (a) | 25 | 50 | 75 | 495 |
| | | | | | | B | 33.4 | | 10 | 35 | 45 | 350 |
| | | | | | | C | 35.2 | | 14 | 42 | 56 | 415 |
| A356.0-T7 to 6061-T6 | 4043 | None | 25.2 | (a) | 11.8 | A | 39.0 | (a) | 37 | 41 | 78 | 410 |
| | | | | | | B | 34.0 | | 16 | 40 | 56 | 395 |

Specimens per Fig. A1.8. Each line represents average results of tests of duplicate specimens for one individual lot of material. (a) Joint yield strength not determined; ratio of tear strength to yield strength not available

# CHAPTER 7

# Fracture Toughness

THROUGH THE WORK of A.A. Griffith (Ref 38), G.R. Irwin (Ref 39), and the ASTM Committee E-24 on Fracture Testing of High-Strength Metallic Materials, now ASTM Committee E9 (Ref 40, 41, and many others), about 19 ASTM Standard Test methods, including E 399 (Ref 9), are available for the determination of fracture toughness parameters that relate the load-carrying capacity of structural members stressed in tension to the size of cracks, flaws, or design discontinuities that may be present in the stress field. These parameters, primarily the stress-intensity factor, $K$, and the strain-energy release rate, $G$, are more useful to the designer than those measures of toughness that provide only a relative merit rating of materials, such as notch-tensile and tear tests. $K$ and $G$ characterize the potential fracture conditions in terms that permit structural designers to design resistance to unstable crack growth and catastrophic fracture into a structure, even with materials that are relatively low in toughness, including those sometimes described as brittle.

It is appropriate in such a survey of the fracture characteristics of aluminum alloys to very briefly review the fracture mechanics theory, describe the test procedures most often used to determine critical values of those fracture parameters, present representative data for aluminum alloys, and illustrate some of the ways the data might be used. It is beyond the scope of this book to go deeply into the science of fracture mechanics or to describe the wide range of analytical techniques now employed in using fracture mechanics in design.

The limited applicability of linear elastic fracture mechanics to most aluminum alloys, that is, other than the high-strength heat-treatable alloys, must be emphasized. Since the analysis is based upon the assumption that unstable crack growth develops in elastically stressed material, the fracture-toughness approach is applicable primarily to relatively high-strength materials with relatively low ductility. The type of behavior assumed in the development of the fracture-mechanics concepts is essentially nonexistent in the majority of aluminum alloys. Nevertheless, it is useful to overview the approach, provide representative data for those alloys for which the analysis is useful, and illustrate ways of estimating the fracture toughness of the tougher alloys.

## 7.1 Theory

Consider a large panel (representing a structure) stressed in tension uniformly and elastically in one direction in the plane of the panel, and containing a through-the-thickness crack that is 1) perpendicular to the direction of stress and 2) small with respect to the size of the panel, as shown schematically in Fig. 7.1. Although it is assumed that the panel is stressed uniformly, it is recognized that 1) at the tips of the crack, the stress is greater than the average stress, and 2) within regions immediately above and below the crack, the stress is less than the average stress and may be considered to be zero.

As the uniform gross stress increases, so does the stored elastic strain energy in the specimen that is available to propagate the crack. The elastic energy in the region immediately surrounding the crack becomes a "crack driving force," which is generally defined in terms of the elastic strain-energy release rate, $G$, and is related to $K$, the stress-intensity factor describing the stress field local to the crack tip. This crack driving force is opposed by the resistance of the material to crack extension, which also increases with stress and maintains an equilibrium. When the stress increases to the point that the rate of increase of the crack driving force

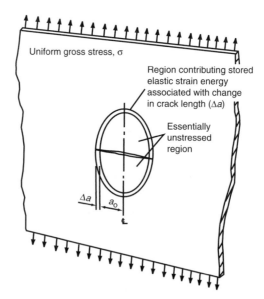

Uniform gross stress, σ

Region contributing stored elastic strain energy associated with change in crack length (Δa)

Essentially unstressed region

Δa  $a_o$

**Fig. 7.1**  Schematic drawing of large elastically stressed panel containing a crack. Uniform gross stress, σ

with respect to crack length is equal to the rate of increase of the resistance, unstable crack growth ensues (Ref 41). In terms of a slowly growing crack under constant or increasing stress, when the elastic strain energy released by a minute increment of crack length, $\Delta a$, is sufficient to develop a new increment of crack length, $\Delta a$, the crack will become self-propagating.

By examining the conditions at the "critical" situation, that is, when the crack growth becomes unstable, a measure of the "critical" strain-energy release rate, $G_c$, and stress-intensity factor, $K_c$, can be established empirically.

From Griffith's work (Ref 38), Irwin (Ref 39) suggested that, in very large systems involving brittle materials, the critical strain-energy release rate, (i.e., the rate at the onset of unstable crack growth) is related to stress and crack length by:

$$G_c = \frac{K_c^2}{E} = \frac{\pi \sigma_c^2 a_c}{E} \qquad \text{(Eq 1)}$$

where $G_c$ is critical strain-energy release rate, in.-lb/in.²; $K_c$ is critical stress-intensity factor, psi/in.; $\sigma_c$ is gross-section stress at the onset of unstable crack growth, psi; $2a_c$ is total crack length at the onset of unstable crack growth, in.; and $E$ is modulus of elasticity, psi.

This relationship between stress and crack length must be modified to take into account the facts that a) the dimensions of test panels are finite and may not always be considered large with respect to the crack size, and b) most materials are not perfectly brittle, so that an appreciable amount of plastic deformation takes place at the tips of the crack. The considerations of finite dimensions lead to:

$$G_c = \frac{K_c^2}{E} = \frac{\sigma_c^2}{E}\left[W\tan\left(\frac{\pi a_c}{W}\right)\right] \qquad \text{(Eq 2)}$$

where $W$ equals the width of the panel. The effect of the plastic deformation at the crack tip is to increase the "effective" length of the crack ($2a$, in the equations) by the size of the plastic zone at the tip of the crack, namely:

$$a_c = a'_c + \frac{EG_c}{2\pi\sigma_{ys}^2} = a_c + \frac{K_c^2}{2\pi\sigma_{ys}^2} \qquad \text{(Eq 3)}$$

where $a_c$ is the physical size of the crack and $\sigma_{ys}$ is the tensile yield strength of the material.

The complete relationship, then, is:

$$G_c = \frac{K_c^2}{E} = \frac{\sigma_c^w}{E}\left[W\tan\left(\frac{\pi a_c}{W} + \frac{EG_c}{2W\sigma_{ys}^2}\right)\right] \qquad \text{(Eq 4)}$$

The relationship of Eq 4 to Eq 1 may be seen if the width of the panel, $W$, in Eq 4 is allowed to become large, so that the angle becomes small and the tangent of the angle can be considered equal to the angle:

$$G_c = \frac{K_c^{\,2}}{E} = \frac{\sigma_c^2}{E}\left(\pi a_c + \frac{EG_c}{2\sigma_{ys}^2}\right)$$

(Eq 5)

and if a very brittle material is assumed, so that $G_c$ is very small and $EG_c$ is small compared with the square of the yield strength:

$$\frac{EG_c}{2\sigma_{ys}^2} \sim 0$$

then

$$G_c = \frac{K_c^{\,2}}{E} = \frac{\pi\sigma_c^2 a_c}{E}$$

(Eq 6)

Many other analyses have been developed for other stress-flaw size situations (part-through cracks, edge cracks, etc.), but it is beyond the scope of this treatment to review them here (Ref 41, 42, et al.).

It may be noted that the material thickness, $t$, does not enter into these equations except in relationship to load and stress; however, it should not be assumed that thickness is unimportant. Whether plane-stress or plane-strain conditions prevail is dependent primarily upon thickness. It has been well established empirically that $G_c$, and hence $K_c$, vary with material thickness in a pattern similar to that shown schematically in Fig. 7.2.

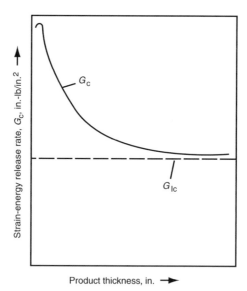

**Fig. 7.2** Schematic representation of influence of thickness on critical strain-energy release rate, $G_c$. Also indicative of pattern for critical stress intensity factor, $K_c$

As the plate thickness increases and plane-strain conditions become dominant, the value of $G$ approaches a minimum value, designated $G_{Ic}$ and called the plane-strain strain-energy release rate. ($K_{Ic}$ is the associated plane-strain stress-intensity factor.)

For a center-cracked panel, the type of specimen representing the conditions in Fig. 7.1, the parameters are expressed as:

$$G_{Ic} = \frac{K_{Ic}^2}{E}(1 - \mu^2) = \frac{\sigma_{Ic}^2}{E}(1 - \mu^2)W \tan\left(\frac{\pi a_o}{W} + \frac{K_{Ic}^2}{6W\sigma_{ys}^2}\right) \qquad \text{(Eq 7)}$$

where $2a_o$ is the total original crack length, in.; $a_{Ic}$ is the gross stress at the initiation of slow crack growth, psi; and $A$ is Poisson's ratio, which is 0.33 for aluminum alloys. The other terms are as defined previously. (Note the use of 6 in the denominator of the plastic zone size correction factor in place of 2, which takes into account the fact that the plastic zone is smaller under plane-strain conditions.)

For many alloys, plane-strain conditions may be difficult to achieve. So in order to measure plane-strain fracture toughness, it is necessary to approach or approximate the conditions well enough as to provide a good representation of plane-strain fracture. For example, it has been established that the conditions for fracture under plane-strain conditions are essentially the same as those existing at the initial burst of crack extension in statically loaded specimens, referred to early on as "pop-in" (Ref 16). If there is such a burst of unstable crack growth, values of $G_I$ and $K_{Ic}$ can be calculated with Eq 6, with $G_{Ic}$ being the stress at pop-in. The initial pop-in will still have reasonably represented unstable plane-strain crack growth even if the crack is subsequently arrested by virtue of the ability of the material to develop a shear fracture (shear lip on the fracture surface) and, thus, change the mode of fracture from plane strain to some mixture of plane strain and plane stress (mixed mode).

The significance of $G_{Ic}$ and $K_{Ic}$ is similar to that of $G_c$ and $K_c$ in that they are parameters relating critical gross stress and crack (or flaw) size when the stress conditions are those of plane strain. They have more general applicability, however, in that they represent the lowest level of stress at which unstable crack growth can take place in material of any thickness under any type of stress (static or fatigue) and, hence, represent a conservative (safe) design tool.

It has been found more convenient to handle fracture mechanics problems in terms of the stress-intensity factor, $K$, rather than the strain-energy release rate, $G$. This parameter provides a direct relationship between the gross-section stress and crack length without the involvement of other material properties or strain (through modulus of elasticity). Therefore, most analytical and experimental procedures focus on $K_{Ic}$ and $K_c$ rather than $G_{Ic}$ and $G_c$.

## 7.2 Test Procedures

It is useful in discussing test methods for fracture-toughness testing to look first at the early evolution of test procedures that led to the primary focus on plane-strain fracture-toughness testing and then the later emergence of the importance of refining mixed-mode and plane-strain test methods. While both types of tests were employed throughout the period, the standardization and application of the various procedures followed the plane-strain to plane-stress refocus.

**Early Evolution of Test Methods.** In the early years of fracture-toughness testing, center-cracked specimens of the general design in Fig. A1.9 were most frequently used because they permit the evaluation of the fracture parameters under mixed-mode conditions at the onset of unstable crack growth to fracture, as well as at the plane-strain instability using the pop-in concept. By this latter approach, the initial spurt of crack growth in a relatively large specimen may, under certain conditions, adequately represent the plane-strain instability even though the crack subsequently arrests. By instrumenting the crack opening and establishing an appropriate empirical relationship between crack opening and crack length, the conditions at the initial crack instability may be determined and a calculation of $K_{Ic}$ made. Detailed discussion of the early development of the various test procedures that may be used to evaluate the fracture toughness parameters is given in Ref 41, 42 and other publications of ASTM Committee E-24 (now E-9).

In early tests made at Alcoa Laboratories (Ref 1, 2, 43–48), many of which are reported herein, center-notched specimens of various sizes were tested, ranging in thickness from 0.063 to 1.00 in., and with widths ranging from 3 to 20 in. As better understandings developed, the relatively wider specimens were used more often with center-crack length between 25 and 50% of the total width. Generally, the specimens were fatigue cracked, although some of the earlier mixed mode $K_c$ values were obtained from specimens with sharply machined notches (notch radius equal to or less than 0.0005 in.), and the 1 in. thick × 20 in. wide × 64 in. long specimens (Fig. A1.9b) were not fatigue cracked because of load-capacity requirements (Ref 48).

The center-notched specimens were fatigue cracked by axial-stress loading; the single-edge-notched specimens were fatigue cracked in bending. The maximum nominal fatigue-precracking stresses were equal to or less than 20% of the yield strength of the material. The fatigue cracks were extended at least 1/8 in.

As noted earlier, a compliance-gage technique involving SR-4 electrical-resistance strain gage units was used to obtain an autographic load-deformation curve, from which it was possible to detect the load at pop-in and, for center-cracked specimens, the length of the crack at the onset of subsequent unstable crack growth to fracture. A gage length of two-thirds of the specimen width was generally used. A representative 20 in. wide center-cracked specimen in a 3,000,000 lb Southwark testing machine, under test, is shown Fig. 7.3, where the mounting of the strain gage units to measure crack opening may be seen.

Fatigue-cracked single-edge-notched specimens of the type in Fig A1.10 were also used. With such specimens, values of the plane-strain parameters were determined from the loads at the initial burst of unstable crack growth. In the case of the single-edge-notched, like the center-cracked specimens, the initial burst of crack growth was almost always at a stress less than that at the subsequent onset of unstable crack growth to fracture (i.e., at pop-in).

To ensure that the values of the plane-strain fracture parameters from the center-cracked and single-edge-notched specimens were valid, it was the standard practice to calculate values only for those tests in which significant bursts of unstable crack growth took place, as indicated by a significant pop-in. Examples of load-deformation curves with suitable

**Fig. 7.3** Fracture toughness specimen in 3 million lb testing machine

indications of pop-in are shown in Fig. 7.4. In the majority of instances where these were not present, the data were discarded; in those instances where the pop-in is questionable, the data were so indicated. Data from tests in which the net stress at the onset of unstable crack growth exceeds 0.8 of the tensile yield strength were also generally discarded, because of the likelihood of a greater amount of plastic action than is properly accounted for by the elastic stress analysis and associated corrections.

As study of fracture under plane-strain conditions continued and ASTM standard test methods for plane-strain fracture-toughness testing were more fully developed, notched bend specimens of the type in Fig. A1.11(a) and, eventually, compact tension specimens of the type in Fig. A1.12(a) became almost universally used. ASTM Standard Method E 399 for plane-strain fracture toughness testing emerged as the most widely used set of procedures, and they have been gradually broadened to include a wide range of component and specimen variations. For aluminum alloys in particular, ASTM Methods B 645 and B 646, adding current requirements applicable to aluminum, are used in conjunction with E 399.

It has always been and continues to be a standard feature of fracture toughness testing that it is essential to ensure that a "valid" measurement has been achieved. Reference to the applicable test methods will provide a detailed listing of the individual criteria for validity, but the essential requirements are that specimen size, specimen preparation, and testing conditions are such that the test adequately represents unstable growth of a relatively small crack in a large elastic stress field. Typically, values obtained from a test are labeled $K_Q$, a candidate value of $K_{Ic}$ or $K_c$, until the validity criteria have been checked, and only labeled $K_{Ic}$ or $K_c$ when the criteria are met. In some few cases, values may be labeled something

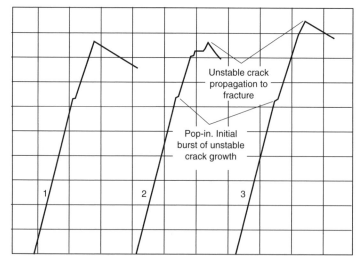

**Fig. 7.4** Typical autographic load-deformation curves from fracture toughness tests

akin to "essentially valid" or reasonably indicative of valid results when deviations from validity are few and very small. Indiscriminate use of this practice is not recommended, and strict adherence to the validity criteria in the standard test methods is prescribed.

An additional aspect of plane-strain fracture toughness test methods that deserves particular mention is that of specimen orientation. Unlike most other tests (though equally true for tear specimens), there are really six standard orientations, not just three, because both the plane of the crack and the direction of crack growth must be considered. The six standard orientations are illustrated in Fig. A1.2 and described in Chapter 2. The definitions of orientations in specimens containing welds are also illustrated. As noted previously, ASTM Method E 399 includes other combinations of component shapes and specimen orientation, with which the more experienced testing organization may find it useful to be familiar.

Yet another important consideration in plane-strain fracture toughness testing in particular is the influence of residual stresses upon the fracture toughness test results, particularly for nonsymmetrical specimens taken from relatively thick and/or metallurgically complex components such as die forgings. Thanks to the work of Bucci and his associates (Ref 49, 50), this influence is now relatively well understood and may be taken into account in testing. See section 7.8 for more detailed discussion of this factor.

**Mixed-Mode and Plane-Stress Fracture Toughness Testing.** With the gradual recognition of the fact that relatively few structures provided the combinations of thickness and limited plastic-zone development that would provide the constraint needed to be properly described as plane strain in nature, attention returned to the need for standardized methods of defining fracture under mixed-mode and plane-stress conditions. This was accomplished for the aluminum industry by standardization of center-notched panel testing of the type described earlier, to ensure adequate width and initial crack dimensions to ensure consistent and relatively geometrically independent measures of $K_c$ in ASTM Standard B 646. It also led to more focus and standardization of methods for measurement of crack resistance curves, ASTM Standard E 561.

Crack resistance curves, or $R$-curves, are continuous records of stress-intensity factor, $K$, as a function of crack extension, as illustrated schematically in Fig. 7.5(a). They are generated by recording the conditions while driving a crack by increasing the stress intensity, usually in a wide center-cracked panel of the type in Fig. A1.12. Measurements of the crack length are made as in the center-cracked fracture toughness tests, that is, by instrumenting the specimen to record crack-opening displacement, which was then interpreted via a compliance calibration relating crack-opening displacement to actual crack length. $R$-curves generated in this manner may be used to analyze the potential for crack-growth instability by

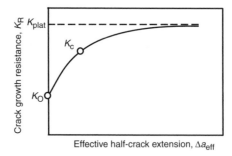

**Fig. 7.5(a)** Schematic of typical $R$ curve. $K_0$, stress-intensity factor corresponding to initial crack extension; $K_c$, critical stress-intensity factor (point of fracture instability); $K_{plat}$, plateau stress-intensity factor; $K_R$, crack-extension resistance; $\Delta a_{eff}$, effective crack extension increment

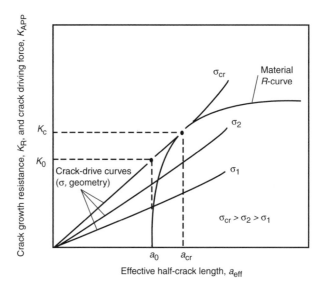

**Fig. 7.5(b)** Schematic of typical $R$-curve, illustrating overlay of crack driving force curves, including tangency indicative of instability when crack driving force equals crack resistance

overlaying crack-driving-force curves based upon the expected design conditions, and looking for the point of tangency with the $R$-curve, as illustrated in Fig. 7.5(b).

Representative crack resistance curves for some aluminum alloys are presented subsequently along with the $K_{Ic}$ and $K_c$ measurements.

## 7.3 $K_{Ic}$ and $K_c$ Data

Values of the critical stress-intensity factors from tests of aluminum alloys are presented in the following tables at the end of this Chapter:

**Individual test results**

Table 7.1      $K_{Ic}$ and $K_c$ values from tests of center-notched
                      specimens of sheet and thin plate (Fig. A1.9 a, b)

Table 7.2      $K_{Ic}$ and $K_c$ values from tests of center-notched
                      specimens of plate, 1 in. thick or more (Fig A1.9b)

Table 7.3      $K_{Ic}$ values from tests of single-edge-notched
                      specimens (Fig. A1.10)

Table 7.4      $K_{Ic}$ values from tests of notched bend and compact
                      tension specimens (Fig. A1.11, A1.12, respectively)

Table 7.5      Representative summary of $K_{Ic}$ values from tests of
                      compact tension specimens (Fig. A1.13)
                      from industry database

**Published typical values**

Table 7.6      Published typical $K_{Ic}$ and $K_c$ values for wrought
                      aluminum alloys

**Published specific minumum values**

Table 7.7      Published specified minimum $K_{Ic}$ values for wrought
                      aluminum alloys

Table 7.8      Published specified minimum $K_{Ic}$ values for wrought
                      aluminum alloys

(Note: No typical or minimum values of fracture toughness have been published for aluminum alloy castings or for welds in wrought or cast alloys.)

These data are presented in terms of the stress-intensity factor, $K$ ($K_c$ or $K_{Ic}$); values of the strain-energy release rate, $G$ ($G_c$ or $G_{Ic}$), may be calculated with Eq 1 or 6.

# 7.4 Discussion of $K_{Ic}$ and $K_c$ Data

**Plane-Strain Fracture Toughness, $K_{Ic}$.** Review of the data in Tables 7.1 through 7.4 illustrates that values of $K_{Ic}$ determined from specimens of various types and sizes are fairly consistent so long as the validity conditions are carefully maintained, especially those regarding specimen size in terms of the plastic-zone size, that is, that specimen thickness, $B$, and crack length, $a$, are equal to or greater than 2.5 $(K_{Ic}/\sigma_{ys})^2$. Rather large variations in $K_{Ic}$ will be observed for several quite logical reasons, and it is well to note these:

- Variations in $K_{Ic}$ values are to be expected for different specimen orientations, that is, situations in which different patterns of crack growth are developed with respect to the microstructure. Values are usually highest when stress is applied in the longitudinal direction

(L-T or L-S) when crack growth is across grain-flow patterns; values are usually lowest when stress is applied normal to the thickness, when crack growth is in the plane of major grain flow (S-L or S-T).

- Variations in $K_{Ic}$ values are to be expected for different products (sheet, plate, forgings, extrusions, etc) and for different thicknesses of the same product when the metallurgical structures produced by the fabrication procedures vary.
- Greater scatter is to be expected in data from fracture-toughness tests than in data from ordinary tensile or compressive tests because of the greater number of variables and the uncertain nature of some of them (e.g., fatigue-cracking stress, shape of crack front, number of cycles of loading, etc.). The problem is even greater in measurements of $K_c$ than for $K_{Ic}$ since the crack length at the onset of rapid fracture must be established (see next paragraph).

It is appropriate to note that notched bend and compact tension specimens are now considered the most useful and reliable for determining values of $K_{Ic}$, and are the focus of ASTM Standard Test Method E 399. Measurements from center-cracked panels are least reliable because they depend in large part on the degree of clarity of the initial burst of crack growth, often disguised for aluminum alloys because of the ability of most to plastically deform in the presence of stress raisers.

It is also useful to note that specimen-size studies for relatively tough alloys such as 2219-T851 (Ref 47) have indicated that more consistent and reliable values are obtained when specimen thickness, $B$, and crack length, $a$, are equal to or greater than 5 $(K_{Ic}/\sigma_{ys})^2$ rather than the standard of 2.5; and that when insufficient thickness is available to obtain valid values, reasonable estimates of such values can be obtained if $a$ is maintained at 5 times that of the plastic-zone-size factor.

While there was some doubt on the matter in the early days of fracture testing, it is now clear that fatigue precracking is an important prerequisite to useful measures of $K_{Ic}$; values determined without precracking of the specimens would be considered approximations at best, likely to be 5 to 10% higher than the correct values.

**Plane Stress and Mixed-Mode Fracture Toughness $K_c$.** The data in Tables 7.1 and 7.2 for alloys such as 7075-T6 and 7075-T651 illustrate the fact that the critical stress-intensity factor, $K_c$, tends to decrease with increase in thickness in the manner illustrated schematically for $G_c$ in Fig. 7.2, discussed previously. The decrease in the value of $K_c$ with increase in thickness reflects the transition from the conditions of plane stress through a mixture of modes moving toward plane-strain conditions and approaching, asymptotically, a minimum value approximately equal to $K_{Ic}$. The rate of decrease will differ for different alloys and tempers, and also for testing direction. For thicknesses up to about

1 in., the $K_c$ values for 7075-T651 plate in the longitudinal direction (L-T) are as high or higher than those for relatively thin 7075-T6 sheet, but at greater thicknesses they decrease toward the plane-strain value; in the transverse direction (T-L), $K_c$ decreases much more rapidly with thickness.

It should be noted that values of $K_c$ are more variable than $K_{Ic}$ values, at least partly because there may be larger specimen size effects. It is for this reason, among others, that no ASTM standards have ever been developed for measurement of $K_c$ but rather have focused on the measurement of crack resistance curves (see section 7.7).

Fatigue precracking, used as a means for developing the initial flaw in fracture toughness specimens, is not necessarily an important factor in determining the critical stress-intensity factor, $K_c$ under mixed-mode or plane-stress conditions. Under such conditions there is an appreciable amount of slow crack growth, and the stress condition at the tip of this crack is practically the same regardless of the original crack starter.

# 7.5 Industry $K_{Ic}$ Database, ALFRAC

As noted previously, an industry-wide effort (Ref 51) was made to build a database of $K_{Ic}$ values from regular testing of plant production lots of the higher-toughness aluminum alloys, most notably of those alloys such as 2124-T851 and 7475-T7351 for which fracture-toughness guarantees were to be specified. Great care was taken to detail the factors considered in determining the validity of the data in accordance with the ASTM E 399 standards applicable at the time, and the specific reasons for any individual test result not considered fully valid were documented.

Thousands of test results were compiled, and it is beyond the scope of this book to include all of those results herein. An example of the types of data and the analyses carried out is given in the example in Table 7.5. In many cases, these results were later augmented by others, and the total became the basis of the online database known as ALFRAC, distributed for several years online and searchable via the Scientific and Technical Information Network, STN International. (Access to STN International in North America is provided by Chemical Abstracts Service, a division of the American Chemical Society, Columbus, OH.) One of the unique and most useful aspects of this database is the inclusion therein of all of the validity criteria for each test, so that even those results not entirely valid per ASTM Standard E 399 may be judged based upon the reasons for and degree of noncompliance.

The typical and specified minimum values of $K_{Ic}$ and $K_c$ discussed subsequently came from analyses of data compilations such as those in the ALFRAC database.

## 7.6 Typical and Specified Minimum Values of $K_{Ic}$ and $K_c$ Fracture Toughness

**Typical Values.** Based upon testing at a number of laboratories, MIL-HDBK-5, the design handbook for the Aerospace Industry (Ref 52) has, for a number of years, published typical values of plane-strain fracture-toughness, $K_{Ic}$, for a number of wrought high-strength aluminum alloys. These, combined with those from recent publications by the aluminum industry summarized in Ref 2, are presented in Table 7.6.

**Specified Minimum Values.** In addition, the aluminum industry, through the Aluminum Association, Inc. in cooperation with the aerospace industry standards organizations, has developed specified minimum values of $K_{Ic}$ and $K_c$ for high-toughness alloys such as 2124, 7050, and 7475, developed especially for use in fracture-critical applications (Ref 2 and 53–56). These values are presented in Tables 7.7 and 7.8 for $K_{Ic}$ and $K_c$, respectively. In these cases, the statistical basis used has been the same as that used for other specified minimum values for the aluminum industry and for MIL-HDBK-5, namely, the values will be equaled or exceeded by 99% of production lots with 95% confidence.

## 7.7 Crack-Resistance Curves

Representative crack-resistance curves ($R$-curves) for several high-toughness aluminum alloys are presented in Fig. 7.6 and 7.7 (Ref 2, 57, 58). Crack-resistance curve testing has not reached the stage where statistically significant curves can be presented; rather, the curves, including those in Fig. 7.7 and 7.8, are for individual lots of material that are representative of commercial production of the alloys and tempers for which data are presented.

Included among the alloys and tempers for which crack-resistance curves are presented are:

| | |
|---|---|
| Fig. 7.6 | 2024-T3 and 2524-T3 sheet |
| Fig. 7.7 | 7075-T6 and 7475-T6 sheet, and 7075-T651 and 7075-T7351, and 7475-T651, 7475-T7651, and 7475-T7351 plate |

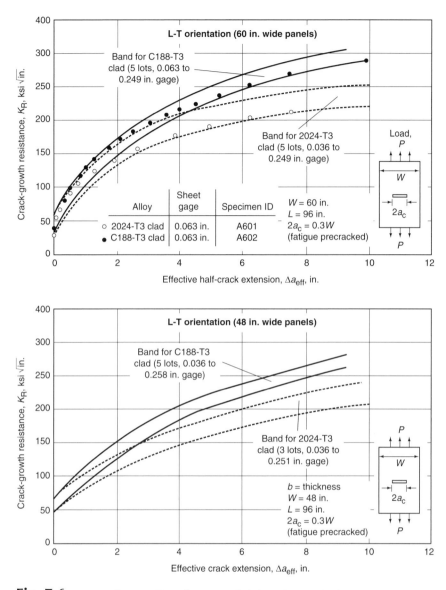

**Fig. 7.6** *R*-curves for 2024-T3 and 2524-T3 clad sheet. Source: Boeing

Each of these sets of *R*-curves illustrates the benefits of the composition and production-process control used in making the higher-toughness version of each alloy type (2524 vs. 2024, and 7475 vs. 7075), as discussed further in Chapter 11.

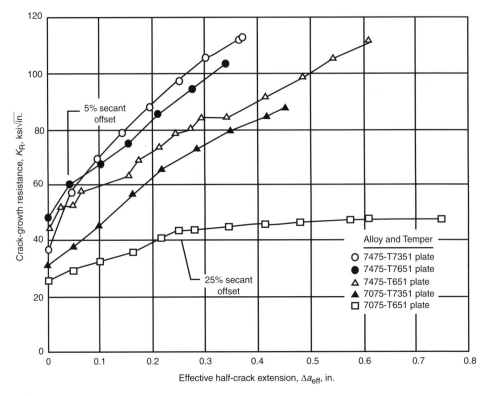

**Fig. 7.7**  R-curves for 7475-T7351, 7475-T7651, 7475-T651, and 7075-T7351, 7075-T651 plate. L-T orientation. Thickness is 0.50 in.; width is 4.00 in.

# 7.8 Use of Fracture-Toughness Data

While it is beyond the scope of this book to go deeply into fracture-mechanics design concepts, it is appropriate to describe them briefly and note that they have considerable value in the design of high-performance aircraft or aerospace structures, where high strength-to-weight ratios are essential, and where the initiation and propagation of cracks in regions of high tensile stress must be avoided or adequately taken into account. They are also useful in critical tankage design for other fields, such as for the containment and transportation of liquified gases, where failure might lead to catastrophic losses of property and possibly life. It is improbable that they will ever be needed in the design of a broad range of civil-engineering structures (bridges, buildings, etc.) or in most chemical process equipment, because the materials usually used are generally so tough that this method of stress analysis is not applicable.

The fracture-toughness approach to design may be considered ultraconservative (ultrasafe) by some designers, even for high-strength alloys and tempers, Indeed, if internal discontinuities, shear or weld cracks, and fatigue cracks can be avoided in structures, the allowable stress value may safely exceed that based on the $K_{Ic}$ values of the fracture-toughness approach. However, in most cases, there are *minimum* limits beyond which the size of discontinuities and cracks cannot be detected by practical production and inspection procedures and eliminated, so it is prudent to consider that they might be present. Further, in applications involving cyclic, that is, fatigue loading of any type, some consideration must be given to the consequences of fatigue-crack initiation and growth from any such flaws already present.

As indicated previously, the stress-intensity factor, $K$, is, from the designer's viewpoint, a parameter relating fracture stress to the critical size of flaw, design detail, or discontinuity, having sharpness of the ends equal to that of a crack. Representative curves showing the relationship of gross fracture stress and critical crack of flaw size for mixed-mode fracture of 0.063 in. aluminum alloy sheet are shown in Fig. 7.8, and those for plane-strain fracture (independent of thickness) are illustrated in Fig. 7.9. These curves are based on the average values of $K_c$ and $K_{Ic}$ from tests of center-notched specimens in Tables 7.1 and 7.2. They were developed for members of infinite width with Eq 5 and its equivalent for plane-strain conditions rewritten in the form:

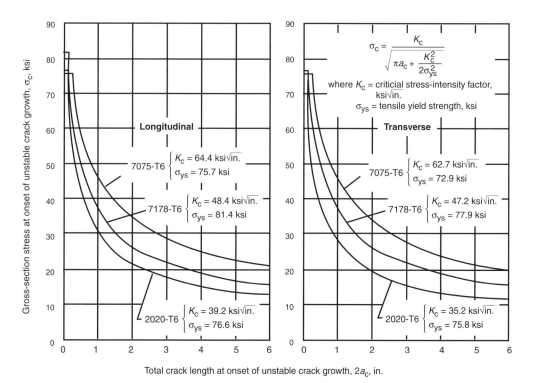

**Fig. 7.8** Gross-section stress at onset of rapid fracture vs. crack length—infinitely wide panels, 0.063 in. sheet

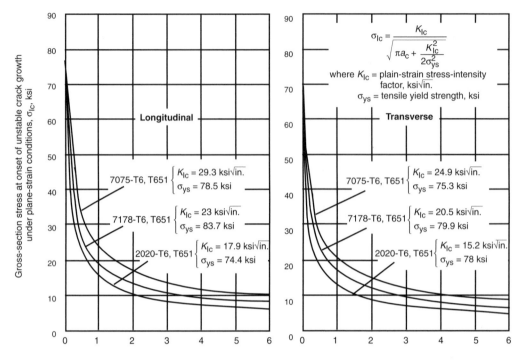

Total crack length at onset of unstable crack growth under plane-strain conditions, $2a_o$, in.

**Fig. 7.9** Gross-section stress at initiation of slow crack growth or rapid crack propagation under plane-strain conditions vs. crack length—infinitely wide panels

(Eq 8)

$$\sigma_c = \frac{K_c}{\sqrt{\pi a_c + \dfrac{K_c^2}{2\sigma_{ys}^2}}}$$

The curves are cut off at the yield strength of the materials, since failure of the structure with flaws smaller than that at the cut-off point would be by general yielding, and the principles of fracture mechanics would not be applicable.

Specifically, fracture toughness data, such as those in Tables 7.6 and 7.7 as well as those from more sophisticated tests developed in recent years, are used for the following purposes:

- Alloy selection
  a. By merit rating based on values of $K_c$ and/or $K_{Ic}$
  b. By determining residual load-carrying capacity with due regard for initial size of the discontinuity, the rate of the fatigue-crack propagation, and the design life of the structure
- Design of new structures
  a. By establishing the design stress for a given component consistent with maximum expected crack length

b. By establishing limiting crack length for a component on the basis of a given operating stress

c. By establishing inspection criteria (including thoroughness and frequency) consistent with the potential initial crack size and the expected rate of fatigue-crack propagation

• Evaluation of existing structures

a. By estimating residual strength and tolerance for additional loading

b. By estimating residual life consistent with observed crack length, rate of fatigue-crack propagation, and critical crack length

It is important to recognize that values of "flaws" or "crack" size, as referred to previously, must take into account any design discontinuities to which the real flaw or crack are adjacent or from which they grow. For example, a $\frac{3}{16}$ in. rivet hole, with a $\frac{1}{8}$ in. fatigue crack growing out of one side, constitutes a total flaw size or discontinuity $\frac{5}{16}$ in. in length.

Brief comment is given on the approach to three general design problems:

• Design for large cracks ($2a$ greater than $2t$)
• Design for small flaws or cracks ($2a$ less than $2t$)
• Design for part-through cracks

Internal residual stresses from production of the component are, of course, also a factor in such designs, and they must be considered over and above the analyses given subsequently.

**Design for Large Cracks ($2a$ greater than $t$).** In a typical situation, the designer of a structure in which unstable crack growth must be considered has a material, a tentative design stress, and an estimate of the width and thickness of the member. Before the designer can complete the fracture analysis, the maximum size of discontinuity that might exist in the structure after fabrication and inspection and the size to which that crack might grow during service before it is detected.

With this information and curves of the type shown in Fig. 7.8 and 7.9, the designer can determine whether or not the previous choice of material and tentative design stress is satisfactory. These evaluations are made on the basis of the original design and for the most severe set of expected circumstances, that is, after development of fatigue cracks. The process might be as illustrated in the following example.

A designer has selected 0.063 in. 7075-T6 sheet for an application that requires a design stress of 35 ksi (transverse). The inspection department ensures that even in areas hidden from easy view, a crack 1 in. or more in length would be detected, but one 0.75 in. long might not be. In this application, thorough inspections are to be made at intervals of $Y$ hours.

Reference to Fig. 7.8 indicates that unstable crack propagation in 7075-T6 sheet stressed to 35 ksi in the transverse direction would not be expected until the crack length reached 1.8 in.; with a 1.0 in. crack, the stress could safely be as high as 45 ksi. Therefore, the original design would appear to be safe from the aspect of unstable crack growth.

On the other hand, the data available on fatigue crack propagation indicate that if a crack 0.75 in. long existed in the original structure, it would grow to 1.9 in. in length in less than $Y$ hours, the time of the next inspection. If this takes place without reduction of the applied stress, catastrophic failure would be expected (Fig. 7.8). Hence, the time between inspections must be shortened, the stress must be lowered, or the material must be changed to one with greater fracture toughness.

**Design for Small Flaws or Cracks (2a less than 2t).** Situations arise in which the designer can be certain of restricting the size of flaws or cracks to a length equal to or less than the thickness of the material, and may only wish to know if it is safe to base the design on the full yield strength. If the maximum anticipated flaw size is to the left of the point of cut-off in Fig. 7.8 or 7.9, or if the critical stress calculated by Eq 3 for the anticipated flaw size is greater than the yield strength, design on the basis of the yield strength is safe.

A related approach is to restrict the use of a material to situations where it can sustain a stress equal to the yield strength in the presence of cracks equal in length to at least twice the thickness ($2a_c \geq 2t$), without developing rapid crack propagation (sometimes referred to as the $\beta = 2\pi$ criterion). This is another way of requiring that the use of a material be restricted to thicknesses less than one-half the crack size associated with the cut-off point in the $\sigma_c$ versus $2a$ curves; for example, for 7075-T6 in Fig. 7.10, the greatest thickness allowed would be one-half of 0.22 in., or 0.11 in. The basis of this criterion for crack length is the observation that a part-through crack developing from the inside surface of a pressure vessel has a nearly semicircular shape. Thus, the crack may be expected to be $2t$ long at the inside surface when it first reaches the outside surface, where it can be detected visually or by leakage. Thus, the requirement for tolerance of a $2t$ crack theoretically assures the possibility of visual or leakage detection of the crack before catastrophic fracture. Experience has shown that this criterion is often conservative (safe) for aluminum alloys.

With this approach, it is useful to know the $a_{c,2t}$ fracture stresses for infinite panels, computed with the equation:

$$\sigma_{c,2t} = \frac{K_c}{\sqrt{\pi t + \dfrac{K_c^2}{2\sigma_{ys}^2}}}$$

This equation can be developed from Eq 7 by setting $2a$ equal to $2t$. When $\sigma_{c,2t}$ exceeds $\sigma_{ys}$, it is safe to base the design on the full yield strength of the material.

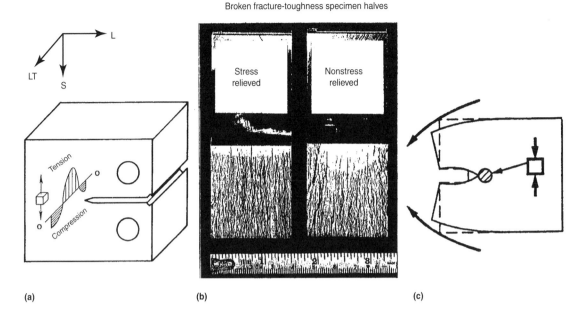

**Fig. 7.10** Illustrations of potential residual stresses in fracture toughness specimens. (a) Potential residual stress pattern. (b) Effect of significant residual stresses on fatigue crack front curvature. (c) Likely influence of residual stresses during testing of specimens

**Design for Part-Through Cracks.** When part-through cracks are involved, the problem may be handled as follows:

1. The situation is first analyzed in terms of $K_{Ic}$ with Fig. 7.9. The value of $K$ required for compatibility with the anticipated length and design stress is either calculated with the equation or determined from the position of the point representing these values in Fig. 7.9. If the calculated value of $K$ exceeds $K_{Ic}$ for the material involved, or if the plotted point falls above the applicable line in Fig. 7.9, the part-through crack must be considered as being likely to grow rapidly through the member and to at least $2t$. If the required value of $K$ is less than $K_{Ic}$ for the material, or if the plotted point falls below the applicable line in Fig. 7.9, no additional growth would be expected unless there was fatigue action or an unexpected increase in stress.

2. If the crack is likely to grow through the thickness, the new situation must then be examined in terms of $K_c$ with Fig. 7.8. The new crack length (equal to $2t$) and design stress are used to calculate a new required value of $K$ for the through-the-thickness crack or to locate a point in Fig. 7.8. If the new value of $K$ exceeds $K_c$, or the point falls on or above the line in Fig. 7.8, rapid crack propagation is now a probability, and corrective measures of the type described previously are called for. If the new value of $K$ is appreciably less than $K_c$ (or if the point falls below the line in Fig 7.8), rapid crack propagation is not an immediate possibility, although any growth of the crack by fatigue must still be considered, as in the previous example.

## 7.9 Discussions of Individual Alloys

Discussion of the relative performance of specific alloys and tempers, especially those shown to have excellent properties for use in fracture-critical components, are covered under alloy and metallurgical considerations in Chapter 11.

## 7.10 Understanding the Effect of Residual Stresses on Fracture Toughness Values

The potential role of residual stresses in complicating fracture toughness measurements and interpretation is sufficiently important that this Chapter closes by emphasizing the need to understand these effects. We are indebted to the fine work of Dr. R.J. Bucci and his associates at Alcoa Laboratories (Ref 49, 50) for developing this understanding in their efforts to interpret and deal with the variability of data observed, especially in plane-strain fracture toughness measurements from asymmetrical specimens taken from relatively thick and complex parts, such as forging or those machined from thick plate.

Fig. 7.10(a) illustrates the residual stress state commonly found in L-T and T-L compact fracture toughness specimens following their machining from thick plate or forgings. Also illustrated (Fig. 7.10b) is one important result of such residual stresses: the excess curvature of the fatigue crack front following precracking of the specimen. The same residual stress pattern may also result in the equivalent of clamping forces holding the compact specimen arms closed (Fig. 10c), which in turn may result in an artificial elevation of the resultant $K_{Ic}$ value when such a specimen is tested. The net result of this type of artifact is likely to be abnormally great scatter in test results from lot to lot and, therefore, somewhat ironically, lower design values because of the effect of the great scatter on the statistical analysis.

Bucci and his associates have provided two approaches to minimizing the influence of residual stresses on fracture-toughness measurements:

- Minimize the residual stresses in the specimen prior to the $K_{Ic}$ measurement by making the specimen thickness as small as possible while meeting other validity criteria, and by fatigue precracking at a fatigue-stress ratio of +0.7 rather than +0.1, as has historically been the standard. These practices will not only minimize the residual stresses but also have the benefit of providing straighter fatigue crack fronts in the specimens.
- Use a post-test correction method to estimate the fracture toughness, $K_c$ or $K_{Ic}$, that would have been obtained had the test specimen been free of residual stress. The need for this correction is usually suggested by substantial curvature in the early part of the load-displacement curve.

The details of these methods and appropriate supporting information are included in Ref 50 and are being incorporated in current ASTM standard methods for fracture-toughness testing (Ref 59).

**Table 7.1(a)   Results of fracture toughness tests of thin, center-cracked panels of aluminum alloy sheet and plate—longitudinal (L-T) orientation**

| Alloy and temper | Thickness, in. | Specimen width, in. | Ultimate tensile strength (UTS), ksi | Tensile yield strength (TYS), ksi | Elongation in 2 in., % | At pop-in | | | | At unstable crack growth | | | |
|---|---|---|---|---|---|---|---|---|---|---|---|---|---|
| | | | | | | Crack length, 2a, in. | Gross stress ($\sigma_G$), ksi | $\sigma_N/$ $\sigma_{ys}$(a) | $K_{Ic}$, ksi $\sqrt{in.}$ | Crack length, 2a, in. | Gross stress, ($\sigma_G$), ksi | $\sigma_N/$ $\sigma_{ys}$(a) | $K_c$, ksi $\sqrt{in.}$ |
| 2014-T6 | 0.063 | 16 | 72.8 | 67.6 | 11.2 | ... | ... | ... | ... | (b) | (b) | (b) | 65.0 |
| 2014-T651 | 0.250 | 4 | 69.7 | 64.3 | 11.6 | 1.40 | 17.9 | 0.43 | 28.2 | 2.28 | 32.0 | 1.16 | (c) |
| | | 4 | 67.8 | 62.2 | 11.0 | 1.40 | 14.9 | 0.37 | 23.5 | 2.31 | 32.0 | 1.22 | (c) |
| | | 3 | 70.3 | 65.0 | 11.0 | 1.16 | 18.0 | 0.45 | 27.2 | 1.76 | 30.4 | 1.14 | (c) |
| 2020-T6(d) | 0.063 | 16 | 81.4 | 77.2 | 8.0 | ... | ... | ... | ... | (b) | (b) | (b) | 37.4 |
| | | 2 | 80.2 | 75.9 | 7.8 | ... | ... | ... | ... | 0.85 | 30.3 | 0.70 | 41.0 |
| 2020-T6 (Alclad)(d) | 0.125 | 3 | 73.8 | 68.6 | 8.0 | 1.20 | 11.6 | 0.28 | 17.2 | 1.51 | 21.4 | 0.63 | 37.1 |
| 2020-T651 | 0.250 | 3 | 81.6 | 77.4 | 8.5 | 1.18 | 12.4 | 0.26 | 18.2 | 1.45 | 14.8 | 0.37 | 25.0 |
| | | 4 | 81.6 | 77.4 | 8.5 | 1.48 | 10.8 | 0.22 | 17.6 | 1.66 | 12.7 | 0.28 | 22.2 |
| 2024-T351 | 0.500 | 15 | ... | ... | ... | ... | ... | ... | ... | ... | ... | ... | ... |
| 2024-T81 | 0.125 | 3 | 71.3 | 65.2 | 9.0 | 1.08 | 17.6 | 0.42 | 24.4 | 1.68 | 33.8 | 1.16 | (e) |
| | | 3 | 71.0 | 64.9 | 9.0 | 1.05 | 19.7 | 0.47 | 27.0 | 1.64 | 33.4 | 1.12 | (e) |
| | | 3 | 69.6 | 62.2 | 8.1 | 1.08 | 19.4 | 0.48 | 26.7 | 1.74 | 33.2 | 1.28 | (e) |
| 2024-T351 | 0.250 | 4 | 71.8 | 65.2 | 10.0 | 1.41 | 16.3 | 0.39 | 25.8 | 2.28 | 30.0 | 1.08 | (e) |
| | | 4 | 73.0 | 66.4 | 8.8 | 1.44 | 16.2 | 0.38 | 26.0 | 2.21 | 31.3 | 1.05 | (e) |
| | 1.375 | 3 | 71.6 | 65.0 | 12.3 | 1.09 | 18.6 | 0.40 | 26.0 | 1.61 | 29.0 | 0.96 | (e) |
| | | 3 | 71.4 | 66.0 | 13.3 | 1.06 | 18.2 | 0.39 | 25.0 | 1.50 | 28.2 | 0.88 | (e) |
| | | 3 | 71.4 | 65.8 | 13.3 | 1.10 | 18.2 | 0.40 | 25.5 | 1.52 | 29.4 | 0.90 | (e) |
| 2024-T86 | 0.063 | 2 | 77.2 | 72.9 | 6.5 | ... | ... | ... | ... | 0.86 | 35.2 | 0.85 | 52.1(c) |
| | | 16 | 77.2 | 72.9 | 6.5 | ... | ... | ... | ... | (b) | (b) | (b) | 50.9 |
| | | 2 | 72.7 | 68.0 | 6.5 | ... | ... | ... | ... | 0.84 | 37.4 | 0.89 | 53.5(c) |
| | | 2 | 77.0 | 72.5 | 6.2 | ... | ... | ... | ... | 0.85 | 36.8 | 0.90 | 55.0(c) |
| 2219-T87 | 0.250 | 3 | 69.3 | 57.6 | 10.5 | 1.14 | 19.2 | 0.54 | 27.9 | 1.94 | 32.0 | 1.35 | (e) |
| | | 4 | ... | 57.6 | | 1.50 | 17.4 | 0.48 | 28.8 | 2.21 | 33.2 | 1.36 | (e) |
| | | 4 | 69.4 | 56.0 | 10.8 | 1.43 | 18.1 | 0.50 | 29.0 | 2.57 | 31.0 | 1.55 | (e) |
| | | 4 | 68.0 | 55.6 | 11.0 | 1.44 | 16.8 | 0.47 | 27.0 | 2.47 | 30.1 | 1.44 | (e) |
| 7075-T6 | 0.063 | 16 | 82.6 | 75.7 | 10.5 | ... | ... | ... | ... | (b) | (b) | (b) | 64.4 |
| | 0.125 | 3 | 82.8 | 77.6 | 11.2 | ... | ... | ... | ... | 1.44 | 40.8 | 1.00 | 67.4(c) |
| | | 3 | 80.7 | 73.2 | 11.5 | 1.06 | 18.8 | 0.40 | 25.8 | 1.56 | 37.4 | 1.06 | (e) |
| | | 3 | 84.7 | 78.0 | 11.0 | 1.15 | 19.6 | 0.41 | 27.9 | 1.46 | 33.8 | 0.84 | 57.9 |
| 7075-T651 | 0.250 | 4 | 83.0 | 77.3 | 14.5 | 1.45 | 15.8 | 0.32 | 25.4 | 2.11 | 26.2 | 0.72 | 54.8 |
| | | 4 | 85.0 | 78.8 | 13.5 | 1.44 | 18.8 | 0.37 | 30.1 | 1.98 | 33.8 | 0.85 | 67.0(c) |
| | | 4 | 83.9 | 78.2 | 13.0 | 1.39 | 20.8 | 0.40 | 32.8 | 1.96 | 32.3 | 0.81 | 65.6 |
| | 0.500 | 15 | ... | ... | ... | ... | ... | ... | ... | ... | ... | ... | ... |
| 7079-T6(d) | 0.063 | 16 | 72.4 | 63.4 | 11.0 | ... | ... | ... | ... | (b) | (b) | (b) | 103.0 |
| | 0.125 | 3 | 81.1 | 75.6 | 11.5 | ... | ... | ... | ... | 1.48 | 40.3 | 1.01 | 68.5 |
| 7079-T651(d) | 0.250 | 4 | 80.2 | 74.7 | 11.0 | 1.20 | 17.4 | 0.33 | 25.4 | 1.63 | 26.6 | 0.78 | 48.8 |
| 7178-T6 | 0.063 | 16 | 88.6 | 80.5 | 11.5 | ... | ... | ... | ... | (b) | (b) | (b) | 47.7 |
| | | 2 | 89.4 | 82.4 | 11.5 | ... | ... | ... | ... | 0.81 | 35.7 | 0.79 | 49.0 |
| | | 3 | 89.4 | 82.4 | 11.5 | ... | ... | ... | ... | 1.18 | 31.5 | 0.59 | 44.0 |
| | 0.125 | 3 | 89.6 | 83.5 | 12.6 | 1.07 | 16.7 | 0.31 | 22.8 | 1.36 | 27.6 | 0.62 | 44.7 |
| | 0.125 | 3 | 90.0 | 83.6 | 12.2 | 1.09 | 17.6 | 0.33 | 24,. | 1.37 | 31.1 | 0.69 | 50.5 |
| 7178-T651 | 0.250 | 4 | 88.7 | 84.3 | 13.0 | 1.42 | 13.5 | 0.25 | 21.5 | 2.29 | 21.4 | 0.25 | 48.1 |

Specimens per Fig. A1.9. Each line of data represents the average of duplicate or triplicate tests of one lot of material. For tensile yield strengths, offset is 0.2%. (a) $\sigma_N$ /$\sigma_{ys}$ is ratio of net-section stress at instability to tensile yield strength. (b) Average of five tests with different center-crack lengths. (c) Conditions close to general yielding, but value considered indicative. (d) Obsolete alloy. (e) General yielding indicated by ratio $\sigma_N$ /$\sigma_{ys}$ ; no value calculated

**Table 7.1(b)  Results of fracture toughness tests of thin, center-cracked panels of aluminum alloy sheet and plate—transverse (T-L) orientation**

| Alloy and temper | Thickness, in. | Specimen width, in. | Ultimate tensile strength (TYS), ksi | Tensile yield strength (TYS), ksi | Elongation in 2 in., % | At pop-in | | | | At unstable crack growth | | | |
|---|---|---|---|---|---|---|---|---|---|---|---|---|---|
| | | | | | | Crack length, $2a$, in. | Gross stress $(\sigma_G)$, ksi | $\sigma_N/\sigma_{ys}$(a) | $K_{Ic}$, ksi $\sqrt{\text{in.}}$ | Crack length, $2a$, in. | Gross stress, $(\sigma_G)$, ksi | $\sigma_N/\sigma_{ys}$(a) | $K_c$, ksi $\sqrt{\text{in.}}$ |
| 2014-T6 | 0.063 | 16 | 72.6 | 65.6 | 9.8 | ... | ... | ... | ... | (b) | (b) | (b) | 57.8 |
| 2014-T651 | 0.250 | 4 | 70.0 | 62.2 | 10.2 | 1.41 | 15.1 | 0.37 | 24.5 | 2.14 | 25.9 | 0.90 | 64.3(c) |
| | | 4 | 68.4 | 60.7 | 10.0 | 1.42 | 14.3 | 0.36 | 22.7 | 2.11 | 26.2 | 0.91 | 54.7(c) |
| | | 3 | 69.6 | 62.8 | 10.5 | 1.14 | 15.0 | 0.39 | 27.2 | 1.61 | 24.3 | 0.84 | 65.0(c) |
| 2020-T6(c) | 0.063 | 16 | 81.8 | 75.8 | 7.0 | ... | ... | ... | ... | (b) | (b) | (b) | 34.2 |
| | | 2 | 81.1 | 75.8 | 7.5 | ... | ... | ... | ... | 0.87 | 26.5 | 0.62 | 36.1 |
| 2020-T6 (Alclad)(c) | 0.125 | 3 | 73.5 | 68.4 | 6.8 | 1.11 | 11.9 | 0.28 | 16.6 | 1.24 | 15.5 | 0.41 | 24.7 |
| 2020-T651 | 0.250 | 3 | 83.1 | 78.0 | 6.0 | 1.12 | 10.7 | 0.22 | 15.2 | 1.14 | 10.7 | 0.22 | 15.3 |
| | | 4 | 83.1 | 78.0 | 6.0 | ... | ... | ... | ... | 1.32 | 13.5 | 0.26 | 20.4 |
| 2024-T351 | 0.500 | 15 | 68.7 | 48.6 | 19.5 | 4.98 | 15.0 | 0.47 | 46.4 | 8.25 | 23.7 | 1.10 | (d) |
| 2024-T81 | 0.125 | 3 | 72.5 | 66.4 | 8.0 | 1.05 | 17.6 | 0.41 | 24.2 | 1.49 | 30.6 | 0.92 | 50.0(c) |
| | | 3 | 71.7 | 66.0 | 8.2 | 1.07 | 17.0 | 0.40 | 23.4 | 1.54 | 30 | 0.93 | 53.1(c) |
| | | 3 | 70.5 | 64.1 | 7.4 | 1.06 | 17.8 | 0.43 | 24.7 | 1.60 | 29.7 | 0.99 | 54.2(c) |
| 2024-T351 | 0.250 | 4 | 72.0 | 66.2 | 8.0 | 1.44 | 13.0 | 0.31 | 20.7 | 1.94 | 23.3 | 0.68 | 45.5 |
| | | 4 | 72.4 | 65.8 | 7.5 | 1.42 | 15.7 | 0.37 | 25.0 | 1.94 | 27 | 0.80 | 52.6 |
| | 1.375 | 3 | 71.1 | 65.5 | 12.0 | 1.11 | 15.6 | 0.39 | 22.1 | 1.54 | 24.8 | 0.78 | 44.6 |
| | | 3 | 70.2 | 64.4 | 10.7 | 1.12 | 15.4 | 0.39 | 21.8 | 1.47 | 25.5 | 0.78 | 42.5 |
| | | 3 | 71.2 | 65.4 | 9.3 | 1.00 | 16.9 | 0.41 | 23.0 | 1.34 | 26.9 | 0.74 | 42.9 |
| 2024-T86 | 0.063 | 2 | 75.8 | 71.3 | 6.2 | ... | ... | ... | ... | 0.88 | 31.8 | 0.80 | 45.5 |
| | | 16 | 75.8 | 71.3 | 6.2 | ... | ... | ... | ... | (b) | (b) | (b) | 46.4 |
| | | 2 | 72.4 | 67.6 | 6.0 | ... | ... | ... | ... | 0.86 | 32.7 | 0.79 | 46.0 |
| | | 2 | 76.4 | 71.6 | 6.0 | ... | ... | ... | ... | 0.87 | 33.3 | 0.82 | 46.3 |
| 2219-T87 | 0.250 | 3 | 70.2 | 57.2 | 10.5 | 1.14 | 19.4 | 0.53 | 272.0 | 1.71 | 31.8 | 1.28 | (d) |
| | | 4 | ... | 57.2 | ... | 1.42 | 15.2 | 0.42 | 244.0 | 2.17 | 29 | 1.11 | (d) |
| | | 4 | 69.5 | 55.9 | 10.0 | 1.46 | 16.3 | 0.46 | 265.0 | 2.31 | 26.6 | 1.13 | (d) |
| | | 4 | 69.0 | 55.5 | 10.7 | 1.44 | 16.7 | 0.47 | 26.8T | 2.24 | 27.8 | 1.12 | (d) |
| 7075-T6 | 0.063 | 16 | 82.2 | 72.9 | 10.5 | ... | ... | ... | ... | (b) | (b) | (b) | 62.3 |
| | 0.125 | 3 | 82.8 | 74.1 | 11.5 | ... | ... | ... | ... | 1.24 | 36.2 | 0.83 | 58.2 |
| | | 3 | 82.7 | 72.9 | 11.0 | 1.11 | 18.8 | 0.41 | 26.6 | 1.37 | 32.5 | 0.82 | 52.7 |
| | | 3 | 86.9 | 77.0 | 10.5 | 1.11 | 16.1 | 0.34 | 23.4 | 1.48 | 27 | 0.69 | 46.3 |
| 7075-T651 | 0.250 | 4 | 84.8 | 74.2 | 13.0 | 1.28 | 16.2 | 0.38 | 24.8 | 1.67 | 22.4 | 0.68 | 40.5 |
| | | 4 | 85.8 | 75.4 | 13.2 | 1.44 | 15.1 | 0.31 | 24.2 | 2.05 | 25.7 | 0.70 | 52.4 |
| | | 4 | 84.0 | 72.0 | 13.0 | 1.46 | 16.9 | 0.37 | 27.3 | 2.18 | 25.6 | 0.78 | 54.9 |
| | 0.500 | 15 | 87.0 | 77.2 | 12.2 | ... | ... | ... | ... | 6.75 | 23.1 | 0.33 | 33.4 |
| 7079-T6(c) | 0.063 | 16 | 73.0 | 62.7 | 11.0 | ... | ... | ... | ... | (b) | (b) | (b) | 82.6 |
| | 0.125 | 3 | ... | 73.5 | 11.3 | ... | ... | ... | ... | 1.42 | 37 | 0.94 | 60.5(d) |
| 7079-T651(c) | 0.250 | 4 | 81.0 | 72.6 | 11.5 | 1.18 | 14.8 | 0.33 | 21.7 | 1.53 | 20.8 | 0.58 | 34.3 |
| 7178-T6 | 0.063 | 16 | 88.5 | 78.0 | 11.2 | ... | ... | ... | ... | (b) | (b) | (b) | 46.2 |
| | | 2 | 88.0 | 77.8 | 11.2 | ... | ... | ... | ... | 0.80 | 36.5 | 0.77 | 48.2 |
| | | 3 | ... | 78.8 | ... | ... | ... | ... | ... | 1.21 | 28.6 | 0.63 | 43.6 |
| | 0.125 | 3 | 89.8 | 77.4 | 12.8 | 1.06 | 14.5 | 0.30 | 21.0 | 1.36 | 23.5 | 0.55 | 39.2 |
| | 0.125 | 3 | 91.4 | 79.2 | 12.5 | 1.09 | 15.1 | 0.30 | 22.0 | 1.34 | 23.9 | 0.55 | 38.1 |
| 7178-T651 | 0.250 | 4 | 90.1 | 80.4 | 12.5 | 1.44 | 11.5 | 0.22 | 18.4 | 2.06 | 15.2 | 0.39 | 27.5 |

Specimens per Fig. A1.9. Each line of data represents the average of duplicate or triplicate tests of one lot of material. For tensile yield strengths, offset is 0.2%. (a) $\sigma_N/\sigma_{ys}$ is ratio of net-section stress at instability to tensile yield strength. (b) Average of five tests with different center-crack lengths. (c) Obsolete alloy. (d) Conditions close to general yielding, but value considered indicative

**Table 7.2(a)  Results of fracture toughness tests of 1 × 20 in. center slotted panels of aluminum alloy sheet and plate center cracked specimens—longitudinal (L-T) orientation**

| Alloy and temper | Thickness, in. | Lot | Ultimate tensile strength (UTS), ksi | Tensile yield strength (TYS), ksi | Elongation in 2 in., % | At pop-in Gross stress (σG), ksi | σN/σys(a) | KIc, ksi√in. | At unstable crack growth Crack length, 2a, in. | Gross stress, (σG), ksi | Net Stress at fracture σN, ksi | σN/σys(a) | Kc, ksi√in. |
|---|---|---|---|---|---|---|---|---|---|---|---|---|---|
| 2020-T651 (b) | 1.375 | A | 83.2 | 77.5 | 6.0 | 5.8 | 0.07 | 20.5 | 8.32 | 6.8 | 11.7 | 0.15 | 26.6 |
| | | B | 81.6 | 76.1 | 5.8 | 6.1 | 0.08 | 21.3 | 9.15 | 7.4 | 13.7 | 0.18 | 31.0 |
| | | C | 81.9 | 76.3 | 6.0 | 6.1 | 0.08 | 21.4 | 8.20 | 7.5 | 12.8 | 0.17 | 29.3 |
| 2024-T351 | 1.375 | A | 72.7 | 58.2 | 13.7 | 13.7 | 0.24 | 48.2 | 10.00 | 23.0 | 46.0 | 0.79 | 102.7(c) |
| 2024-T851 | 1.375 | A | 72.0 | 65.8 | 7.8 | 7.6 | 0.12 | 26.5 | 9.40 | 10.8 | 20.4 | 0.31 | 46.0 |
| | | B | 70.7 | 65.6 | 8.5 | 7.2 | 0.11 | 26.6 | 9.77 | 11.4 | 22.2 | 0.34 | 50.0 |
| | | C | 72.0 | 66.1 | 8.0 | 7.6 | 0.11 | 25.3 | 9.24 | 10.0 | 18.7 | 0.24 | 42.4 |
| 2219-T851 | 1.375 | A | 66.8 | 51.0 | 10.2 | 11.0 | 0.22 | 42.7 | 9.80 | 16.0 | 31.6 | 0.62 | 70.9(c) |
| | | B | 66.4 | 50.6 | 10.2 | 12.0 | 0.24 | 44.3 | 10.20 | 19.2 | 29.6 | 0.78 | 89.9(c) |
| | | C | 66.6 | 52.0 | 11.0 | 12.4 | 0.24 | 39.3 | 10.90 | 18.4 | 28.4 | 0.80 | 86.6(c) |
| 7001-T75 (b) | 1.375 | A | 81.8 | 72.2 | 9.5 | 6.9 | 0.10 | 24.2 | 8.40 | 8.8 | 15.1 | 0.21 | 34.6 |
| | | B | 80.6 | 70.6 | 9.5 | 6.7 | 0.09 | 23.3 | 8.46 | 9.3 | 16.1 | 0.22 | 36.7 |
| | | C | 80.6 | 70.6 | 9.5 | 7.0 | 0.10 | 24.5 | 8.10 | 8.1 | 13.6 | 0.20 | 31.6 |
| 7005-T6351 | 1.375 | A | 54.2 | 47.2 | 17.0 | 14.4 | 0.31 | 51.1 | 12.00 | 30.2 | 75.5 | 1.60 | (d) |
| 7075-T651 | 1.375 | A | 83.9 | 76.6 | 15.5 | 8.6 | 0.11 | 30.2 | 10.80 | 17.3 | 37.5 | 0.49 | 82.3 |
| | | B | 86.3 | 80.3 | 14.2 | 8.8 | 0.11 | 30.9 | 11.00 | 13.6 | 30.4 | 0.38 | 65.5 |
| | | C | 86.0 | 78.5 | 15.2 | 8.9 | 0.11 | 31.1 | 10.60 | 13.8 | 29.5 | 0.38 | 64.7 |
| 7075-T7351 | 1.375 | A | 76.7 | 66.3 | 12.0 | 10.2 | 0.15 | 35.8 | 9.54 | 19.8 | 37.9 | 0.57 | 85.2 |
| | | C | 70.8 | 59.1 | 12.5 | 10.8 | 0.18 | 38.2 | 10.00 | 23.6 | 47.1 | 0.80 | 105.4 |
| 7079-T651 (b) | 1.375 | A | 84.0 | 77.6 | 11.5 | 9.2 | 0.12 | 32.1 | 10.14 | 15.6 | 31.4 | 0.41 | 69.8 |
| | | B | 82.8 | 76.0 | 11.0 | 8.4 | 0.11 | 30.0 | 10.22 | 12.6 | 25.8 | 0.34 | 57.4 |
| | | C | 82.2 | 75.2 | 11.2 | 7.7 | 0.10 | 30.2 | 9.68 | 12.9 | 25.1 | 0.34 | 56.4 |
| 7178-T7651 | 1.375 | A | 80.2 | 71.2 | 10.2 | 8.4 | 0.12 | 29.5 | 8.60 | 11.8 | 21.0 | 0.29 | 47.1 |

Each line of data represents the average of duplicate or triplicate tests of one lot of material. Specimens per Fig. A1.9(b), 1 in. thick, with 7.00 in. long, machined-sharp center slots, with slot-tip radii <0.0005 in. For tensile yield strengths, offset is 0.2%. (a) $\sigma_G/\sigma_{ys}$ is ratio of gross-section stress at pop-in to tensile yield strength and $\sigma_N/\sigma_{ys}$ is the ratio of net-section stress at fracture instability to tensile yield strength. (b) Obsolete alloy. (c) Conditions close to general yielding, but value considered indicative. (d) Conditions of general yielding; no value calculated

**Table 7.2(b)  Results of fracture toughness tests of 1 × 20 in. center-slotted panels of aluminum alloy sheet and plate center cracked specimens—transverse (L-T) orientation**

| Alloy and temper | Thickness, in. | Lot | Ultimate tensile strength (UTS), ksi | Tensile yield strength (TYS), ksi | Elongation in 2 in., % | At pop-in Gross stress (σG), ksi | σG/σys(a) | KIc, ksi√in. | At unstable crack growth Crack length, 2a, in. | Gross stress, (σG), ksi | Net Stress at fracture σN, ksi | σNσys(a) | Kc, ksi√in. |
|---|---|---|---|---|---|---|---|---|---|---|---|---|---|
| 2020-T651 (b) | 1.375 | A | 82.4 | 78.4 | 1.8 | 5.2 | 0.07 | 19.2 | 7.00 | 5.2 | 8.0 | 0.10 | 19.2 |
| | | B | 82.2 | 77.5 | 2.6 | 5.4 | 0.07 | 18.5 | 7.00 | 5.4 | 8.3 | 0.11 | 18.9 |
| | | C | 82.2 | 77.4 | 2.4 | 5.6 | 0.07 | 19.5 | 7.00 | 5.6 | 8.6 | 0.11 | 19.6 |
| 2024-T351 | 1.375 | A | 72.4 | 52.0 | 16.5 | 11.8 | 0.23 | 41.7 | 10.10 | 19.9 | 40.4 | 0.77 | 89.7 |
| 2024-T851 | 1.375 | A | 71.1 | 65.5 | 12.0 | 6.3 | 0.10 | 21.3 | 9.10 | 8.1 | 15.1 | 0.23 | 34.2 |
| | | B | 70.2 | 64.4 | 10.7 | 6.4 | 0.10 | 23.0 | 9.13 | 7.4 | 13.4 | 0.20 | 32.2 |
| | | C | 71.2 | 65.4 | 9.3 | 6.0 | 0.09 | 22.3 | 8.84 | 7.9 | 14.0 | 0.22 | 30.5 |
| 2219-T851 | 1.375 | A | 66.0 | 50.8 | 12.0 | 10.6 | 0.21 | 40.0 | 9.88 | 13.3 | 26.0 | 0.52 | 58.5 |
| | | B | 65.6 | 51.2 | 10.7 | 11.2 | 0.22 | 33.1 | 9.91 | 14.8 | 29.6 | 0.57 | 65.6 |
| | | C | 65.8 | 49.3 | 9.3 | 9.8 | 0.20 | 38.2 | 9.72 | 14.9 | 29.3 | 0.59 | 65.2 |
| 7001-T75 (b) | 1.375 | A | 80.8 | 71.3 | 8.8 | 6.5 | 0.09 | 22.8 | 7.71 | 6.6 | 10.2 | 0.15 | 24.7 |
| | | B | 79.9 | 69.6 | 9.0 | 6.4 | 0.09 | 22.6 | 7.90 | 7.2 | 11.0 | 0.17 | 27.2 |
| | | C | 80.5 | 70.6 | 8.8 | 6.1 | 0.09 | 21.3 | 8.04 | 6.4 | 10.7 | 0.15 | 24.5 |
| 7005-T6351 | 1.375 | A | 53.3 | 46.5 | 16.2 | 13.7 | 0.29 | 48.4 | 12.00 | 28.5 | 71.1 | 1.53 | (c) |
| 7075-T651 | 1.375 | A | 82.9 | 73.6 | 16.3 | 7.8 | 0.11 | 27.1 | 8.66 | 9.0 | 15.8 | 0.22 | 36.2 |
| | | B | 85.6 | 77.4 | 14.0 | 7.5 | 0.10 | 26.1 | 8.13 | 8.0 | 13.7 | 0.22 | 30.9 |
| | | C | 86.6 | 76.0 | 16.5 | 7.9 | 0.10 | 27.8 | 8.86 | 8.4 | 15.2 | 0.20 | 34.4 |
| 7075-T7351 | 1.375 | A | 74.9 | 64.6 | 10.5 | 8.8 | 0.14 | 30.8 | 8.82 | 11.0 | 19.5 | 0.31 | 44.6 |
| | | C | 70.1 | 58.5 | 11.0 | 9.8 | 0.17 | 34.4 | 8.71 | 14.4 | 26.4 | 0.48 | 60.4 |
| 7079-T651 (b) | 1.375 | A | 83.1 | 74.2 | 11.2 | 7.8 | 0.11 | 27.4 | 9.03 | 9.0 | 16.5 | 0.23 | 37.2 |
| | | B | 82.5 | 72.8 | 11.2 | 7.6 | 0.10 | 26.7 | 9.30 | 8.4 | 15.8 | 0.22 | 35.7 |
| | | C | 82.8 | 72.6 | 11.2 | 7.7 | 0.11 | 27.0 | 9.16 | 8.6 | 15.9 | 0.22 | 36.2 |
| 7178-T7651 | 1.375 | A | 79.8 | 70.5 | 10.3 | 6.8 | 0.10 | 23.9 | 8.12 | 8.3 | 12.7 | 0.20 | 31.9 |

Each line of data represents the average of duplicate or triplicate tests of one lot of material. Specimens per Fig. A1.9(b), 1-in. thick, with 7.00-in. long, machined-sharp center slots, with slot-tip radii <0.0005 in. For tensile yield strengths, offset is 0.2%. (a) $\sigma_G/\sigma_{ys}$ is ratio of gross-section stress at pop-in to tensile yield strength, and $\sigma_N/\sigma_{ys}$ is the ratio of net-section stress at fracture instability to tensile yield strength. (b) Obsolete. (c) Conditions of general yielding; no value calculated

**Table 7.3(a)   Results of fracture toughness tests of aluminum alloy sheet and plate, single-edge-cracked specimens—longitudinal (L-T) orientation**

| Alloy and temper | Thickness, in. | Ultimate tensile strength (UTS), ksi | Tensile yield strength (TYS), ksi | Elongation in 2 in. or 4D, % | Specimen width, in. | Crack length, 2a, in. | Critical net stress, $\sigma_N$, ksi | $\sigma_N/\sigma_{ys}$(a) | $K_{Ic}$, ksi $\sqrt{\text{in.}}$ |
|---|---|---|---|---|---|---|---|---|---|
| 2014-T651 | 0.250 | 69.7 | 64.3 | 11.6 | 1.50 | 0.56 | 55.8 | 0.86 | 35.8(b) |
|  | 0.500 | 69.1 | 64.6 | 11.7 | 2.25 | 0.67 | 34.0 | 0.53 | 26.8 |
| 2020-T651(c) | 0.250 | 81.6 | 77.4 | 8.5 | 1.50 | 0.58 | 36.0 | 0.46 | 22.2 |
| 2024-T851 | 0.250 | 71.8 | 65.2 | 10.0 | 1.50 | 0.57 | 41.6 | 0.64 | 26.0 |
|  | 0.250 | 73.0 | 66.4 | 8.8 | 1.50 | 0.57 | 47.6 | 0.72 | 30.4 |
|  | 1.375 | 71.4 | 65.8 | 13.3 | 3.00 | 1.00 | 29.0 | 0.44 | 26.2 |
| 7039-T63 | 0.500 | 71.1 | 61.8 | 14.0 | 3.00 | 0.98 | 48.2 | 0.78 | 44.6(b) |
| 7039-T6351 | 1.000 | 65.2 | 56.2 | 14.0 | 3.00 | 0.94 | 47.0 | 0.83 | 44.0(b) |
| 7039-T6 | 1.000 | ... | ... | ... | ... | ... | ... | ... | ... |
| 7075-T651 | 0.250 | 83.9 | 78.2 | 13.0 | 1.50 | 0.56 | 51.9 | 0.66 | 32.7 |
| 7079-T651(c) | 0.250 | 80.2 | 74.7 | 11.0 | 1.50 | 0.56 | 43.1 | 0.58 | 27.0 |
| X7106-T6351(c) | 1.000 | ... | ... | ... | ... | ... | ... | ... | ... |
| X7139-T6351(c) | 0.500 | 70.4 | 62.5 | 15.0 | 3.00 | 0.94 | 51.3 | 0.82 | 48.0(b) |
|  | 0.750 | 69.9 | 61.7 | 13.0 | 3.00 | 0.98 | 50.8 | 0.82 | 47.2(b) |
|  | 1.000 | 67.1 | 57.8 | 14.2 | 3.00 | 1.01 | 42.2 | 0.82 | 44.0(b) |
| 7178-T651 | 0.250 | 88.7 | 84.3 | 13.0 | 1.50 | 0.25 | 45.6 | 0.54 | 26.6 |
|  | 0.625 | ... | ... | ... | ... | ... | ... | ... | ... |

Specimens per Fig. A1.10. Each line of data represents the average of duplicate or triplicate tests of one lot of material. For tensile yield strengths, offset is 0.2%. (a) $\sigma_N/\sigma_{ys}$ is ratio of net-section stress at instability to tensile yield strength. (b) Not valid by present criteria; excess deviation from linearity prior to instability. (c) Obsolete alloy

**Table 7.3(b)   Results of fracture toughness tests of aluminum alloy sheet and plate, single-edge-cracked specimens—transverse (T-L) orientation**

| Alloy and temper | Thickness, in. | Ultimate tensile strength (UTS), ksi | Tensile yield strength (TYS), ksi | Elongation in 2 in. or 4D, % | Specimen width, in. | Crack length, 2a, in. | Critical net stress, $(\sigma_N)$, ksi | $\sigma_N/\sigma_{ys}$(a) | $K_{Ic}$, ksi $\sqrt{\text{in.}}$ |
|---|---|---|---|---|---|---|---|---|---|
| 2014-T651 | 0.250 | ... | ... | ... | ... | ... | ... | ... | ... |
|  | 0.500 | 69.4 | 62.6 | 10.3 | 2.25 | 0.71 | 31.4 | 0.50 | 24.6 |
| 2020-T651(c) | 0.250 | ... | ... | ... | ... | ... | ... | ... | ... |
| 2024-T851 | 0.250 | 72.0 | 66.2 | 8.0 | 1.50 | 0.81 | 40.3 | 0.61 | 19.5 |
|  | 0.250 | 72.4 | 65.8 | 7.5 | 1.50 | 0.55 | 45.2 | 0.70 | 28.9 |
|  | 1.375 | 70.2 | 64.8 | 10.7 | 3.00 | 1.00 | 24.4 | 0.38 | 21.9 |
| 7039-T63 | 0.500 | 71.0 | 62.8 | 13.5 | 3.00 | 0.98 | 40.5 | 0.64 | 37.0(b) |
| 7039-T6351 | 1.000 | 64.0 | 55.2 | 13.0 | 3.00 | 1.00 | 39.0 | 0.71 | 35.9(b) |
| 7039-T6 | 1.000 | 73.6 | 67.0 | 12.6 | 3.00 | 1.03 | 33.2 | 0.50 | 30.5(b) |
| 7075-T651 | 0.250 | 84.0 | 72.0 | 13.0 | 1.50 | 0.56 | 48.8 | 0.68 | 30.9 |
| 7079-T651(c) | 0.250 | 81.0 | 72.6 | 11.5 | 1.50 | 0.55 | 39.4 | 0.54 | 24.8 |
| X7106-T6351(c) | 1.000 | 64.5 | 56.2 | 14.2 | 3.00 | 1.00 | 42.8 | 0.76 | 39.1(b) |
| X7139-T6351(c) | 0.500 | 68.6 | 60.3 | 14.0 | 3.00 | 1.01 | 44.5 | 0.74 | 40.8(b) |
|  | 0.750 | 69.0 | 60.4 | 12.2 | 3.00 | 0.96 | 44.0 | 0.73 | 40.6(b) |
|  | 1.000 | 66.6 | 57.0 | 13.0 | 3.00 | 1.00 | 38.3 | 0.68 | 35.1(b) |
| 7178-T651 | 0.250 | ... | ... | ... | ... | ... | ... | ... | ... |
|  | 0.625 | 89.1 | 82.9 | 12.8 | 3.00 | 1.06 | 22.4 | 0.27 | 19.9 |

Specimens per Fig. A1.10. Each line of data represents the average of duplicate or triplicate tests of one lot of material. For tensile yield strengths, offset is 0.2%. (a) $\sigma_N/\sigma_{ys}$ is ratio of net-section stress at instability to tensile yield strength. (b) Not valid by present criteria; excess deviation from linearity prior to instability. (c) Obsolete alloy

**Table 7.4  Results of fracture toughness tests of aluminum alloy plate and of welds in plate-notched bend (NB) and compact tension (CT) specimens**

| Alloy and temper | Filler alloy | Thickness, in. | Ultimate tensile strength (UTS), ksi | Tensile yield strength (TYS), ksi | Elongation in 2 in. or 4D, % | Type of specimen | Specimen orientation Fig. A1.2 | Specimen width W, in. | Initial crack length, 2a, in. | Maximum nominal net stress, ksi | Specimen strength ratio, Rsb or Rsc(a) | $K_Q$, ksi √in. | $K_{max}$, ksi √in. | Valid $K_{Ic}$, ksi √in. |
|---|---|---|---|---|---|---|---|---|---|---|---|---|---|---|
| **Unwelded plate** | | | | | | | | | | | | | | |
| 2014-T651 | None | 1.000 | 72.0 | 65.8 | 9.2 | NB | T-L | 1.00 | 0.99 | ... | ... | 21,200 | ... | Yes |
| 2024-T651 | None | 1.375 | 70.8 | 64.4 | 7.2 | NB | T-L | 3.00 | 1.51 | ... | ... | 20,300 | ... | Yes |
| 5083-O | None | 7.000 | 45.0 | 20.8 | 18.8 | NB | T-S | 3.00 | 3.71 | 27.0 | 1.30 | ... | 53,300 | No(b) |
| 5083-O | None | 7.700 | 38.0 | 17.5 | 24.0 | NB | T-L | 7.70 | 4.12 | 22.7 | 1.30 | ... | 44,300 | No(b) |
|  |  |  |  |  |  | NB | T-S | 7.70 | 4.20 | ... | ... | ... | 48,000 | No(b) |
|  |  |  | 45.0 | 19.7 | 24.2 | NB | L-S | 6.00 | 3.14 | ... | ... | ... | 48,600 | No(b) |
|  |  |  | 38.0 | 17.5 | 24.0 | NB | T-S | 6.00 | 3.26 | ... | ... | ... | 41,200 | No(b) |
|  |  |  | 35.6 | 16.8 | 10.0 | CT | S-L | 6.00 | 3.21 | ... | ... | ... | 36,200 | No(b) |
| 5083-H321 | None | 3.000 | 48.4 | 35.9 | 15.0 | NB | L-T | 6.00 | 3.00 | 46.6 | 1.30 | 32,200 | 39,100 | No(b) |
|  |  |  | 48.5 | 34.0 | 15.0 | NB | T-L | 6.00 | 3.00 | 38.7 | 1.14 | 32,300 | 39,100 | No(b) |
|  |  |  | 44.7 | 29.6 | 9.0 | CT | S-L | 3.00 | 1.50 | 29.0 | 0.98 | 21,700 | 23,300 | No(b) |
| 5086-H32 | None | 3.000 | 43.9 | 32.3 | 17.0 | NB | L-T | 6.00 | 3.00 | 51.6 | 1.60 | 30,100 | 37,800 | No(b) |
|  |  |  | 42.0 | 28.8 | 18.0 | NB | T-L | 6.00 | 3.00 | 44.3 | 1.54 | 34,500 | 45,300 | No(b) |
|  |  |  | 39.3 | 26.3 | 12.0 | CT | S-L | 3.00 | 1.50 | 33.1 | 1.26 | 23,300 | 25,200 | No(b) |
| 6061-T6 | None | 1.500 | 45.1 | 41.9 | 15.0 | NB | L-T | 3.00 | 1.48 | 34.7 | 0.83 | 26,500 | 26,200 | Yes |
| 6061-T6 | None | 3.000 | 45.1 | 41.9 | 15.0 | NB | L-T | 6.00 | 3.00 | 29 | 0.70 | 26,200 | 31,500 | No(b) |
|  |  |  | 46.4 | 41.5 | 12.5 | NB | T-L | 6.00 | 3.00 | 27.7 | 0.70 | 26,900 | 21,800 | No(b) |
|  |  |  | 45.8 | 39.6 | 10.0 | CT | S-L | 3.00 | 1.50 | 48.7 | 0.92 | 21,300 | 48,400 | Yes |
| 7005-T6351 | None | 3.000 | 59.6 | 53.0 | 12.0 | NB | L-T | 6.00 | 3.00 | 40.6 | 0.79 | 46,700 | 42,300 | Yes |
|  |  |  | 58.5 | 51.5 | 14.0 | NB | T-L | 6.00 | 3.00 | 36.1 | 0.76 | 40,000 | 28,800 | Yes |
|  |  |  | 56.4 | 47.5 | 7.0 | CT | S-L | 3.00 | 1.50 | ... | ... | 27,600 | ... | Yes |
| 7075-T651 | None | 1.375 | 86.1 | 77.7 | 10.8 | NB | T-L | 3.00 | 1.54 | ... | ... | 46,700 | ... | Yes |
| 7075-T7351 | None | 1.375 | 68.2 | 56.8 | 12.0 | NB | T-L | 3.00 | 1.53 | ... | ... | 46,700 | ... | Yes |
| 7079-T651 | None | 1.375 | 82.5 | 72.8 | 11.2 | NB | T-L | 3.00 | 1.64 | ... | ... | 46,700 | ... | Yes |
| **Welded plate** | | | | | | | | | | | | | | |
| 5083-O | 5183 | 7.000 | 43.7 | 25.0 | 16.2 | NB | CNT | 7.00 | 3.50 | ... | ... | ... | 46,600 | No(b) |
| 5083-O | 5183 | 7.000 | 38.4 | 22.8 | 12.7 | NB | FNT | 7.00 | 3.48 | ... | ... | ... | 50,300 | No(b) |
|  | 5183 | 7.700 | 35.8 | 22.5 | 6.5 | NB | FNT | 7.70 | 3.64 | ... | ... | ... | 49,200 | No(b) |
|  | 5183 | 7.700 |  |  |  | NB | FNT | 7.70 | 3.77 | ... | ... | ... | 49,800 | No(b) |
| 5083-O | 5183 | 7.700 | 43.8 | 24.5 | 23.5 | CT | CPT | 6.00 | 3.55 | ... | ... | ... | 58,000 | No(b) |
|  | 5356 | 7.700 | 41.1 | 24.4 | 15.5 | CT | CTP | 6.00 | 2.92 | ... | ... | ... | 35,800 | No(b) |
|  | 5556 | 7.700 | 39.9 | 19.7 | 14.0 | CT | FNT | 6.00 | 3.04 | ... | ... | ... | 22,200 | No(b) |
| 5083-H321 | 5183 | 3.000 | 43.2 | 24.0 | ... | NB | CTL | 6.00 | 2.51 | 37.2 | 1.55 | 27,800 | 43,200 | No(b) |
|  | 5356 | 3.000 | 40.9 | 24.0 | ... | NB | CTL | 6.00 | 2.53 | 36.2 | 1.51 | 24,300 | 41,600 | No(b) |
|  | 5556 | 3.000 | 42.1 | 24.0 | ... | NB | CTL | 6.00 | 2.65 | 42.9 | 1.79 | 31,300 | 31,800 | No(b) |
| 6061-T6 | 4043AW | 3.000 | 30.5 | 15.0 | ... | NB | CTL | 6.00 | 2.72 | 28.3 | 1.89 | 19,300 | 30,500 | No(b) |
|  | 4043 HTAW | 3.000 | 37.8 | 35.0 | ... | NB | CTL | 6.00 | 3.60 | 31.5 | 0.90 | 23,000 | 26,500 | No(b) |
|  | 5356AW | 3.000 | 34.8 | 20.0 | ... | NB | CTL | 6.00 | 2.75 | 42.0 | 2.10 | 26,500 | 46,600 | No(b) |
|  | 5356 HTAW | 3.000 | 33.7 | 20.0 | ... | NB | CTL | 6.00 | 3.58 | 30.0 | 1.50 | 22,000 | 29,400 | No(b) |
| 7005-T6351 | 5039AW | 3.000 | 46.4 | 32.0 | ... | NB | CTL | 6.00 | 3.05 | 45.1 | 1.41 | 28,200 | 48,800 | No(b) |
|  | 5039 HTAW | 3.000 | 46.5 | 35.0 | ... | NB | CTL | 6.00 | 3.35 | 31.5 | 0.90 | 29,200 | 32,500 | 29,200 |
|  | 5356AW | 3.000 | 40.2 | 30.0 | ... | NB | CTL | 6.00 | 2.77 | 43.5 | 1.45 | 29,900 | 48,700 | No(b) |
|  | 5356 HTAW | 3.000 | 38.4 | 30.0 | ... | NB | CTL | 6.00 | 3.73 | 54.0 | 1.80 | 21,800 | 51,700 | No(b) |

Specimens per Fig. A1.11 (NB) and A1.12 (CT); each line of data represents the average of four tests of one lot of material. For tensile yield strengths, offset is 0.2%. $K_Q$ = candidate value of $K_{Ic}$. (a) Rsb or Rsc = $\sigma_n/\sigma_{ys}$ = ratio of maximum net-section stress to tensile yield strength. (b) Not valid by present criteria; excessive plasticity and/or insufficient thickness for plane-strain conditions

**Table 7.5  Representative summary of plane-strain fracture toughness test data for 7475-T7351 plate**

| Alloy and temper | Typical plane-strain fracture toughness, $K_{Ic}$ | | |
|---|---|---|---|
| | L-T, ksi $\sqrt{in.}$ | T-L, ksi $\sqrt{in.}$ | S-T, ksi $\sqrt{in.}$ |
| Average value | 49.2 | 40.9 | 32.4 |
| Actual minimum value | 39.5 | 34.9 | 28.0 |
| Number of tests | 137.0 | 81.0 | 24.0 |
| Standard deviation | 4.5 | 3.9 | 1.4 |
| Skewness value | +0.2 | +0.8 | −0.4 |
| A value, normal distribution | 37.4 | 30.2 | 27.9 |
| A value, using skewness(a) | 38.1 | 32.9 | 27.5 |
| B value, normal distribution | 42.5 | 34.8 | 29.8 |
| B value, using skewness(a) | 42.6 | 35.3 | 30.3 |

From Aluminum Industry Database, ALFRAC. All tests with compact tension specimens per ASTM Standard E399 (of type in Fig. A1.12a). All data included from valid tests per ASTM Standard E 399. (a) Based upon Person's Type III function

**Table 7.6  Published typical $K_{Ic}$ and $K_c$ values for aluminum alloys**

| Alloy and temper | Product form | Typical plane-strain fracture toughness, $K_{Ic}$ | | | |
|---|---|---|---|---|---|
| | | L-T, ksi $\sqrt{in.}$ (MPa $\sqrt{m}$) | T-L, ksi $\sqrt{in.}$ (MPa $\sqrt{m}$) | S-T, ksi $\sqrt{in.}$ (MPa $\sqrt{m}$) | Reference |
| 2090-T81 | Plate | 35 (38) | ... | ... | 53 |
| 2014-T651 | Plate | 22 (24) | 20 (22) | 17 (19) | 2 |
| 2024-T351 | Plate | 33 (36) | 30 (33) | 24 (26) | 2 |
| 2024-T851 | Plate | 22 (24) | 21 (23) | 16 (18) | 2 |
| 2024-T852 | Hand forging | 26 (29) | 19 (21) | 16 (18) | 2 |
| 2124-T851 | Plate | 29 (32) | 23 (25) | 22 (24) | 2 |
| 2219-T851 | Plate | 35 (39) | 33 (36) | 23 (25) | 2, 53 |
| 2219-T87 | Plate | 25 (27) | 22 (24) | ... | 53 |
| 2219-T852 | Hand forging | 39 (43) | 27 (30) | 25 (27) | 53 |
| 2419-T851 | Plate | 39 (43) | 35 (38) | 27 (30) | 53 |
| 6013-T651 | Plate | 38 (42) | ... | ... | 53 |
| 6061-T651 | Plate | 35 (39) | ... | ... | 53 |
| 7050-T73651,T7451 | Plate | 32 (35) | 27 (30) | 26 (29) | 2 |
| 7050-T7651 | Plate | 31 (34) | 28 (31) | 24 (26) | 53 |
| 7050-T74,T7452 | Die forging | 33 (36) | 23 (25) | 23 (25) | 2 |
| 7050-T73652,T7452 | Hand forging | 33 (36) | 21 (23) | 20 (22) | 2 |
| 7050-T73510,T73511 | Extrusion | 41 (45) | 29 (32) | 24 (26) | 2 |
| 7050-T76510,T76511 | Extrusion | 37 (41) | 26 (29) | 22 (24) | 2 |
| 7055-T7751 | Plate | 26 (29) | 24 (26) | ... | 53 |
| 7055-T7751X | Extrusion | 30 (33) | 25 (27) | ... | 53 |
| 7150-T651 | Plate | 27 (30) | 24 (26) | | 53 |
| 7150-T7751 | Plate | 27 (30) | 24 (26) | | 53 |
| 7150-T6510,T6511 | Extrusion | 29 (32) | 23 (25) | ... | 53 |
| 7149-T73 | Die forging | 31 (34) | 22 (24) | 22 (24) | 2 |
| 7075-T651 | Plate | 26 (29) | 23 (25) | 18 (20) | 2, 53 |

(continued)

## Table 7.6 (continued)

| Alloy and temper | Product form | Typical plane-strain fracture toughness, $K_{Ic}$ | | | |
|---|---|---|---|---|---|
| | | L-T, ksi $\sqrt{in.}$ (MPa $\sqrt{m}$) | T-L, ksi $\sqrt{in.}$ (MPa $\sqrt{m}$) | S-T, ksi $\sqrt{in.}$ (MPa $\sqrt{m}$) | Reference |
| 7075-T7351 | Plate | 29 (32) | 26 (290) | 19 (21) | 2 |
| 7075-T7651 | Plate | 27 (30) | 22 (24) | 18 (20) | 2, 53 |
| 7075-T7352 | Die forging | 29 (32) | 27 (30) | 26 (29) | 2 |
| 7075-T73,T7351 | Hand forging | 34 (37) | 26 (29) | 21 (23) | 2 |
| 7075-T6510,T6511 | Extrusion | 28 (31) | 24 (26) | 19 (21) | 2 |
| 7075-T73510,T76511 | Extrusion | 32 (35) | 26 (29) | 20 (22) | 2 |
| 7175-T736,T73651,T74,T7451 | Die forging | 30 (33) | 26 (29) | 26 (29) | 2 |
| 7175-T736,T73652,T74,T7452 | Hand forging | 34 (37) | 27 (30) | 24 (26) | 2, 53 |
| 7175-T73510,T73511 | Extrusion | 36 (400) | 31 (34) | ... | 53 |
| 7475-T651 | Plate | 39 (43) | 34 (37) | 29 (32) | 2 |
| 7475-T7351 | Plate | 50 (55) | 41 (45) | 33 (36) | 2 |
| 7475-T7651 | Plate | 39 (43) | 35 (39) | 28 (31) | 2 |

| Alloy and temper | Product form | Typical plane-stress fracture toughness, $K_c$ | | | |
|---|---|---|---|---|---|
| | | L-T, ksi $\sqrt{in.}$ (MPa $\sqrt{m}$) | T-L, ksi $\sqrt{in.}$ (MPa $\sqrt{m}$) | S-T, ksi $\sqrt{in.}$ (MPa $\sqrt{m}$) | Reference |
| 2090-T83 | Sheet | 40 (44) | ... | ... | 53 |
| 2090-T84 | Sheet | 65 (71) | ... | ... | 53 |
| 2024-T3 | Sheet | | 128 (141) | ... | 53 |
| Alclad 2024-T351 | Sheet | ... | 128 (140) | ... | 56 |
| Alclad 2524-T351 | Sheet | ... | 158 (175) | ... | 56 |
| 6013-T6 | Sheet | 134 (147) | 133 (146) | ... | 53 |
| 6061-T6 | Sheet | 73 (80) | 70 (77) | ... | 53 |
| 7055-T7751 | Plate | 85 (93) | 42 (46) | ... | 53 |
| 7150-T651 | Plate | 95 (104) | 60 (66) | ... | 53 |
| 7170-T77511 | Extrusion | 95 (104) | 60 (66) | ... | 53 |

## Table 7.7 Published minimum values of plane-strain fracture toughness for aluminum alloys

| Alloy and temper | Product form | Minimum plane–strain fracture toughness, $K_{Ic}$ | | | | | | | | | |
|---|---|---|---|---|---|---|---|---|---|---|---|
| | | Thickness, in. | L-T, ksi$\sqrt{in.}$ | T-L, ksi$\sqrt{in.}$ | S-L, ksi$\sqrt{in.}$ | Reference | Thickness, in. | L-T, MPa$\sqrt{m}$ | T-L, MPa$\sqrt{m}$ | S-L, MPa$\sqrt{m}$ | Reference |
| 2014–T651 | Plate | >0.499 | 19 | 18 | ... | 52(a) | >12.5 | 19 | 18 | ... | (c) |
| 2014–T652 | Hand forging | >0.499 | 24 | 18 | ... | 52(a) | >12.5 | 24 | 18 | ... | (c) |
| 2024–T351 | Plate | >0.999 | 27 | ... | ... | 52(a) | >25 | 27 | ... | ... | (c) |
| 2024–T851 | Plate | >0.499 | 15 | ... | ... | 52(a) | >12.5 | 15 | ... | ... | (c) |
| 2124–T8151 | Plate | 1.500–2.000 | 26 | 23 | 13 | 13 | >40.0–50.0 | 29 | 25 | 21 | 13 |
| | | 2.001–4.000 | 26 | 22 | 20 | 13 | >50.0–100.0 | 29 | 24 | 22 | 13 |
| | | 4.001–6.000 | 25 | 21 | 21 | 13 | >100.0–160.0 | 27 | 23 | 23 | 13 |
| 2124–T851 | Plate | 1.500–6.000 | 24 | 20 | 18 | 12(b) | >35.0–155.0 | 26 | 22 | 20 | 12(b) |
| 2219–T851 | Plate | >0.999 | 33 | 29 | 22 | 52(a) | >25 | 33 | 29 | 22 | (c) |
| 7149–T73511 | Extrusion | Up through 5.000 | 26 | 19 | 19 | 13 | >19.0 | 29 | 21 | 21 | 13 |
| 7050–T7451 | Plate | 1.000–2.000 | 29 | 25 | ... | 12(b) | >25.0–50.0 | 32 | 27 | ... | 12(b) |
| | | 2.001–3.000 | 27 | 24 | 21 | 12(b) | >50.0–80.0 | 30 | 26 | 23 | 12(b) |
| | | 3.001–4.000 | 26 | 23 | 21 | 12(b) | >80.0–100.0 | 28 | 25 | 23 | 12(b) |
| | | 4.001–5.000 | 25 | 22 | 21 | 12(b) | >100.0–130.0 | 27 | 24 | 23 | 12(b) |
| | | 5.001–6.000 | 24 | 22 | 21 | 12(b) | >130.0–150.0 | 26 | 24 | 23 | 12(b) |
| 7050–T7651 | Plate | 1.000–2.000 | 26 | 24 | ... | 12(b) | >25.0–50.0 | 28 | 26 | ... | 12(b) |
| | | 2.001–3.000 | 24 | 23 | 20 | 12(b) | >50.0–80.0 | 26 | 25 | 22 | 12(b) |

(continued)

(a) For those alloys for which near 100 lots or more were analyzed. Note: MIL-HDBK-5 values are not guaranteed values. (b) Values from Ref 12 are specification limits. (c) Calculated using standard metric conversion from standard values.

**Table 7.7 (continued)**

| Alloy and temper | Product form | Thickness, in. | L-T, ksi $\sqrt{in.}$ | T-L, ksi $\sqrt{in.}$ | S-L, ksi $\sqrt{in.}$ | Reference | Thickness, in. | L-T, MPa $\sqrt{m}$ | T-L, MPa $\sqrt{m}$ | S-L, MPa $\sqrt{m}$ | Reference |
|---|---|---|---|---|---|---|---|---|---|---|---|
| | | | | | | | Minimum plane–strain fracture toughness, $K_{Ic}$ | | | | |
| 7050–T7452 | Die forging | 0.750–3.500 | 27 | 19 | 19 | 13 | >19.0–90.0 | 30 | 21 | 21 | 13 |
| | | 3.501–7.000 | 25 | 19 | 19 | 13 | >90.0–160.0 | 27 | 21 | 21 | 13 |
| 7050–T76510, T76511 | Extrusion | Up through 5.000, 32 in.² max | 28 | ... | ... | 13 | Up to 120.0 20,000 mm² | 31 | ... | ... | 13 |
| 7150–T6151 | Plate | 0.750–1.500 | 22 | ... | ... | 13 | >20.0–40.0 | 24 | ... | ... | 13 |
| 7150–T651 | Plate | 0.750–1.500 | 22 | ... | ... | 13 | >20.0–40.0 | 24 | ... | ... | 13 |
| 7150–T7751 | Plate | ~1.000 | 22 | ... | ... | 53 | ~25.0 | 24 | ... | ... | (c) |
| 7150-T61510, T61511 | Extrusion | 0.750–1.500 | 21 | ... | ... | 13 | >20.0–40.0 | 23 | ... | ... | 13 |
| 7075–T651 | Plate | >0.499 | 26 | 22 | 20 | 52(a) | >12.5 | 28 | 24 | 22 | (c) |
| 7075–T7651 | Plate | >0.499 | 29 | 23 | 20 | 52(a) | >12.5 | 32 | 25 | 22 | (c) |
| 7475–T651 | Plate | 0.250–1.000 | 30 | 28 | ... | 13 | 0.250–1.000 | 33 | 31 | ... | 13 |
| | | 1.250–1.500 | 30 | 28 | ... | 12(b) | >30.0–35.9 | 33 | 31 | ... | 12(b) |
| 7475–T7351 | Plate | 1.250–2.499 | 40 | 33 | ... | 12(b) | >30.0–65.0 | 44 | 36 | ... | 12(b) |
| | | 2.599–4.000 | 40 | 33 | 25 | 12(b) | >65.0–100.0 | 44 | 36 | 27 | 12(b) |
| 7475–T7651 | Plate | 0.250–1.000 | 33 | 30 | ... | 13 | >25.0–30.0 | 36 | 33 | ... | 13 |
| | | 1.250–1.500 | 33 | 30 | ... | 12(b) | >30.0–35.0 | 36 | 33 | ... | 12(b) |

(a) For those alloys for which near 100 lots or more were analyzed. Note: MIL-HDBK-5 values are not guaranteed values. (b) Values from Ref 12 are specification limits. (c) Calculated using standard metric conversion from standard values.

**Table 7.8  Published specified minimum values of plane-stress fracture toughness, $K_c$, for aluminum alloys**

| Alloy and temper | Product form | Thickness, in. | L-T, ksi $\sqrt{in.}$ | T-L, ksi $\sqrt{in.}$ | Thickness, mm | L-T, MPa $\sqrt{m}$ | T-L, MPa $\sqrt{m}$ | Reference |
|---|---|---|---|---|---|---|---|---|
| | | | | Plane stress fracture toughness, $K_c$ | | | | |
| 7475-T61 | Sheet | 0.040–0.125 | ... | 75 | >1.00–3.20 | ... | 82 | 12(a) |
| | | 0.126–0.249 | 60 | 60 | >3.20–6.30 | 66 | 66 | 12(a) |
| Alclad 7475-T61 | Sheet | 0.040–0.125 | ... | 75 | >1.00–3.20 | ... | 82 | 13 |
| | | 0.126–0.249 | ... | 60 | >3.20–6.30 | ... | 66 | 13 |
| 7475-T761 | Sheet | 0.040–0.125 | ... | 87 | >1.00–3.20 | 110 | 96 | 12(a) |
| | | 0.126–0.249 | ... | 80 | >3.20–6.30 | ... | 88 | 12(a) |
| Alclad 7475-T761 | Sheet | 0.040–0.125 | 100 | 87 | >1.00–3.20 | 110 | 96 | 13 |
| | | 0.126–0.249 | ... | 80 | >3.20–6.30 | ... | 88 | 13 |

(a) Values from Ref 12 are specification limits.

CHAPTER **8**

# Interrelation of Fracture Characteristics

IT WAS NOTED in Chapter 3 that elongation and reduction in area from the tensile test are broad indicators of ductility for certain purposes, but that there are no consistent and reliable correlations between these properties and the more definitive toughness parameters, including notch toughness, tear resistance, and fracture toughness. One illustration of this was the rather broad relationship between elongation and unit propagation energy in Fig. 6.7; it is further illustrated in Fig. 8.1, showing notch-yield ratio against both elongation and reduction in area from a series of tests in which each were measured for the same lots (Ref 19). Again, both show a broad correlation, but not of tightness adequate for correlative purposes.

On the other hand, there are fairly well-defined and useful correlations between both notch-yield ratio (NYR) and unit propagation energy (UPE)

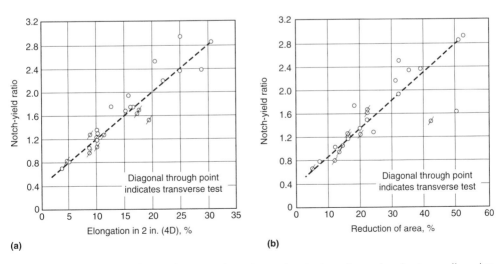

**Fig. 8.1** Notch-yield ratio in relation to elongation and reduction of area for aluminum alloy plate. Notch-yield ratio is notch tensile strength/tensile yield strength. Notched specimens, Fig. A1.7(a)

and the fracture-toughness parameters, $K_c$ and $K_{Ic}$ (Ref 1, 24, 36, 37). For example, NYR and UPE correlate well with $K_c$ from the same lots of material, as illustrated in Fig. 8.2 and 8.3. The relationship between UPE and $K_{Ic}$ has been refined over the years for predictive purposes, as illustrated in Fig. 8.4; this relationship is sufficiently well defined that in situations where fully valid $K_{Ic}$ values cannot be determined or when the greater expense of the more complicated tests must be avoided, tear test results can be used to estimate plane-strain fracture toughness values.

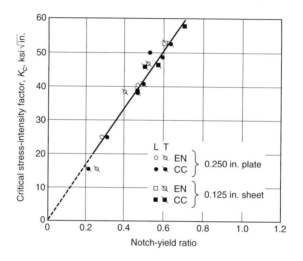

**Fig. 8.2** Critical stress-intensity factor, $K_c$ vs. notch-yield ratio (edge-notched specimen) for aluminum alloy and plate. EN, edge notched, Fig. A1.5; CC, center cracked, Fig. A1.6. Notch-yield ratio is notch tensile strength/tensile yield strength.

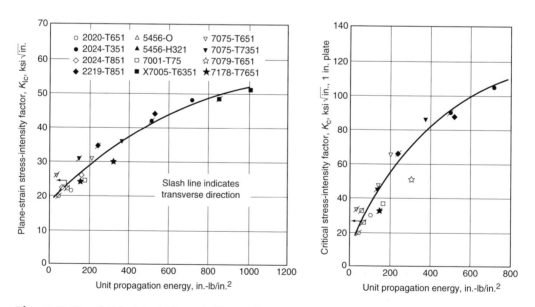

**Fig. 8.3** $K_{Ic}$ and $K_c$ for 1 in. thick panels (Fig. A1.9b) vs. unit propagation energy from tear tests for aluminum alloy plates

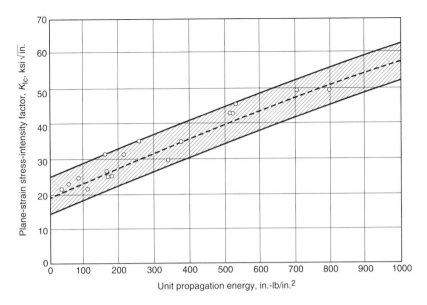

**Fig. 8.4** Relationship between plane-strain fracture toughness and unit propagation energy from tear tests for aluminum alloy products

These correlations are not surprising since the notch tensile, tear, and fracture toughness tests were all designed to measure the same material behavior from different perspectives: the ability to resist crack development and/or growth by plastic deformation at the site of severe stress raisers, including preexisting cracks. The fracture toughness test permits a calculation of the amount of stored elastic strain energy required to produce unstable crack growth. The tear test is a direct measurement of the external energy that is required to produce the crack growth; this is most useful when the energy is normalized based on crack growth area, as with the UPE.

Thus, while the fracture toughness test has the limitation that the specimens must be large enough to provide plane-strain conditions and enough recoverable elastic strain energy to produce unstable crack growth in an elastic stress field (a severe limitation for tough aluminum alloys, requiring massive specimens, if, indeed, the condition can ever be achieved), the tear test is limited only by the capacity of the source of external loading. Therefore, the tear test has been quite useful in screening tests for alloy development (Ref 19, 37), which is discussed in more detail in Chapter 11. In addition, it can even be used by extrapolation to estimate the fracture toughness of those materials that could rarely, if ever, be measured in a manner meeting all validity requirements.

Additional use has been made of these correlations through the use of notch-tensile testing as quality control for fracture toughness for those alloys where fracture toughness values are included in purchase specifications, as illustrated in Fig. 8.5 and 8.6 for 2124-T851 and 7475-T7351, respectively (Ref 60). In these cases, the relatively less-expensive notch-tensile test is sometimes used in plant production testing and fracture

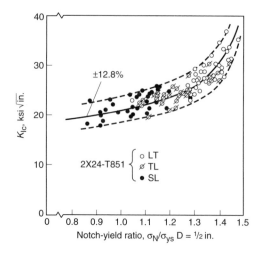

**Fig. 8.5** Correlation of plane-strain fracture toughness and notch-yield ratio (specimens per Fig. A1.7a) for 2024 and 2124 plate

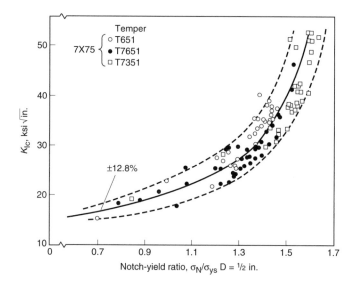

**Fig. 8.6** Correlation of plane-strain fracture toughness with notch-yield ratio (specimens per Fig. A1.7a) for 7075 and 7475 plate

toughness testing is used only for those lots for which meeting the appropriate specifications is in doubt. At higher toughness levels especially, the correlation is weaker and the amount of retesting required by this approach may be unacceptable.

Several other interesting relationships have been observed (Ref 2, 19) that enable the results of notch-tensile and tear tests to be used to estimate behavior under dynamic/fatigue loading:

- The relationship between the NYR and the ratio of notch-fatigue strength to the tensile yield strength appears useful for estimating fatigue life in the presence of notches from notch-tensile tests, as illustrated in Fig. 8.7. In this illustration, the notch fatigue strengths are for sharply notched rotating beam fatigue specimens. The value of this relationship can be rationalized on the basis that both tests measure in different ways the ability of materials to resist crack initiation and propagation in the presence of severe stress raisers.

- The relationship between tear resistance, as measured by UPE, and fatigue-crack growth rate is sufficiently well defined, as in Fig. 8.8, to potentially be useful for estimating the growth rate in terms of

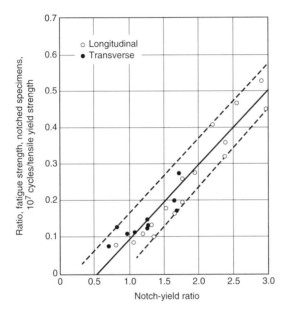

**Fig. 8.7** Relationship between ratio of fatigue strength of notched specimens to tensile yield strength and notch-yield ratio for aluminum alloy plate. Notch-yield ratio is notch tensile strength/tensile yield strength. Fatigue specimens were R.R. Moore rotating beam specimens; notched tensile and fatigue specimens were notched as in Fig. A1.7.

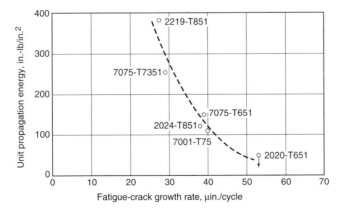

**Fig. 8.8** Relationship between unit propagation energy and fatigue-crack growth rate, where $K_{max}$ is 15 ksi$\sqrt{in.}$ and stress ratio is + 0.33

stress-intensity factor in cases where fatigue-crack growth rate measurements are not available.

For the record, there does not appear to be any relationship between any measures of fracture toughness and resistance to stress-corrosion cracking (Ref 2, 19). This is well illustrated by looking at the relationship between plane-strain fracture toughness, $K_{Ic}$, and the threshold stress-intensity factor for the initiation of stress-corrosion crack growth from tests of pre-cracked specimens, $K_{ith}$, in Fig. 8.9.

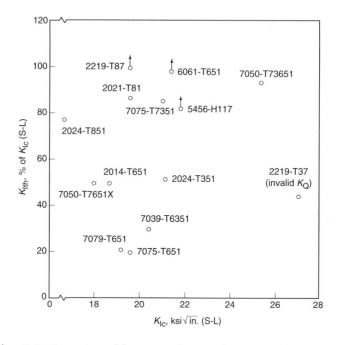

**Fig. 8.9** Comparison of fracture toughness and stress-corrosion resistance for some aluminum alloys. Stress-corrosion data are from ring-loaded 1/2 to 3/4 in. thick specimens in salt dichromate acetate-corrodent formula: 0.6M (3 1/2%) NaCl + 0.02M $Na_2Cr_2O_7$ + 0.07M $NaC_2H_3O_2$ at a pH of 4. $K_{ith}$, threshold stress intensity for stress-corrosion crack growth. $K_Q$, candidate value of $K_{Ic}$, invalid in instance shown

# CHAPTER 9

# Toughness at Subzero and Elevated Temperatures

THE NOTCH-TENSILE, tear, and fracture toughness tests described previously have been widely and effectively used to determine the effect of both subzero and relatively high temperatures on the toughness of aluminum alloys. A number of aluminum alloys, both wrought and cast, and welds in both wrought and cast alloys have been tested over a wide range of temperatures (Ref 25–35, 61–64), and the results are included in the following tables at the end of this Chapter.

Key parameters from these tests are plotted in the following figures as a function of test temperature and, in the case of elevated temperatures, the time at temperature:

| | |
|---|---|
| Fig. 9.1 | Notch-yield ratio: wrought alloy sheet; 1 in. wide, edge-notched specimens |
| Fig. 9.2 | Notch-yield ratio; wrought alloy plate; ½ in. diam specimens |
| Fig. 9.3 | Notch-yield ratio; welds in wrought alloy sheet; 1 in. wide, edge-notched specimens |
| Fig. 9.4 | Notch-yield ratio; cast alloys; ½ in. diam specimens |
| Fig. 9.4(a) | Sand-cast alloys |
| Fig. 9.4(b) | Permanent-mold cast alloys |
| Fig. 9.4(c) | Premium-strength sand-cast alloys |
| Fig. 9.5 | Notch-yield ratio; welds in wrought and cast alloys; ½ in. diam specimens |
| Fig. 9.6(a) | Unit propagation energy; wrought alloy sheet (−320 °F to RT) |
| Fig 9.6(b) | Unit propagation energy; wrought alloy sheet (−320 °F to 400 °F) |
| Fig. 9.7 | Unit propagation energy; welds in wrought alloy plate (−320 °F to RT) |
| Fig. 9.8 | Fracture toughness; (−320 °F to RT) |

For all of the test results included in these tables and figures, the sub-zero temperatures were achieved by immersing the specimens in cryostats containing liquefied gases as follows:

- For −112 °F, liquefied petroleum gas
- For −320 °F, liquefied nitrogen
- For − 423 °F, liquefied hydrogen
- For −452 °F, liquefied helium

In all such cases, tensile yield strengths and crack-opening displacements were measured by the use of extensometers incorporated into strain-transfer devices mounted directly on the specimens.

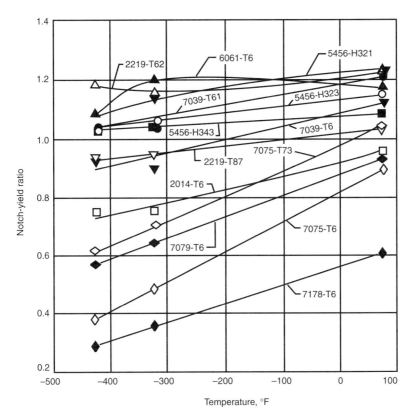

**Fig. 9.1** Notch-yield ratios (notch tensile strength/tensile yield strength) for ⅛ in. aluminum alloy sheet (average for longitudinal and transverse directions) at various temperatures. Specimens per Fig. A1.4(a)

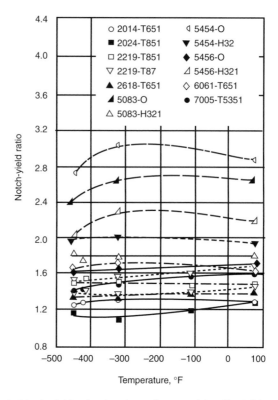

**Fig. 9.2** Notch-yield ratios (notch tensile strength/tensile yield strength) for plate at various temperatures. Specimens per Fig. A1.7(a)

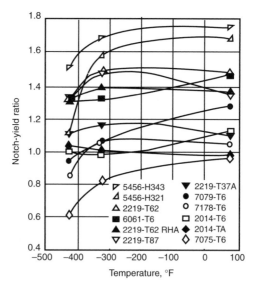

**Fig. 9.3** Notch-yield ratios (notch tensile strength/tensile yield strength) for welds in ⅛ in. aluminum alloy sheet at various temperatures (as-welded, unless noted otherwise; average for longitudinal and transverse directions). A, aged after welding; RHA, reheated and aged after welding. Specimens per Fig. A1.4(b)

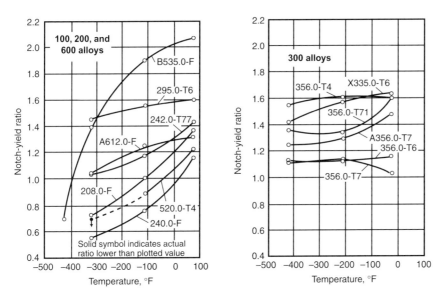

**Fig. 9.4(a)**  Notch-yield ratio for sand cast aluminum alloy slabs at various temperatures. Specimens per Fig. A1.7(a)

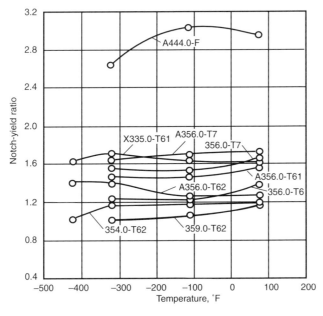

**Fig. 9.4(b)**  Notch-yield ratio for permanent mold cast aluminum alloy slabs at various temperatures. Specimens per Fig. A1.7(a)

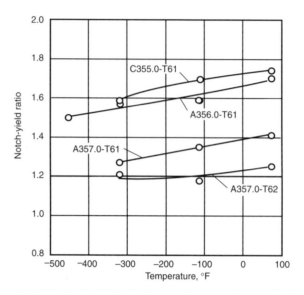

**Fig. 9.4(c)** Notch-yield ratio for premium strength cast aluminum alloy slabs at various temperatures. Specimens per Fig. A1.7(a)

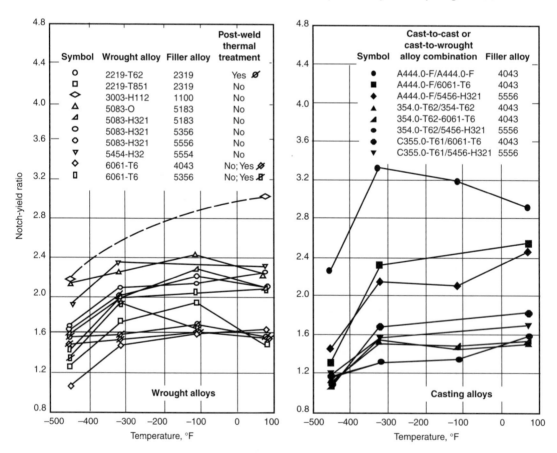

**Fig. 9.5** Notch-yield ratio (notch-tensile strength/joint yield strength) for groove welds in wrought and casting alloys at various temperatures. Specimens per Fig. A1.7(b)

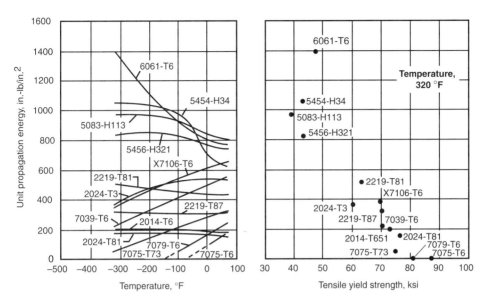

**Fig. 9.6(a)**  Tear resistance of sheet and plate of aluminum alloys at various temperatures (transverse direction). Specimens per Fig. A1.8

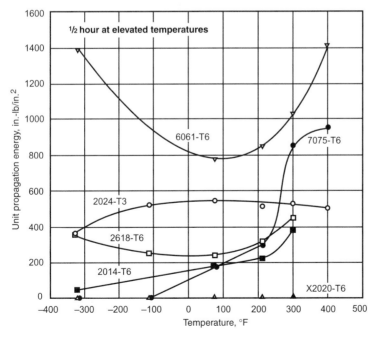

**Fig. 9.6(b)**  Tear resistance of aluminum alloy sheet at high temperatures. Specimens per Fig. A1.8

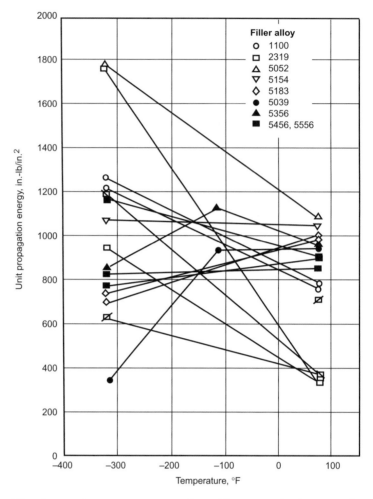

**Fig. 9.7** Unit propagation energy for welds in wrought aluminum alloy plate at various temperatures. Specimens per Fig. A1.8

## 9.1 Wrought Alloys at Subzero Temperatures

As the data for notch-yield ratio (NYR) (Fig. 9.1 and 9.2), unit propagation energy (UPE) (Fig. 9.6), and fracture toughness, $K_{Ic}$ (Fig. 9.8) indicate, the toughness of most 2xxx, 5xxx, and 6xxx wrought alloys remains about the same or decreases gradually as temperature decreases below room temperature (RT), even down to temperatures as low as –423 °F (Fig. 9.1) or –452 °F (Fig. 9.2). For the 7xxx wrought alloys, the values of NYR and UPE decrease more significantly with decrease in temperature, though even for these alloys, some toughness parameters like $K_{Ic}$ (Fig. 9.8) remain about the same at subzero temperatures as at RT.

Looking for indications of the most desirable alloys for cryogenic service, it is helpful to look at the low-temperature behavior as a function of tensile yield strength level at the more extreme temperatures. For example,

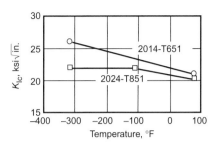

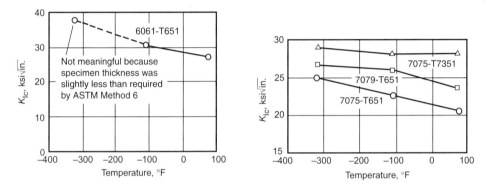

**Fig. 9.8**  Plane-strain fracture toughness of aluminum alloy plate at various temperatures. Specimens per Fig. A1.11(a), A1.12(a)

NYR is plotted as a function of tensile yield strength for sharply notched sheet-type specimens at –423 °F in Fig. 9.9 and for notched round specimens at –452 °F in Fig. 9.10; unit propagation energy is plotted as a function of yield strength at –320 °F in the right-hand part of Fig. 9.6(a). In general, the wrought alloy data fit fairly tight relationships indicating a trade-off of toughness with increasing strength. Based upon the data at –452 °F (Fig. 9.10), the 5*xxx* alloys in the annealed (O) temper have the highest overall toughness indices. Among the higher-strength combinations, 2219 in various tempers and 6061-T6 generally perform well.

In particular, the toughness of the 5083 and most other 5*xxx* alloys in the annealed (O) temper is outstanding at subzero temperatures (Ref 45, 61–64). This has been further confirmed by the testing of very large, thick notch bend and compact tension fracture toughness specimens (Fig. A1.11b, A1.12b, c) at temperatures as low as –320 °F, the results of which are included in Table 9.11. It is appropriate to note that even with fracture-toughness specimens as thick as 8 in., plane-strain conditions were still not encountered with 5083-O, and fully plastic tearing fracture was observed even at the lower temperature. From the combination of notch tensile, tear, and fracture toughness tests of 7.0 and 7.7 in. thick 5083-O plate and of 5183 welds in that 5083-O plate, as illustrated in the summary that follows, $K_{Ic}$ can conservatively be estimated to be approximately 45 to 50 ksi $\sqrt{\text{in}}$. in the L-T and T-L orientations, and $K_c$ can be conservatively

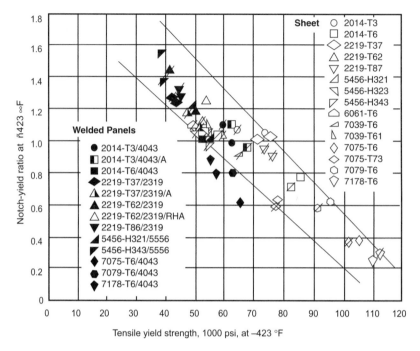

**Fig. 9.9** Notch-yield ratio vs. tensile yield strength for $\frac{1}{8}$ in. aluminum alloy sheet at –423 °F. Specimens per Fig. A1.4(a)

estimated to be at least 100 ksi $\sqrt{\text{in.}}$ (Ref 62, 63). That $K_c$ value leads to the stress-versus-flaw size diagram in Fig. 9.11 and the finding that even at stresses up to the tensile yield strength of the material at –320 °F, cracks with lengths in excess of 12 in. would not lead to unstable crack growth.

It was on the basis of such information that thick 5083-O plate with 5183 welds was chosen for the shipboard transportation of liquefied natural gas held in 125 ft diameter tanks at –260 °F, as illustrated by the cross section in Fig. 9.12. Assembly of the highly stressed girth supports required welded 5083-O plate as thick as 7.7 in., and the majority of the tank walls ranged from 2 to 4 in. in thickness. The added confidence required for the safety of this approach was gained by large-scale fracture toughness tests (Fig. A1.11b and A1.12b, c) that permitted estimates of both $K_c$ or $K_{Ic}$ for thick plate at temperatures as low as –320 °F, described previously. While even in the largest scale tests conducted, no unstable crack growth conditions were ever experienced for this alloy and weld combination, the results of these tests combined with the extrapolative techniques using notch-tensile and tear tests led to estimates of the toughness of the material that showed that even at its tensile yield strength, very thick 5083-O plate and 5183 welds in the walls of the tanks will support part-through cracks of any depth and through cracks more than 12 in. in length without unstable crack growth. "Leak-before-break" is assured by the analysis.

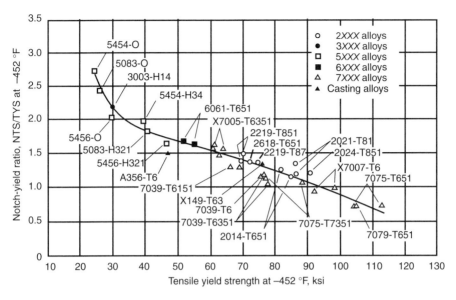

**Fig. 9.10** Notch-yield ratio vs. tensile yield strength for aluminum alloys at –452 °F. Specimens per Fig. A1.7(a)

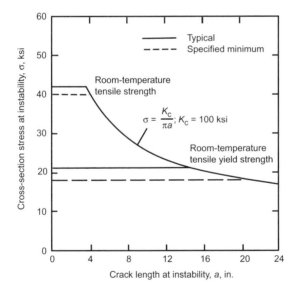

**Fig. 9.11** Estimated (conservative) fracture stress vs. flaw size relationship for 5083-O plate and 5183 welds

These findings are supported by totally independent studies, one by the Naval Research Laboratories (Ref 65) involving dynamic tear tests, and the other by Battelle Columbus (Ref 66) involving burst tests of pressure vessels at –220 °F. From the Naval Research Laboratories tests, the investigators concluded that the critical flaw sizes for 5083/5183 were "huge" and that there is "no need to calculate critical flaw sizes." From the Battelle burst tests, $K_c$ values in the range from 125 to 165 ksi $\sqrt{\text{in}}$. were

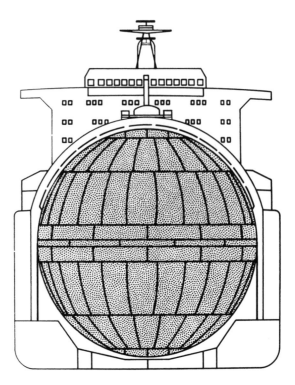

**Fig. 9.12** Cross section of 125 ft diam tank for shipboard transportation of liquefied natural gas. Tank is fabricated of 5083-O plate welded with 5183 filler alloy

estimated for the 5083/5183 vessels, totally supporting the patterns indicated by the smaller scale notch-tensile and tear tests as well as the large-scale fracture toughness tests.

## 9.2 Wrought Alloys at Elevated Temperatures

As would be expected, the toughnesses of most aluminum alloys at temperatures above RT, as represented by UPE (Fig. 9.6b), are higher than at RT with the difference increasing with temperature and, generally, with time at temperature. Among the alloys tested, no great increase was noted for only 2024-T3, but that is to be expected because 2024-T3 age hardens with high-temperature exposure; however, even in the aged condition after exposure at 300 and 400 °F, the UPE for 2024-T3 was as high as at RT.

Elevated temperature exposure does not appear to pose any serious toughness problems for aluminum alloys, though the effects of age hardening of susceptible tempers of heat treatable alloys (e.g., T3, T4 types) must always be considered.

## 9.3 Cast Alloys at Subzero Temperatures

Among the cast alloys (Fig. 9.4a, b, and c), the 3xx.x alloys consistently retain their toughness at subzero temperatures, even to –423 and –452 °F. Alloy 444.0-F (Fig. 9.4a), with its relatively low yield strength, showed an exceptionally high NYR, at or above 2.5, even at –320 °F. From the notch-tensile data in Fig. 9.4(c), it is also clear that A356.0-T6 performed quite well even at –452 °F. Other casting alloys, notably the 2xx.0 and 5xx.0 series, generally show relatively lower toughness at the lower temperatures, or more rapidly decreasing toughness with decrease in temperature.

When the notch-yield ratios for cast alloys are viewed on the basis of yield strength level (Fig. 9.13), A444.0-F still exhibits exceptionally high toughness. Among the higher strength alloys, the premium strength cast alloys (i.e., those cast with special attention to chill rates in critical regions) have the most consistently superior strength toughness combination, similar to the case at RT. Permanent mold castings yield performance close to the premium strength castings, and in fact, A356.0 permanent mold castings essentially match the performance of the premium strength castings. The sand castings generally exhibit the poorest performance.

When selecting cast alloys for cryogenic service, it seems especially important to pay careful attention to the casting process as well as the alloy itself; high-quality casting processes involving higher chill rates in areas that will experience the most critical service exposure yield superior combinations of strength and toughness.

## 9.4 Welds at Subzero Temperatures

For welds in wrought alloys, (Fig. 9.5 for NYR and 9.7 for UPE), both NYRs and UPEs tend to remain about the same or decrease very gradually as temperature decreases below RT, at least to –320 °F. Among the exceptions were (a) those joints made with 1100, 5052, and most with 2319 filler alloy, which have UPEs substantially higher at –320 °F than at RT, and (b) 5039 welds in 7005, for which the UPE at –320 °F was well below the values at RT and –112 °F.

At temperatures below –320 °F, there is a greater tendency for NYRs to decrease below the RT value, even for the toughest filler alloys such as 3003 and 5183. However, for all alloys except 4043 and 5039, NYRs were about 1.3 or higher and UPEs were above 600 in.-lb/in.$^2$ at all temperatures for which values were measured.

From the plot in Fig. 9.9 of NYR versus TYS (tensile yield strength) for aluminum alloy sheet at –423 °F, it is clear that in general, welds in 2xxx, 5xxx, and 6xxx aluminum alloy sheet provide about the same combination

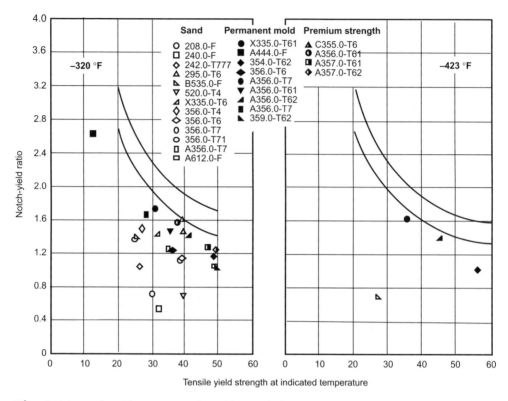

**Fig. 9.13** Notch-yield ratio vs. tensile yield strength for cast aluminum alloys at –320 and –423 °F. Specimens per Fig. A1.7(a)

of strength and toughness at that temperature as does unwelded sheet. However, welds in the high-strength 7*xxx* alloys fall below the band of other data, indicating poorer toughness at a given strength level. As for the case at RT, welds made with filler alloys 2319 and 5556 provide higher toughness levels at subzero temperatures than do welds made with 4043 filler alloy.

These same trends exist for welds in wrought and cast alloys –452 °F as illustrated in Fig. 9.14. This plot also illustrates that postweld heat treatment of welded 2219-T62 plate provides a superior combination of strength and toughness. Both 5183 welds in 5083-O and postweld heat treated 6061-T6 welds performed relatively well at –452 °F.

For welds in cast alloys (Fig. 9.5, right), both 4043 and 5556 welds exhibited NYRs about the same at temperatures down to –320 °F as at RT. However, at lower temperatures, both filler alloys exhibited some significant reduction; in all cases, NYRs were above 1.0 at –452 °F.

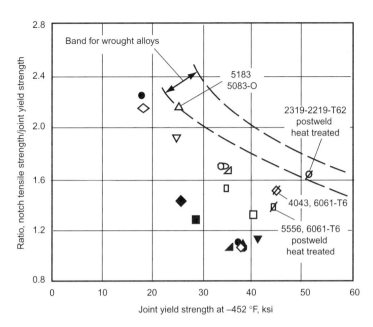

**Fig. 9.14** Joint yield strength vs. notch-yield ratios for groove welds in wrought and cast aluminum alloys at –452 °F. Specimens per Fig. A1.7(b). Open symbols, wrought, as-welded; slash symbols, wrought, postweld heat treated; solid symbols, casting, as-welded. See Fig. 9.5 for symbol identification.

**Table 9.1(a)** Results of tensile tests of smooth and notched 1 in. wide, edge-notched sheet-type tensile specimens from 0.125 in. sheet at subzero temperatures, longitudinal

| Alloy and temper | Test temperature, °F | Ultimate tensile strength (UTS), ksi | Tensile yield strength (TYS), ksi | Elongation in 2 in., % | Notch tensile strength (NTS), ksi | NTS/TS | NTS/YS |
|---|---|---|---|---|---|---|---|
| 2014-T3 | RT | 66.0 | 46.7 | 20.4 | 53.2 | 0.81 | 1.14 |
|  | −320 | 85.4 | 61.1 | 30.8 | 66.8 | 0.78 | 1.09 |
|  | −423 | 106.8 | 73.3 | 21.9 | 76.7 | 0.72 | 1.05 |
| 2014-T6 | RT | 70.2 | 65.8 | 10.3 | 65.3 | 0.93 | 0.99 |
|  | −320 | 87.2 | 78.7 | 13.5 | 61.4 | 0.70 | 0.78 |
|  | −423 | 100.6 | 85.5 | 10.5 | 65.8 | 0.65 | 0.77 |
| 2219-T37 | RT | 57.1 | 48.3 | 15.9 | 53.6 | 0.94 | 1.11 |
|  | −320 | 76.0 | 61.1 | 32.2 | 66.7 | 0.88 | 1.09 |
|  | −423 | 99.4 | 74.3 | 19.8 | 76.2 | 0.77 | 1.03 |
| 2219-T62 | RT | 58.3 | 39.0 | 11.0 | 47.9 | 0.82 | 1.23 |
|  | −320 | 74.1 | 50.3 | 14.5 | 58.1 | 0.78 | 1.16 |
|  | −423 | 92.2 | 54.0 | 14.0 | 67.4 | 0.73 | 1.25 |
| 2219-T87 | RT | 68.2 | 57.5 | 11.6 | 61.0 | 0.89 | 1.06 |
|  | −320 | 84.1 | 68.2 | 12.3 | 67.1 | 0.80 | 0.98 |
|  | −423 | 99.1 | 73.4 | 16.4 | 69.5 | 0.70 | 0.95 |
| 5456-H321 | RT | 57.4 | 39.5 | 14.5 | 47.6 | 0.83 | 1.21 |
|  | −320 | 76.6 | 46.7 | 26.8 | 52.7 | 0.69 | 1.13 |
|  | −423 | 96.4 | 52.6 | 21.7 | 56.7 | 0.59 | 1.08 |
| 5456-H323 | RT | 56.5 | 42.1 | 11.2 | 46.9 | 0.83 | 1.11 |
|  | −320 | 76.4 | 49.0 | 23.3 | 50.8 | 0.66 | 1.04 |
|  | −423 | 93.0 | 53.5 | 13.3 | 56.0 | 0.60 | 1.05 |
| 5456-H343 | RT | 58.2 | 46.1 | 8.3 | 48.7 | 0.84 | 1.06 |
|  | −320 | 77.1 | 51.4 | 18.5 | 53.3 | 0.69 | 1.04 |
|  | −423 | 82.3 | 56.8 | ... | 59.0 | 0.72 | 1.04 |
| 6061-T6 | RT | 44.9 | 40.8 | 13.8 | 46.2 | 1.03 | 1.14 |
|  | −320 | 61.0 | 48.6 | 23.3 | 55.3 | 0.91 | 1.03 |
|  | −423 | 75.8 | 55.3 | 18.3 | 57.0 | 0.75 | 2.00 |
| 7075-T6 | RT | 82.1 | 74.4 | 11.2 | 68.4 | 0.83 | 0.92 |
|  | −320 | 100.2 | 91.0 | 14.3 | 42.5 | 0.42 | 0.47 |
|  | −423 | 113.2 | 105.2 | 6.3 | 40.4 | 0.36 | 0.38 |
| 7075-T73 | RT | 73.0 | 61.8 | 12.8 | 65.4 | 0.90 | 1.06 |
|  | −320 | 91.9 | 74.8 | 13.8 | 55.0 | 0.60 | 0.74 |
|  | −423 | 106.3 | 77.6 | 15.5 | 49.1 | 0.46 | 0.63 |
| 7079-T6(a) | RT | 77.5 | 72.3 | 11.5 | 70.8 | 0.91 | 0.98 |
|  | −320 | 94.8 | 85.6 | 14.3 | 60.5 | 0.64 | 0.71 |
|  | −423 | 114.9 | 95.5 | 15.8 | 59.6 | 0.46 | 0.62 |
| 7178-T6 | RT | 90.0 | 83.6 | 12.2 | 51.8 | 0.58 | 0.62 |
|  | −320 | 107.7 | 99.6 | 9.5 | 35.5 | 0.33 | 0.36 |
|  | −423 | 123.1 | 111.6 | 8.2 | 31.5 | 0.26 | 0.28 |

Specimens per Fig. A1.4(a). Each line represents average of three tests for a single lot of material. For yield strengths, offset is 0.2%. RT, room temperature. (a) Obsolete alloy

**Table 9.1(b)  Results of tensile tests of smooth and notched 1 in. wide, edge-notched sheet-type tensile specimens from 0.125 in. sheet at subzero temperatures, transverse**

| Alloy and temper | Test temperature, °F | Ultimate tensile strength (UTS), ksi | Tensile yield strength (TYS), ksi | Elongation in 2 in., % | Notch tensile strength (NTS), ksi | NTS/TS | NTS/YS |
|---|---|---|---|---|---|---|---|
| 2014-T3 | RT | 66.0 | 40.8 | 20.5 | 50.3 | 0.76 | 1.23 |
| | –320 | 84.2 | 52.2 | 26.8 | 63.4 | 0.75 | 1.21 |
| | –423 | 104.5 | 64.4 | 19.5 | 69.0 | 0.66 | 1.07 |
| 2014-T6 | RT | 73.1 | 66.2 | 11.8 | 58.5 | 0.80 | 0.88 |
| | –320 | 88.6 | 76.8 | 12.5 | 53.1 | 0.60 | 0.69 |
| | –423 | 95.1 | 82.2 | 7.6 | 58.4 | 0.61 | 0.71 |
| 2219-T37 | RT | 60.3 | 46.7 | 13.3 | 54.9 | 0.91 | 1.18 |
| | –320 | 78.9 | 57.6 | 24.6 | 66.3 | 0.84 | 1.15 |
| | –423 | 103.8 | 71.0 | 12.0 | 74.0 | 0.72 | 1.04 |
| 2219-T62 | RT | 57.4 | 38.2 | 12.0 | 46.1 | 0.80 | 1.21 |
| | –320 | 74.3 | 49.3 | 14.0 | 56.0 | 0.75 | 1.14 |
| | –423 | 90.6 | 52.4 | 16.0 | 58.2 | 0.64 | 1.11 |
| 2219-T87 | RT | 70.1 | 58.4 | 10.8 | 57.4 | 0.82 | 0.98 |
| | –320 | 87.1 | 69.6 | 13.7 | 63.5 | 0.73 | 0.91 |
| | –423 | 103.0 | 75.8 | 15.2 | 68.0 | 0.66 | 0.90 |
| 5456-H321 | RT | 57.9 | 39.6 | 17.5 | 48.2 | 0.83 | 1.22 |
| | –320 | 72.5 | 46.7 | 23.5 | 53.1 | 0.73 | 1.14 |
| | –423 | 88.9 | 54.1 | 14.8 | 56.4 | 0.63 | 1.04 |
| 5456-H323 | RT | 56.6 | 40.1 | 13.5 | 46.7 | 0.83 | 1.16 |
| | –320 | 74.1 | 46.2 | 21.7 | 48.4 | 0.65 | 1.05 |
| | –423 | 88.3 | 50.3 | 13.4 | 52.5 | 0.57 | 1.04 |
| 5456-H343 | RT | 59.7 | 43.0 | 12.6 | 47.6 | 0.80 | 1.11 |
| | –320 | 75.4 | 49.6 | 18.3 | 51.4 | 0.68 | 1.04 |
| | –423 | 85.1 | 54.7 | 19.5 | 54.7 | 0.64 | 1.00 |
| 6061-T6 | RT | 44.3 | 38.0 | 14.0 | 45.6 | 1.03 | 1.20 |
| | –320 | 59.2 | 44.0 | 23.5 | 54.5 | 0.92 | 1.24 |
| | –423 | 68.3 | 50.5 | 14.3 | 55.3 | 0.81 | 1.10 |
| 7039-T6 | RT | 63.0 | 54.2 | 11.0 | 60.2 | 0.96 | 1.11 |
| | –320 | 80.8 | 64.3 | 14.0 | 56.8 | 0.70 | 0.88 |
| | –423 | 92.7 | 65.5 | 17.5 | 60.4 | 0.65 | 0.92 |
| 7039-T61 | RT | 60.5 | 46.0 | 14.2 | 55.2 | 0.91 | 1.20 |
| | –320 | 77.6 | 55.8 | 23.5 | 57.8 | 0.74 | 1.04 |
| | –423 | 91.8 | 59.7 | 18.5 | 60.6 | 0.66 | 1.03 |
| 7075-T6 | RT | 83.7 | 71.8 | 13.0 | 60.5 | 0.72 | 0.84 |
| | –320 | 101.0 | 86.7 | 7.0 | 41.9 | 0.41 | 0.48 |
| | –423 | 116.5 | 101.5 | 6.2 | 37.4 | 0.32 | 0.37 |
| 7075-T73 | RT | 74.6 | 61.1 | 10.8 | 62.4 | 0.84 | 1.02 |
| | –320 | 95.6 | 74.0 | 12.3 | 48.7 | 0.51 | 0.66 |
| | –423 | 110.0 | 77.0 | 11.8 | 45.5 | 0.41 | 0.59 |
| 7079-T6(a) | RT | 79.3 | 71.4 | 11.3 | 61.5 | 0.78 | 0.86 |
| | –320 | 95.2 | 81.0 | 12.8 | 43.7 | 0.46 | 0.54 |
| | –423 | 112.5 | 90.9 | 9.5 | 43.4 | 0.39 | 0.48 |
| 7178-T6 | RT | 91.4 | 79.2 | 12.5 | 46.9 | 0.51 | 0.59 |
| | –320 | 110.3 | 93.8 | 5.8 | 33.7 | 0.31 | 0.36 |
| | –423 | 130.0 | 112.2 | 4.2 | 32.0 | 0.25 | 0.29 |

Specimens per Fig. A1.4. Each line represents average of three tests for a single lot of material. For yield strengths, offset is 0.2%. RT, room temperature. (a) Obsolete alloy

**Table 9.2(a)  Results of tensile tests of smooth and notched 0.5 in. diam, round specimens from aluminum alloys at subzero temperatures, longitudinal**

| Alloy and temper | Test temperature, °F | Ultimate tensile strength (UTS), ksi | Tensile yield strength (TYS), ksi | Elongation in 2 in., % | Reduction of area, % | Notch tensile strength (NTS), ksi | NTS/TS | NTS/YS |
|---|---|---|---|---|---|---|---|---|
| 2014-T651 | RT | 69.0 | 63.5 | 10.2 | 24 | 82.8 | 1.20 | 1.30 |
| | −320 | 84.0 | 76.1 | 12.0 | 22 | 99.6 | 1.19 | 1.31 |
| | −423 | 95.6 | 80.2 | 15.0 | 23 | 10.3 | 1.08 | 1.28 |
| | −452 | 95.4 | 81.8 | 12.8 | 20 | 102.2 | 1.07 | 1.25 |
| 2021-T8151 | RT | 74.4 | 66.6 | 8.0 | 8 | 84.6 | 1.14 | 1.27 |
| | −112 | 79.4 | 69.5 | 9.5 | 17 | 95.8 | 1.21 | 1.38 |
| | −320 | 90.2 | 78.9 | 11.0 | 19 | 106.1 | 1.18 | 1.34 |
| | −423 | 101.6 | 86.1 | 12.5 | 20 | 116.4 | 1.15 | 1.35 |
| 2024-T851 | RT | 72.0 | 65.8 | 7.8 | 17 | 83.8 | 1.17 | 1.28 |
| | −112 | 77.8 | 71.3 | 6.0 | 14 | 84.2 | 1.08 | 1.18 |
| | −320 | 99.8 | 83.3 | 7.7 | 13 | 90.4 | 0.91 | 1.09 |
| | −452 | 104.4 | 90.7 | 9.5 | 14 | 106.3 | 1.02 | 1.17 |
| 2219-T851 | RT | 67.6 | 53.8 | 11.0 | 27 | 79.4 | 1.12 | 1.48 |
| | −112 | 71.4 | 57.6 | 11.5 | 28 | 84.3 | 1.18 | 1.46 |
| | −320 | 82.5 | 63.8 | 13.8 | 30 | 94.5 | 1.15 | 1.48 |
| | −423 | 95.6 | 68.8 | 16.0 | 28 | 103.5 | 1.08 | 1.5 |
| | −452 | 95.7 | 70.3 | 15.0 | 26 | 102.1 | 1.06 | 1.48 |
| 2219-T87 | RT | 67.4 | 56.2 | 11.8 | 28 | 82.3 | 1.22 | 1.46 |
| | −112 | 72.0 | 59.9 | 12.0 | 28 | 82.8 | 1.15 | 1.38 |
| | −320 | 83.5 | 67.0 | 14.0 | 28 | 91.5 | 1.09 | 1.37 |
| | −423 | 98.6 | 72.4 | 15.2 | 21 | 102.5 | 1.04 | 1.42 |
| | −452 | 97.8 | 74.2 | 15.2 | 23 | 100.2 | 1.03 | 1.37 |
| 2618-T651 | RT | 62.4 | 57.6 | 10.8 | 32 | 81.2 | 1.3 | 1.41 |
| | −112 | 68.2 | 62.5 | 10.7 | 27 | 87.1 | 1.28 | 1.39 |
| | −320 | 78.0 | 68.7 | 13.3 | 26 | 92.0 | 1.18 | 1.34 |
| | −452 | 87.6 | 72.3 | 15.0 | 23 | 98.7 | 1.13 | 1.37 |
| 3003-H14 | RT | 22.9 | 21.1 | 16.8 | 68 | ... | ... | ... |
| | −452 | 58.1 | 30.1 | 32.0 | 49 | 65.1 | 1.0 | 2.16 |
| 5083-O | RT | 46.8 | 20.4 | 19.5 | 26 | 54.0 | 1.16 | 2.65 |
| | −320 | 63.0 | 23.0 | 34.0 | 34 | 61.0 | 0.97 | 2.65 |
| | −423 | 85.2 | 25.2 | 32.00 | 24 | 59.3 | 0.70 | 2.36 |
| | −452 | 80.8 | 25.8 | 32.0 | 33 | 62.3 | 0.77 | 2.42 |
| 5083-H321 | RT | 48.6 | 34.1 | 15.0 | 23 | 61.1 | 1.26 | 1.80 |
| | −320 | 66.1 | 39.7 | 31.5 | 33 | 70.4 | 1.06 | 1.77 |
| | −423 | 90.0 | 41.8 | 30.0 | 24 | 72.8 | 0.81 | 1.74 |
| | −452 | 85.8 | 40.5 | 29.0 | 33 | 73.7 | 0.86 | 1.82 |
| 5454-O | RT | 35.8 | 16.7 | 24.5 | 48 | 48.1 | 1.34 | 2.88 |
| | −320 | 54.3 | 19.4 | 39.5 | 49 | 60.2 | 1.11 | 3.04 |
| | −452 | 73.9 | 24.1 | 34.3 | 35 | 65.6 | 0.89 | 2.72 |
| 5454-H32 | RT | 40.9 | 28.9 | 15.7 | 32 | 56.2 | 1.37 | 1.94 |
| | −320 | 61.1 | 34.5 | 32.0 | 40 | 69.2 | 1.13 | 2.00 |
| | −452 | 82.3 | 39.4 | 28.6 | 31 | 77.7 | 0.94 | 1.97 |
| 5456-O | RT | 49.0 | 23.2 | 21.8 | 31 | 50.9 | 1.04 | 2.19 |
| | −320 | 66.0 | 26.1 | 34.5 | 35 | 59.6 | 0.89 | 2.29 |
| | −452 | 84.4 | 29.5 | 30.7 | 24 | 60.9 | 0.72 | 2.02 |
| 5456-H321 | RT | 56.3 | 34.5 | 13.5 | 16 | 59.7 | 1.06 | 1.71 |
| | −320 | 73.6 | 40.1 | 27.0 | 28 | 66.2 | 0.90 | 1.65 |
| | −452 | 92.6 | 46.5 | 23.6 | 25 | 75.8 | 0.82 | 1.63 |
| 6061-T651 | RT | 44.9 | 42.2 | 16.5 | 50 | 69.2 | 1.54 | 1.64 |
| | −320 | 58.3 | 48.9 | 23.0 | 48 | 83.4 | 1.43 | 1.71 |
| | −452 | 70.1 | 55.0 | 25.5 | 42 | 89.9 | 1.28 | 1.63 |
| 7005-T5351 | RT | 62.0 | 55.0 | 15.0 | 43 | 86.2 | 1.39 | 1.59 |
| | −112 | 67.8 | 58.6 | 14.0 | 30 | 92.1 | 1.36 | 1.57 |
| | −320 | 83.9 | 67.5 | 17.0 | 27 | 99.1 | 1.18 | 1.47 |
| | −423 | 102.5 | 73.4 | 18.5 | 28 | 107.5 | 1.05 | 1.46 |
| | −452 | 97.6 | 75.6 | 17.0 | 22 | 106.9 | 1.09 | 1.41 |
| 7005-T6351 | RT | 56.8 | 49.4 | 18.0 | 50 | 81.3 | 1.43 | 1.65 |
| | −112 | 65.4 | 52.7 | 16.5 | 41 | 88.4 | 1.35 | 1.68 |
| | −320 | 78.0 | 59.3 | 18.0 | 34 | 98.3 | 1.26 | 1.66 |
| | −452 | 87.9 | 63.9 | 16.5 | 29 | 99.9 | 1.14 | 1.56 |
| 7005-T6351 | RT | 52.6 | 46.0 | 19.8 | 52 | 79.2 | 1.51 | 1.72 |
| | −112 | 61.4 | 50.9 | 18.0 | 44 | 86.4 | 1.41 | 1.70 |
| | −320 | 70.5 | 53.7 | 19.0 | 34 | 93.2 | 1.32 | 1.74 |
| | −452 | 85.6 | 60.9 | 19.5 | 25 | 94.6 | 1.10 | 1.56 |

(continued)

Specimens per Fig. A1.7. Each line represents average of three tests for a single lot of material. For yield strengths, offset is 0.2%. RT, room temperature. (a) Obsolete alloy

**Table 9.2(a)   (continued)**

| Alloy and temper | Test temperature, °F | Ultimate tensile strength (UTS), ksi | Tensile yield strength, (TYS), ksi | Elongation in 2 in., % | Reduction of area, % | Notch tensile strength (NTS), ksi | NTS/TS | NTS/YS |
|---|---|---|---|---|---|---|---|---|
| 7007-T651(a) | RT | 77.0 | 73.1 | 13.0 | 30 | 107.4 | 1.40 | 1.47 |
|  | –112 | 87.5 | 81.3 | 11.2 | 20 | 104.0 | 1.19 | 1.28 |
|  | –320 | 101.4 | 89.3 | 12.8 | 19 | 95.2 | 0.94 | 1.07 |
|  | –452 | 116.1 | 98.1 | 14.2 | 15 | 94.0 | 0.81 | 0.96 |
| 7039-T6151 | RT | 61.9 | 51.8 | 13.5 | 34 | 81.6 | 1.32 | 1.61 |
|  | –320 | 83.1 | 62.9 | 15.0 | 24 | 87.3 | 1.39 | 1.39 |
|  | –452 | 94.2 | 69.0 | 15.5 | 22 | 89.1 | 0.95 | 1.29 |
| 7039-T6351 | RT | 67.2 | 56.5 | 14.5 | 32 | 88.6 | 1.32 | 1.57 |
|  | –112 | 75.4 | 62.1 | 14.5 | 23 | 91.9 | 1.22 | 1.48 |
|  | –320 | 88.4 | 67.8 | 17.0 | 20 | 85.4 | 0.97 | 1.26 |
|  | –452 | 103.1 | 76.4 | 17.5 | 19 | 89.5 | 0.87 | 1.17 |
| 7075-T651 | RT | 88.8 | 80.4 | 9.8 | 14 | 103.6 | 1.17 | 1.29 |
|  | –112 | 95.2 | 88.6 | 10.0 | 11 | 98.4 | 1.03 | 1.11 |
|  | –320 | 110.9 | 100.9 | 9.0 | 10 | 81.8 | 0.74 | 0.80 |
|  | –452 | 120.5 | 112.1 | 8.0 | 9 | 80.2 | 0.67 | 0.72 |
| 7075-T7351 | RT | 76.2 | 66.2 | 10.2 | 22 | 93.4 | 1.23 | 1.41 |
|  | –112 | 83.4 | 72.3 | 10.0 | 17 | 93.0 | 1.12 | 1.29 |
|  | –320 | 98.2 | 82.5 | 10.7 | 14 | 84.4 | 0.86 | 1.02 |
|  | –452 | 110.0 | 88.1 | 11.0 | 12 | 94.0 | 0.85 | 1.07 |
| 7079–T651(a) | RT | 84.2 | 76.8 | 10.0 | 20 | 103.9 | 1.23 | 1.35 |
|  | –452 | 113.9 | 104.9 | 6.5 | 6 | 75.9 | 0.67 | 0.72 |

Specimens per Fig. A1.7. Each line represents average of three tests for a single lot of material. For yield strengths, offset is 0.2%. RT, room temperature. (a) Obsolete alloy

**Table 9.2(b)   Results of tensile tests of smooth and notched 0.5 in. diam, round specimens from aluminum alloys at subzero temperatures, transverse**

| Alloy and temper | Test temperature, °F | Ultimate tensile strength (UTS), ksi | Tensile yield strength, (TYS), ksi | Elongation in 2 in., % | Reduction of area, % | Notch tensile strength (NTS), ksi | NTS/TS | NTS/YS |
|---|---|---|---|---|---|---|---|---|
| 2014-T651 | RT | 69.5 | 62.7 | 8.8 | 16 | 79.8 | 1.15 | 1.27 |
|  | –320 | 85.1 | 74.4 | 9.0 | 12 | 83.1 | 0.98 | 1.12 |
|  | –423 | 96.4 | 77.4 | 11.0 | 15 | ... | ... | ... |
|  | –452 | 96.9 | 84.8 | 10.2 | 12 | 93.7 | 0.97 | 1.15 |
| 2219-T851 | RT | 66.4 | 51.2 | 10.2 | 22 | 77.0 | 1.16 | 1.50 |
|  | –112 | 71.0 | 55.0 | 10.5 | 22 | 81.5 | 1.15 | 1.48 |
|  | –320 | 83.0 | 61.1 | 12.2 | 24 | 90.5 | 1.09 | 1.48 |
|  | –423 | 96.7 | 67.5 | 15.8 | 25 | 96.5 | 1.0 | 1.43 |
|  | –452 | 95.6 | 69.8 | 13.0 | 20 | 96.5 | 1.01 | 1.38 |
| 6061-T651 | RT | 44.9 | 40.4 | 15.2 | 42 | 67.8 | 1.51 | 1.68 |
|  | –320 | 58.8 | 46.6 | 20.5 | 39 | 80.5 | 1.37 | 1.73 |
|  | –452 | 70.4 | 52.6 | 22.8 | 34 | 87.2 | 1.24 | 1.66 |
| 7005-T6351 | RT | 54.8 | 46.0 | 19.8 | 52 | 79.2 | 1.51 | 1.72 |
|  | –112 | 64.1 | 50.9 | 18.0 | 44 | 86.4 | 1.41 | 1.70 |
|  | –320 | 75.0 | 53.7 | 19.0 | 34 | 93.2 | 1.32 | 1.74 |
|  | –452 | 84.6 | 60.9 | 19.5 | 25 | 94.6 | 1.10 | 1.56 |
| 7005-T6351 | RT | 53.3 | 46.4 | 18.0 | 44 | 79.4 | 1.49 | 1.71 |
|  | –112 | 61.9 | 51.2 | 14.8 | 33 | 86.4 | 1.40 | 1.69 |
|  | –320 | 72.8 | 57.1 | 16.5 | 25 | 89.9 | 1.23 | 1.57 |
|  | –452 | 86.3 | 62.7 | 16.5 | 20 | 91.6 | 1.06 | 1.46 |
| 7007-T651(a) | RT | 73.8 | 68.8 | 13.0 | 37 | 100.6 | 1.36 | 1.46 |
|  | –112 | 82.3 | 74.4 | 10.8 | 20 | 86.9 | 1.06 | 1.16 |
|  | –320 | 95.2 | 84.5 | 11.2 | 14 | 81.6 | 0.86 | 0.96 |
|  | –452 | 107.1 | 91.8 | 11.5 | 14 | 85.4 | 0.80 | 0.93 |
| 7039-T6151 | RT | 60.4 | 50.2 | 13.0 | 30 | 80.0 | 1.33 | 1.60 |
|  | –320 | 80.0 | 60.9 | 15.5 | 23 | 80.8 | 1.01 | 1.33 |
|  | –452 | 90.8 | 66.0 | 12.5 | 15 | 85.0 | 0.94 | 1.29 |
| 7039-T6351 | RT | 66.5 | 56.6 | 13.0 | 33 | 88.0 | 1.32 | 1.55 |
|  | –112 | 72.8 | 61.4 | 12.5 | 23 | 84.7 | 1.16 | 1.38 |
|  | –320 | 87.1 | 69.0 | 13.5 | 19 | 75.1 | 0.86 | 1.09 |
|  | –452 | 101.4 | 77.7 | 13.0 | 15 | 80.6 | 0.79 | 1.04 |
| 7075-T651 | RT | 86.6 | 77.4 | 10.0 | 18 | 95.2 | 1.10 | 1.23 |
|  | –112 | 92.8 | 84.6 | 9.5 | 12 | 83.2 | 0.90 | 0.98 |
|  | –320 | 105.4 | 94.2 | 6.0 | 8 | 71.0 | 0.67 | 0.75 |
|  | –452 | ... | 104.0 | ... | ... | 75.3 | ... | 0.72 |

Specimens per Fig. A1.7. Each line represents average of three tests for a single lot of material. For yield strengths, offset is 0.2%. RT, room temperature. (a) Obsolete alloy

**Table 9.3(a)   Results of tensile tests of smooth and notched 1 in. wide, edge-notched sheet-type tensile specimens from welds in 0.125 in. aluminum alloy sheet at subzero temperatures, longitudinal (transverse weld)**

| Parent alloy and temper | Filler alloy | Postweld heat treatment | Test temperature, °F | Ultimate tensile strength (UTS), ksi | Joint yield strength (JYS), ksi | Elongation in 2 in., % | Notch tensile strength (NTS), ksi | NTS/TS | NTS/YS |
|---|---|---|---|---|---|---|---|---|---|
| 2014-T3 | 4043 | None | RT | 50.0 | 41.5 | 2.8 | 47.4 | 0.95 | 1.14 |
| | | | −320 | 61.1 | 50.8 | 2.2 | 57.4 | 0.94 | 1.13 |
| | | | −423 | 65.8 | 62.3 | 0.7 | 61.4 | 0.93 | 0.99 |
| 2014-T3 | 4043 | Aged to T6 | RT | 54.8 | 54.8(a) | (a) | 52.9 | 0.97 | 0.97 |
| | | | −320 | 62.9 | 62.9(a) | (a) | 61.2 | 0.97 | 0.97 |
| | | | −423 | 66.9 | 62.3(a) | (a) | 68.4 | 1.02 | 1.10 |
| 2014-T6 | 4043 | None | RT | 46.3 | 37.8 | 2.8 | 42.9 | 0.93 | 1.13 |
| | | | −320 | 60.9 | 46.8 | 2.0 | 45.7 | 0.75 | 0.98 |
| | | | −423 | 60.0 | 51.0 | 1.2 | 50.9 | 0.85 | 1.00 |
| 2219-T37 | 2319 | None | RT | 41.8 | 28.7 | 4.0 | 36.0 | 0.88 | 1.28 |
| | | | −320 | 58.6 | 35.5 | 6.9 | 49.4 | 0.84 | 1.39 |
| | | | −423 | 58.5 | 43.5 | 2.0 | 54.4 | 0.93 | 1.25 |
| 2219-T37 | 2319 | Aged to T87 | RT | 43.2 | 39.1 | 1.9 | 43.7 | 1.01 | 1.12 |
| | | | −320 | 54.6 | 48.3 | 1.9 | 55.6 | 1.02 | 1.15 |
| | | | −423 | 63.0 | 51.9 | 1.3 | 55.3 | 0.88 | 1.07 |
| 2219-T62 | 2319 | None | RT | 44.0 | 30.5 | 2.7 | 47.3 | 1.08 | 1.55 |
| | | | −320 | 54.3 | 39.7 | 2.3 | 56.5 | 1.04 | 1.42 |
| | | | −423 | 64.0 | 50.0 | 6.0 | 59.2 | 0.92 | 1.18 |
| 2219-T62 | 2319 | Reheat-treated to T62 | RT | 60.5 | 43.5 | 7.5 | 59.5 | 0.98 | 1.37 |
| | | | −320 | 75.2 | 51.8 | 7.5 | 73.0 | 0.97 | 1.41 |
| | | | −423 | 81.6 | 58.5 | 4.0 | 78.7 | 0.96 | 1.35 |
| 2219-T87 | 2319 | None | RT | 45.3 | 32.8 | 2.3 | 41.7 | 0.92 | 1.27 |
| | | | −320 | 59.2 | 39.0 | 3.0 | 53.8 | 0.91 | 1.38 |
| | | | −423 | 65.9 | 45.3 | 2.3 | 57.8 | 0.88 | 1.28 |
| 5456-H321 | 5556 | None | RT | 51.9 | 32.6 | 13.0 | 54.4 | 1.05 | 1.67 |
| | | | −320 | 63.7 | 37.7 | 15.8 | 59.7 | 0.94 | 1.58 |
| | | | −423 | 59.1 | 49.5 | 3.8 | 59.7 | 1.01 | 1.21 |
| 5456-H343 | 5556 | None | RT | 51.9 | 30.7 | 7.0 | 53.2 | 1.03 | 1.73 |
| | | | −320 | 59.1 | 34.1 | 6.5 | 59.0 | 1.00 | 1.73 |
| | | | −423 | 58.8 | 38.6 | 3.0 | 59.9 | 1.02 | 1.55 |
| 6061-T6 | 4043 | None | RT | 32.2 | 23.2 | 5.3 | 34.1 | 1.06 | 1.47 |
| | | | −320 | 47.4 | 28.8 | 10.5 | 38.0 | 0.80 | 1.32 |
| | | | −423 | 65.5 | 32.0 | 10.8 | 41.8 | 0.64 | 1.31 |

Specimens per Fig. A1.4. Each line represents average for three tests for a single lot of material. For yield strengths, offset is 0.2% in 2 in. gage length. RT, room temperature. (a) No joint yield strength or elongation identified

**Table 9.3(b)  Results of tensile tests of smooth and notched 1 in. wide, edge-notched sheet-type tensile specimens from welds in 0.125 in. aluminum alloy sheet at subzero temperatures, transverse (longitudinal weld)**

| Parent alloy and temper | Filler alloy | Postweld heat treatment | Test temperature, °F | Ultimate tensile strength (UTS), ksi | Joint yield strength (JYS), ksi | Elongation in 2 in., % | Notch tensile strength (NTS), ksi | NTS/TS | NTS/YS |
|---|---|---|---|---|---|---|---|---|---|
| | | | | | Transverse (longitudinal weld) | | | | |
| 2014-T3 | 4043 | None | RT | 47.6 | 38.6 | 1.0 | 49.5 | 1.04 | 1.28 |
| | | | −320 | 56.4 | 47.6 | 1.3 | 57.0 | 1.01 | 1.20 |
| | | | −423 | 61.0 | 58.6 | 0.5 | 64.1 | 1.05 | 1.09 |
| 2014-T3 | 4043 | Aged to T6 | RT | 49.3 | 49.3(a) | (a) | 48.9 | 0.99 | 0.99 |
| | | | −320 | 58.8 | 58.8(a) | (a) | 60.9 | 1.04 | 1.04 |
| | | | −423 | 67.6 | 67.6(a) | (a) | 65.2 | 0.96 | 0.96 |
| 2219-T37 | 2319 | None | RT | 42.7 | 27.4 | 4.0 | 38.3 | 0.90 | 1.40 |
| | | | −320 | 56.1 | 36.6 | 4.0 | 49.9 | 0.89 | 1.36 |
| | | | −423 | 55.8 | 42.8 | 1.6 | 54.0 | 0.97 | 1.26 |
| 2219-T37 | 2319 | Aged to T87 | RT | 42.7 | 38.2 | 2.3 | 41.1 | 0.96 | 1.08 |
| | | | −320 | 53.4 | 44.7 | 1.7 | 52.6 | 0.99 | 1.18 |
| | | | −423 | 56.8 | 48.2 | 1.0 | 54.8 | 0.96 | 1.14 |
| 2219-T62 | 2319 | None | RT | 45.2 | 30.8 | 3.5 | 44.9 | 0.99 | 1.46 |
| | | | −320 | 58.1 | 35.6 | 4.7 | 56.7 | 0.98 | 1.00 |
| | | | −423 | 60.0 | 41.5 | 2.0 | 59.8 | 1.00 | 1.44 |
| 2219-T62 | 2319 | Reheat-treated to T62 | RT | 60.2 | 42.8 | 9.2 | 58.0 | 0.96 | 1.36 |
| | | | −320 | 73.8 | 51.1 | 6.2 | 70.0 | 0.95 | 1.37 |
| | | | −423 | 79.4 | 57.1 | 4.3 | 74.4 | 0.94 | 1.30 |
| 2219-T87 | 2319 | None | RT | 44.6 | 31.0 | 2.2 | 44.2 | 0.99 | 1.43 |
| | | | −320 | 58.0 | 34.6 | 4.1 | 55.1 | 0.95 | 1.59 |
| | | | −423 | 62.2 | 44.8 | 2.1 | 59.0 | 0.95 | 1.32 |
| 5456-H321 | 5556 | None | RT | 51.7 | 30.3 | 8.5 | 51.8 | 1.00 | 1.71 |
| | | | −320 | 63.2 | 35.8 | 9.5 | 57.1 | 0.90 | 1.59 |
| | | | −423 | 62.1 | 55.0 | 5.0 | 55.8 | 0.90 | 1.01 |
| 5456-H343 | 5556 | None | RT | 51.8 | 30.1 | 7.3 | 54.0 | 1.04 | 1.79 |
| | | | −320 | 63.7 | 34.3 | 7.2 | 57.5 | 0.90 | 1.68 |
| | | | −423 | 60.9 | 39.2 | 3.5 | 57.5 | 0.94 | 1.47 |

Specimens per Fig. A1.4. Each line represents average for three tests for a single lot of material. For yield strengths, offset is 0.2% in 2 in. gage length. RT, room temperature. (a) No joint yield strength or elongation identified

**Table 9.4  Results of tensile tests of smooth and 0.5 in. diam, notched round specimens from welds in aluminum alloys at subzero temperatures**

| Base alloy and temper | Filler alloy | Postweld thermal treatment | Test temperature, °F | Ultimate tensile strength (UTS), ksi | Tensile yield strength (TYS), ksi | Elongation in 2 in., % | Reduction of area, % | Joint strength efficiency, % | Location of fracture(a) | Notch tensile strength (NTS), ksi | NTS/TS | NTS/YS |
|---|---|---|---|---|---|---|---|---|---|---|---|---|
| 1100-H112 | 1100 | None | RT | 11.6 | 6.1 | 26.5 | (b) | (b) | (b) | 17.8 | 1.53 | 2.92 |
| | | | –320 | 11.6 | 6.1 | 26.5 | (b) | (b) | (b) | 17.8 | 1.53 | 2.92 |
| 2219-T62 | 2319 | Aged to T62 | RT | 57.3 | 40.2 | 7.5 | 7 | 99 | C | 63.7 | 1.11 | 1.58 |
| | | | –112 | 60.4 | 40.5 | 6.5 | 8 | 96 | C | 68.6 | 1.14 | 1.69 |
| | | | –320 | 68.9 | 46.6 | 5.5 | 6 | 94 | C | 74.8 | 1.09 | 1.61 |
| | | | –452 | 72.0 | 51.5 | 3.5 | 5 | 80 | C | 82.8 | 1.15 | 1.61 |
| 2218-T851 | 2319 | None | RT | 32.7 | 26.8 | 2.0 | 5 | 50 | C | 40.7 | 1.24 | 1.52 |
| | | | –112 | 40.8 | 25.0 | 4.0 | 15 | 57 | C | 48.3 | 1.18 | 1.93 |
| | | | –320 | 51.7 | 28.0 | 3.5 | 10 | 62 | C | 48.5 | 0.94 | 1.73 |
| | | | –423 | 59.6 | 40.2 | 2.5 | 10 | 62 | C | 52.7 | 0.88 | 1.31 |
| 3003-H112 | 1100 | None | RT | 16.1 | 7.6 | 24.0 | 67 | 100 | C | 22.7 | 1.41 | 3.00 |
| | | | –112 | 19.3 | 8.3 | 26.5 | 66 | (b) | C | ... | ... | ... |
| | | | –320 | 33.7 | 10.8 | 31.0 | 52 | (b) | C | ... | ... | ... |
| | | | –452 | 51.1 | 18.5 | 28.0 | 25 | (b) | C | 39.8 | 0.78 | 2.15 |
| 5052-H112 | 5052 | None | RT | 29.1 | 13.9 | 18.0 | (b) | (b) | (b) | 32.8 | 1.13 | 2.36 |
| | | | –320 | 45.8 | 16.3 | 25.0 | (b) | (b) | (b) | 45.5 | 0.99 | 2.79 |
| | 5154 | None | RT | 29.2 | 13.7 | 15.0 | (b) | (b) | (b) | 32.1 | 1.10 | 2.34 |
| | | | –320 | 45.9 | 16.5 | 26.0 | (b) | (b) | (b) | 45.7 | 1.00 | 2.76 |
| 5083-O | 5183 | None | RT | 42.5 | 20.1 | 21.5 | (b) | 100 | A | 44.7 | 1.05 | 2.22 |
| | | | –112 | 43.9 | 20.7 | 31.0 | 44 | 100 | C | 49.9 | 1.14 | 2.41 |
| | | | –320 | 58.2 | 22.4 | 19.0 | 20 | 99 | A | 50.1 | 0.86 | 2.24 |
| | | | –452 | 55.3 | 25.2 | 27.0 | 37 | 69 | B | 53.9 | 0.98 | 2.14 |
| 5083-H321 | 5183 | None | RT | 44.2 | 26.0 | 14.0 | 39 | 96 | C | 54.5 | 1.23 | 2.10 |
| | | | –112 | 47.0 | 26.2 | 19.0 | 48 | 98 | A | 59.5 | 1.27 | 2.27 |
| | | | –320 | 64.7 | 31.4 | 19.0 | 23 | 100 | A | 62.4 | 0.96 | 1.98 |
| | | | –452 | 66.1 | 35.7 | 9.0 | 14 | 80 | A | 58.8 | 0.89 | 1.65 |
| | 5356 | None | RT | 41.5 | 24.3 | 13.5 | 47 | 90 | A | 53.8 | 1.30 | 2.22 |
| | | | –112 | 43.9 | 27.0 | 14.5 | 52 | 91 | A | 57.5 | 1.31 | 2.13 |
| | | | –320 | 61.9 | 29.1 | 15.5 | 33 | 97 | A | 60.6 | 0.98 | 2.08 |
| | | | –452 | 66.0 | 34.1 | 9.0 | 17 | 80 | C | 57.7 | 0.87 | 1.69 |
| | 5556 | None | RT | 44.4 | 25.6 | 14.0 | 36 | 97 | A | 53.7 | 1.21 | 2.10 |
| | | | –112 | 46.3 | 26.7 | 18.5 | 46 | 96 | A | 58.1 | 1.26 | 2.19 |
| | | | –320 | 65.3 | 30.6 | 20.5 | 26 | 100 | A | 60.5 | 0.93 | 1.98 |
| | | | –452 | 68.8 | 34.6 | 13.0 | 17.0 | 83 | A | 57.9 | 0.84 | 1.68 |
| 5086-H32 | 5356 | None | RT | 38.5 | 19.1 | 16.0 | (b) | (b) | (b) | 41.4 | 1.07 | 2.17 |
| | | | –320 | 52.8 | 20.3 | 17.0 | (b) | (b) | (b) | 48.4 | 0.92 | 2.38 |
| 5154-H112 | 5154 | None | RT | 32.6 | 14.5 | 17.0 | (b) | (b) | (b) | 34.1 | 1.05 | 2.35 |
| | | | –320 | 48.7 | 16.9 | 27.5 | (b) | (b) | (b) | 43.0 | 0.88 | 2.54 |
| 5454-H32 | 5554 | None | RT | 33.9 | 17.1 | 18.0 | 42 | 85 | A | 39.3 | 1.16 | 2.30 |
| | | | –112 | 36.0 | 17.4 | 22.0 | 47 | 86 | A | ... | ... | ... |
| | | | –320 | 54.7 | 22.5 | 29.0 | 29 | 93 | A | 52.8 | 0.97 | 2.35 |
| | | | –452 | 61.4 | 26.1 | 14.5 | (b) | 90 | A | 47.9 | 0.78 | 1.91 |
| 5456-O | 5456 | None | RT | 43.9 | 21.7 | 13.0 | (b) | (b) | (b) | 40.7 | 0.93 | 1.87 |
| | | | –320 | 57.2 | 24.8 | 18.0 | (b) | (b) | (b) | 48.5 | 0.85 | 1.95 |
| 5456-H321 | 5556 | None | RT | 44.6 | 22.5 | 13.0 | (b) | (b) | (b) | 45.2 | 1.01 | 2.01 |
| | | | –320 | 59.0 | 26.2 | 14.5 | (b) | (b) | (b) | 52.4 | 0.89 | 2.00 |
| 6061-T6 | 4043 | None | RT | 31.0 | 20.9 | 6.0 | 19 | 69 | C | 34.0 | 1.10 | 1.63 |
| | | | –320 | 34.6 | 23.6 | 6.0 | 19 | 71 | A | 38.6 | 1.12 | 1.64 |
| | | | –423 | 44.0 | 25.8 | 5.5 | 12 | 75 | A | 39.6 | 0.90 | 1.54 |
| | | | –452 | 49.1 | 37.6 | 4.5 | 9 | 63 | A | 39.9 | 0.81 | 1.06 |
| | 4043 | None | RT | 26.1 | 15.2 | 12.0 | (b) | (b) | (b) | 27.5 | 1.05 | 1.81 |
| | | | –320 | 38.8 | 18.2 | 7.5 | (b) | (b) | (b) | 33.8 | 0.87 | 1.86 |
| | 4043 | Aged to T6 | RT | 43.3 | 35.9 | 11.0 | 44 | 96 | B | 57.5 | 1.31 | 1.57 |
| | | | –112 | 47.8 | 38.3 | 21.5 | 38 | 98 | B | 61.5 | 1.27 | 1.60 |
| | | | –320 | 57.3 | 42.3 | 16.5 | 12 | 97 | A | 64.8 | 1.13 | 1.53 |
| | | | –452 | 65.6 | 44.8 | 15.0 | 16 | 63 | A | 67.2 | 1.02 | 1.50 |
| | 4043 | HTA | RT | 43.2 | 38.6 | 2.0 | (b) | (b) | (b) | 42.3 | 0.98 | 1.10 |
| | | | –320 | 53.4 | 46.8 | 3.0 | (b) | (b) | (b) | 47.9 | 0.90 | 1.02 |
| | 5356 | None | RT | 32.7 | 22.6 | 8.0 | 31 | 73 | A | 46.9 | 1.44 | 2.07 |
| | | | –112 | 37.1 | 24.7 | 9.0 | 36 | 76 | B | 50.1 | 1.35 | 2.03 |

(continued)

Specimens per Fig. A1.7(b). Each line represents average of two or three tests for one lot of material. For joint yield strength, offset is 0.2%, over a 2 in. gage length. Joint efficiencies based upon typical values for parent alloys. (a) Location of A, through weld; B, ½ to 2½ in. from center of weld, in or near weld heat-affected zone; C, edge of weld. (b) Not recorded

**Table 9.4** (continued)

| Base alloy and temper | Filler alloy | Postweld thermal treatment | Test temperature, °F | Ultimate tensile strength (UTS), ksi | Tensile yield strength (TYS), ksi | Elongation in 2 in., % | Reduction of area, % | Joint strength efficiency, % | Location of fracture(a) | Notch tensile strength (NTS), ksi | NTS/TS | NTS/YS |
|---|---|---|---|---|---|---|---|---|---|---|---|---|
| | | | −320 | 47.0 | 27.3 | 13.5 | 39 | 80 | B | 54.1 | 1.15 | 1.98 |
| | | | −452 | 57.7 | 35.3 | 13.5 | 24 | 84 | A | 53.3 | 0.92 | 1.50 |
| | 5356 | Aged to T6 | RT | 40.5 | 29.3 | 9.5 | 33 | 90 | B | ... | ... | ... |
| | | | −112 | 46.4 | 35.1 | 12.0 | 44 | 95 | A | 57.8 | 1.25 | 1.65 |
| | | | −320 | 57.1 | 33.9 | 20.0 | 29 | 97 | B | 66.4 | 1.16 | 1.96 |
| | | | −452 | 69.1 | 44.5 | 19.0 | 24 | 89 | A | 60.8 | 0.88 | 1.37 |
| 7005-T53 | 5039 | None | RT | 48.3 | 32.2 | 12.2 | (b) | 78 | (b) | 59.0 | 1.83 | 1.85 |
| | | | −112 | 57.0 | 37.8 | 10.0 | (b) | 84 | (b) | 64.7 | 1.71 | 1.71 |
| | | | −320 | 56.8 | 43.3 | 3.5 | (b) | 68 | (b) | 53.7 | 1.24 | 1.24 |
| | | | −452 | 60.0 | 48.4 | 3.5 | (b) | 61 | (b) | 55.5 | 1.15 | 1.15 |
| 7005-T6351 | 5039 | None | RT | 48.4 | 32.3 | 11.5 | (b) | 85 | (b) | 57.6 | 1.78 | 1.78 |
| | | | −112 | 55.2 | 35.4 | 11.0 | (b) | 85 | (b) | 63.6 | 1.80 | 1.81 |
| | | | −320 | 55.8 | 40.8 | 3.8 | (b) | 77 | (b) | 63.4 | 1.55 | 1.55 |
| | | | −452 | 68.8 | 53.9 | 3.5 | (b) | 77 | (b) | 55.8 | 1.04 | 1.04 |
| | 5356 | None | RT | 42.1 | 28.2 | 6.8 | (b) | 74 | (b) | 52.7 | 1.87 | 1.87 |
| | | | −112 | 47.4 | 30.2 | 8.3 | (b) | 73 | (b) | 59.7 | 1.98 | 1.97 |
| | | | −320 | 60.3 | 34.0 | 6.7 | (b) | 77 | (b) | 64.0 | 1.88 | 1.88 |
| | | | −452 | 66.8 | 45.2 | 3.5 | (b) | 76 | (b) | 62.5 | 1.38 | 1.38 |

Specimens per Fig. A1.7(b). Each line represents average of two or three tests for one lot of material. For joint yield strength, offset is 0.2%, over a 2 in. gage length. Joint efficiencies based upon typical values for parent alloys. (a) Location of A, through weld; B, 1/2 to 2 1/2 in. from center of weld, in or near weld heat-affected zone; C, edge of weld. (b) Not recorded

**Table 9.5** Results of tensile tests of smooth and 0.5 in. diam, notched round specimens from aluminum alloy castings at subzero temperatures (former alloy designation in parentheses)

| Alloy and temper | Test temperature, °F | Ultimate tensile strength (UTS), ksi | Tensile yield strength (TYS), ksi | Elongation in 2 in., % | Reduction of area, % | Notch tensile strength (NTS), ksi | NTS/TS | NTS/YS |
|---|---|---|---|---|---|---|---|---|
| **Sand casting** | | | | | | | | |
| 208.0-F | RT | 25.0 | 18.5 | 1.8 | 2 | 25.2 | 1.01 | 1.36 |
| (108-F) | −112 | 26.2 | 21.7 | 2.0(a) | 0(a) | 21.6 | 0.82 | 1.00 |
| | −320 | 30.65 | 30.2 | 1.3 | 0 | 21.7 | 0.71 | 0.72 |
| 240.0-F | RT | 33.8 | 26.0 | 1.4 | 2 | 29.9 | 0.88 | 1.15 |
| (A140-F) | −112 | 32.8 | 22.7 | (b) | (b) | 19.4 | <0.58 | 0.75 |
| | −320 | 36.8 | 32.4 | (b) | (b) | 17.4 | <0.47 | 0.54 |
| 242.0-T77 | RT | 29.8 | 20.4 | 2.1 | 4 | 29.0 | 0.97 | 1.42 |
| (142-T77) | −112 | 32.8 | 22.7 | (b) | (b) | 26.0 | <0.81 | 1.17 |
| | −320 | 32.8 | 26.8 | (b) | (b) | 27.7 | <0.84 | 1.03 |
| 295.0-T6 | RT | 42.0 | 27.1 | 6.4 | 10 | 43.5 | 1.04 | 1.60 |
| (195-T6) | −112 | 45.5 | 32.0 | 6.0 | 5 | 58.0 | 1.08 | 1.45 |
| | −320 | 53.7 | 39.9 | 5.0 | 5 | 58.0 | 1.08 | 1.45 |
| X335.0-T6 | RT | 37.3 | 23.4 | 8.6 | 12 | 38.2 | 1.02 | 1.63 |
| (X335-T6) | −112 | 42.3 | 27.4 | 8.0 | 10 | 43.0 | 1.02 | 1.57 |
| | −320 | 51.6 | 32.1 | 7.6 | 10 | 45.6 | 0.88 | 1.42 |
| 356.0-T4 | RT | 31.1 | 19.8 | 4.4 | 6 | 31.6 | 1.02 | 1.60 |
| (356-T4) | −112 | 36.6 | 23.4 | 4.4 | 6 | 37.6 | 1.04 | 1.61 |
| | −320 | 40.8 | 27.2 | 2.7 | 3 | 42.2 | 1.04 | 1.55 |
| 356.0-T6 | RT | 38.6 | 32.6 | 2.2 | 3 | 37.4 | 0.97 | 1.15 |
| (356-T6) | −112 | 43.1 | 35.8 | 2.7 | 2 | 40.0 | 0.93 | 1.12 |
| | −320 | 47.5 | 39.2 | 2.7 | 2 | 44.0 | 0.93 | 1.13 |
| 356.0-T7 | RT | 37.8 | 33.7 | 1.6 | 2 | 34.5 | 0.91 | 1.02 |
| (356-T7) | −112 | 41.4 | 34.4 | 2.0 | 2 | 38.8 | 0.94 | 1.13 |
| | −320 | 45.1 | 38.8 | 1.3 | 0 | 43.1 | 0.96 | 1.11 |
| 356.0-T71 | RT | 28.8 | 20.2 | 5.0 | ... | 32.0 | 1.11 | 1.59 |
| (356-T71) | −112 | 32.2 | 22.2 | 4.4 | 5 | 29.6 | 0.92 | 1.34 |
| | −320 | 37.4 | 25.3 | 3.0 | 2 | 34.4 | 0.92 | 1.36 |
| A356.0-T7 | RT | 37.1 | 30.5 | 4.4 | 7 | 44.9 | 1.21 | 1.47 |
| (A356-T7) | −112 | 40.0 | 31.7 | 4.4 | 5 | 41.0 | 1.02 | 1.29 |
| | −320 | 45.6 | 35.2 | 3.4 | 4 | 44.0 | 0.96 | 1.25 |

(continued)

Tests of single specimens per Fig. A1.7 at each temperature. For yield strength, offset is 0.2%. RT room temperature. (a) Broke outside middle third. (b) Broke in threads. (c) Broke before reaching 0.2%

**Table 9.5   (continued)**

| Alloy and temper | Test temperature, °F | Ultimate tensile strength (UTS), ksi | Tensile yield strength (TYS), ksi | Elongation in 2 in., % | Reduction of area, % | Notch tensile strength (NTS), ksi | NTS/TS | NTS/YS |
|---|---|---|---|---|---|---|---|---|
| 520.0-F | RT | 34.2 | 31.6 | 2.1 | 2 | 38.4 | 1.12 | 1.22 |
| (220-F) | –112 | 41.6 | 37.4 | 1.3 | 1 | 27.9 | 0.79 | 0.75 |
| | –320 | 39.6 | 39.6(c) | 0.7 | 0 | 27.2 | 0.69 | 0.69 |
| B535.0-F | RT | 41.2 | 21.2 | 12.9 | 13 | 43.8 | 1.06 | 2.07 |
| (B218-F) | –112 | 41.6 | 22.3 | 10.0 | 11 | 42.4 | 1.06 | 1.90 |
| | –320 | 37.3 | 25.5 | 3.7 | 5 | 35.5 | 0.95 | 1.39 |
| | –423 | 30.8 | 28.1 | 0.8 | 1 | 20.0 | 0.65 | 0.71 |
| A612.0-F | RT | 43.1 | 34.8 | 3.2 | 7 | 45.5 | 1.05 | 1.31 |
| (A612-F) | –112 | 45.5 | 41.0 | 2.7 | 2 | 50.8 | 1.12 | 1.24 |
| | –320 | 53.2 | 49.0 | 2.4 | 3 | 51.0 | 0.96 | 1.04 |
| **Permanent-mold casting** | | | | | | | | |
| X335.0-T61 | RT | 40.8 | 28.4 | 8.5 | 13 | 45.7 | 1.12 | 1.61 |
| (X335-T61) | –112 | 40.1 | 29.2 | 4.6 | 7 | 47.2 | 1.18 | 1.62 |
| | –320 | 45.7 | 31.0 | 5.3 | 5 | 53.4 | 1.18 | 1.72 |
| | –423 | 54.0 | 35.8 | 5.0 | ... | 58.3 | 1.08 | 1.63 |
| 354.0-T62 | RT | 50.1 | 45.5 | 1.1 | 3 | 54.2 | 1.08 | 1.19 |
| (354-T62) | –112 | 54.4 | 45.6 | 1.3 | 2 | 53.2 | 0.98 | 1.17 |
| | –320 | 61.0 | 48.7 | 1.3 | 2 | 56.4 | 0.92 | 1.16 |
| | –423 | 60.2 | 56.1 | 0.8 | 1 | 57.2 | 0.95 | 1.02 |
| 356.0-T6 | RT | 36.8 | 31.1 | 1.0 | ... | 43.0 | 1.17 | 1.38 |
| (356-T6) | –112 | 42.0 | 34.1 | 3.7 | 5 | 41.4 | 0.98 | 1.21 |
| | –320 | 45.7 | 36.5 | 3.2 | 4 | 45.0 | 0.98 | 1.23 |
| 356.0-T7 | RT | 28.4 | 21.4 | 4.3 | 7 | 35.3 | 1.24 | 1.65 |
| (356-T7) | –112 | 32.6 | 24.3 | 3.7 | 5 | 37.2 | 1.14 | 1.53 |
| | –320 | 37.3 | 25.6 | 3.0 | 4 | 39.9 | 1.07 | 1.56 |
| A356.0-T61 | RT | 39.4 | 30.8 | 4.3 | ... | 47.8 | 1.21 | 1.55 |
| (A356-T61) | –112 | 41.9 | 32.6 | 3.7 | 6 | 47.7 | 1.14 | 1.46 |
| | –320 | 49.4 | 35.8 | 4.4 | 6 | 52.6 | 1.07 | 1.47 |
| A356.0-T62 | RT | 40.9 | 36.7 | 2.1 | 6 | 46.2 | 1.13 | 1.26 |
| (A356-T62) | –112 | 45.2 | 39.6 | 3.0 | 5 | 49.8 | 1.10 | 1.26 |
| | –320 | 48.6 | 41.4 | 3.0 | 5 | 57.9 | 1.19 | 1.40 |
| | –423 | 5.5 | 45.3 | 3.5 | 3 | 63.3 | 1.15 | 1.40 |
| A356.0-T7 | RT | 28.2 | 21.4 | 5.3 | 9 | 36.9 | 1.31 | 1.72 |
| (A356-T7) | –112 | 35.4 | 25.8 | 5.7 | 8 | 43.9 | 1.24 | 1.70 |
| | –320 | 42.7 | 28.5 | 6.4 | 7 | 47.1 | 1.10 | 1.65 |
| 359.0-T62 | RT | 46.2 | 43.2 | 1.2 | 3 | 49.7 | 1.08 | 1.15 |
| (359-T62) | –112 | 52.4 | 47.3 | 2.0 | 4 | 49.8 | 0.95 | 1.05 |
| | –320 | 57.7 | 49.5 | 1.6 | 4 | 49.8 | 0.86 | 1.01 |
| A444.0-F | RT | 23.2 | 9.7 | 22.2 | 37 | 28.6 | 1.23 | 2.95 |
| (A344-F) | –112 | 26.2 | 10.0 | 19.7 | 24 | 30.4 | 1.16 | 3.04 |
| | –320 | 37.6 | 12.1(a) | 13.3(a) | 12(a) | 32.0 | 0.85 | 2.64 |
| **Premium-strength casting** | | | | | | | | |
| C355.0-T61 | RT | 43.6 | 30.3 | 6.4 | 8 | 52.6 | 1.21 | 1.74 |
| (C355-T61) | –112 | 48.4 | 33.2 | 7.5 | 8 | 56.6 | 1.17 | 1.7 |
| | –320 | 54.4 | 39.4 | 5.4 | 6 | 62.7 | 1.15 | 1.59 |
| A356.0-T61 | RT | 41.6 | 30.2 | 8.8 | 10 | 51.4 | 1.23 | 1.7 |
| (A356-T61) | –112 | 48.2 | 34.8 | 8.9 | 10 | 55.2 | 1.15 | 1.59 |
| | –320 | 51.7 | 38.0 | 4.0 | 4 | 59.8 | 1.15 | 1.57 |
| | –452 | 66.0 | 48.0 | 7.1 | 9 | 71.9 | 1.09 | 1.50 |
| A357.0-T61 | RT | 51.2 | 40.0 | 11.4 | 13 | 56.2 | 1.10 | 1.41 |
| (A357-T61) | –112 | 54.4 | 43.4 | 4.0 | 5 | 58.2 | 1.08 | 1.35 |
| | –320 | 61.5 | 47.0 | 4.0 | 4 | 59.4 | 0.96 | 1.27 |
| A357.0-T62 | RT | 51.2 | 44.4 | 2.5 | 4 | 55.4 | 1.08 | 1.25 |
| (A357-T62) | –112 | 53.1 | 46.7 | 2.1 | 3 | 55.2 | 1.04 | 1.18 |
| | –320 | 62.2 | 49.3 | 2.5 | 4 | 59.7 | 0.96 | 1.21 |

Tests of single specimens per Fig. A1.7 at each temperature. For yield strength, offset is 0.2%. RT room temperature. (a) Broke outside middle third. (b) Broke in threads. (c) Broke before reaching 0.2%

**Table 9.6  Results of tensile tests of smooth and 0.5 in. diam, notched round specimens from welds in aluminum alloy sand castings at subzero temperatures**

| Alloy and temper combination | Filler alloy | Post weld thermal treatment | Test temperature, | Ultimate tensile (UTS), ksi | Joint yield strength (JYS), ksi | Elongation in 2 in., % | Reduction of area, % | Joint strength efficiency, % | Location of fracture (a) | Notch tensile strength (NTS), ksi | NTS/TS | NTS/YS |
|---|---|---|---|---|---|---|---|---|---|---|---|---|
| A444.0-F to | 4043 | None | RT | 23.8 | 9.5 | 12.1 | 22 | 100 | B | 27.5 | 1.15 | 2.90 |
| A444.0-F | | | −112 | 26.1 | 10.0 | 14.3 | 26 | 100 | B | 31.7 | 1.21 | 3.17 |
| | | −320 | 33.5 | 11.5 | 6.4 | 9 | 89 | B | 38.1 | 1.14 | 3.31 |
| | | −452 | 48.6 | 18.0 | 10.0 | 13 | (b) | A | 40.4 | 0.83 | 2.24 |
| A444.0-F to | 4043 | None | RT | 24.0 | 11.4 | 5.7 | 23 | 100 | B | 29.3 | 1.22 | 2.51 |
| 6061-T6 | | | −320 | 34.7 | 14.8 | 7.1 | 9 | 92 | B | 34.1 | 0.98 | 2.30 |
| | | −452 | 45.5 | 28.5 | 7.1 | 8 | (b) | B | 36.9 | 0.81 | 1.29 |
| A444.0-F to | 5556 | None | RT | 24.1 | 12.2 | 12.1 | 27 | 100 | B | 29.5 | 1.22 | 2.42 |
| 5456-H321 | | | −112 | 27.0 | 15.0 | 5.0 | 14 | 100 | B | 31.1 | 1.15 | 2.08 |
| | | −320 | 33.4 | 16.1 | 5.7 | 8 | 89 | C | 34.3 | 1.03 | 2.13 |
| | | −452 | 37.2 | 25.4 | 4.3 | 7 | (b) | C | 36.4 | 0.98 | 1.43 |
| 354.0-T62 to | 4043 | None | RT | 37.8 | 21.5 | 6.4 | 10 | 76 | A | 32.0 | 0.85 | 1.48 |
| 354.0-T6 | | | −112 | 40.4 | 22.9 | 5.7 | 11 | 74 | A | 33.0 | 0.82 | 1.44 |
| | | −320 | 48.9 | 24.1 | 5.0 | 8 | 84 | A | 36.9 | 0.76 | 1.53 |
| | | −452 | 55.0 | 38.3 | 4.3 | 7 | (b) | A | 42.3 | 0.72 | 1.05 |
| 354.0-T62 to | 4043 | None | RT | 30.8 | 19.0 | 9.3 | 39 | 62 | C | 28.7 | 0.93 | 1.51 |
| 6061-T6 | | | −112 | 35.8 | 21.8 | 7.1 | 7 | 66 | A | 31.5 | 0.88 | 1.44 |
| | | −320 | 43.1 | 23.0 | 5.0 | 7 | 71 | A | 34.7 | 0.81 | 1.51 |
| | | −452 | 45.9 | 35.7 | 2.9 | 4 | (b) | A | 37.4 | 0.82 | 1.05 |
| 354.0-T62 to | 5556 | None | RT | 37.7 | 24.6 | 3.6 | 5 | 75 | A | 37.7 | 1.00 | 1.53 |
| 5456-H321 | | | −112 | 42.1 | 27.1 | 3.6 | 6 | 77 | A | 35.7 | 0.85 | 1.42 |
| | | −320 | 47.6 | 30.4 | 3.6 | 5 | 78 | A | 39.5 | 0.83 | 1.30 |
| | | −452 | 47.7 | 37.6 | 2.9 | 3 | (b) | A | 41.3 | 0.87 | 1.10 |
| C355.0-T61 to | 4043 | None | RT | 28.9 | 19.3 | 7.1 | 32 | 66 | C | 34.5 | 1.19 | 1.79 |
| 6061-T6 | | | −320 | 44.4 | 23.3 | 7.9 | 19 | 82 | A | 38.9 | 0.88 | 1.67 |
| | | −452 | 52.3 | 38.6 | 6.4 | 8 | (b) | A | 40.4 | 0.78 | 1.05 |
| C355.0-T61 to | 5556 | None | RT | 35.4 | 24.4 | 3.6 | 5 | 81 | A | 40.5 | 1.15 | 1.66 |
| 5456-H321 | | | −320 | 45.6 | 29.3 | 4.3 | 7 | 84 | C | 45.0 | 0.99 | 1.54 |
| | | −452 | 48.3 | 40.8 | 2.9 | 5 | (b) | C | 45.5 | 0.94 | 1.12 |

Specimens per Fig. A1.7(b). Each line represents the average of duplicate tests on one lot of material. For joint yield strength, offset is 0.2%, over a 2 in. gage length. Joint efficiencies based upon typical values for parent alloys. RT, room temperature. (a) Location of fracture of unnotched specimens: A, through weld; B, approximately 0.5 to 2.5 in. from weld; C, edge of weld. (b) Not recorded; no parent metal tests for comparison

**Table 9.7  Results of tensile and tear tests of aluminum alloy sheet at various temperatures**

| Alloy and temper | Test temperature, °F | Exposure temperature, °F | Time at temperature, h | Ultimate tensile strength (UTS), ksi | Tensile yield strength (TYS) ksi | Elongation in 2 in., % | Tear strength, ksi | Ratio tear strength to yield | Initiate crack, in.-lb | Propagate crack, in.-lb | Total energy in.-lb | Unit propagation (UPE), in lb/in.$^2$ |
|---|---|---|---|---|---|---|---|---|---|---|---|---|
| | | | | | | | | | **Energy required to:** | | | |
| 2014-T6 | −320 | −320 | (a) | 86.0 | 72.8 | 10.0 | 67.0 | 0.92 | 8 | 13 | 21 | 200 |
| | −112 | −112 | (a) | ... | ... | ... | 64.1 | ... | 8 | 13 | 21 | 205 |
| | RT | RT | ... | 73.3 | 63.8 | 35.1 | 62.1 | 0.97 | 8 | 11 | 19 | 172 |
| | RT | 212 | 0.5 | 72.8 | 63.6 | ... | 65.0 | 1.03 | 5 | 14 | 19 | 220 |
| | | | 96 | 73.5 | 65.1 | 13.0 | 66.0 | 1.01 | 11 | 15 | 25 | 228 |
| | RT | 300 | 0.5 | 72.4 | 63.4 | ... | 62.5 | 0.98 | 7 | 12 | 19 | 186 |
| | | | 16 | 73.0 | 65.3 | 5.0 | 61.0 | 0.94 | 5 | 13 | 18 | 206 |
| | | | 96 | 71.9 | 65.7 | 5.5 | 63.4 | 0.96 | 8 | 11 | 21 | 172 |
| | 212 | 212 | 0.5 | 66.1 | 59.3 | 5.2 | 73.0 | 1.23 | 12 | 14 | 26 | 215 |
| | | | 96 | 66.6 | 60.8 | ... | 67.0 | 1.10 | 8 | 14 | 22 | 215 |
| | 300 | 300 | 0.5 | 57.6 | 52.2 | 34.5 | 68.0 | 1.28 | 12 | 24 | 36 | 374 |
| | | | 16 | 58.3 | 53.1 | ... | 67.0 | 1.26 | 11 | 20 | 31 | 313 |
| | | | 96 | 59.5 | 55.0 | 11.5 | 65.5 | 1.19 | 8 | 24 | 32 | 372 |
| | 400 | 400 | 0.5 | ... | ... | ... | 57.7 | ... | 12 | 31 | 43 | 480 |
| 2020-T6(b) | −320 | −320 | (a) | 95.5 | 87.2 | 2.3 | 35.4 | 0.41 | 2 | 0 | 2 | 0 |
| | −112 | −112 | (a) | 87.5 | 80.9 | 5.0 | 39.5 | 0.49 | 2 | 0 | 2 | 0 |
| | RT | RT | ... | 81.1 | 75.8 | 7.5 | 41.6 | 0.55 | 3 | 0 | 3 | 0 |
| | RT | 212 | 0.5 | 81.2 | 75.8 | 7.0 | 41.3 | 0.54 | 3 | 0 | 3 | 0 |
| | | | 480 | 83.3 | 77.1 | 6.5 | 40.5 | 0.52 | 4 | 0 | 4 | 0 |
| | RT | 300 | 0.5 | 81.6 | 76.0 | 7.0 | 48.8 | 0.64 | 3 | 0 | 3 | 0 |
| | | | 96 | 82.1 | 76.8 | 6.0 | 38.2 | 0.50 | 4 | 0 | 4 | 0 |
| | | | 480 | 80.7 | 74.8 | 6.5 | 44.7 | 0.60 | 3 | 0 | 3 | 0 |
| | 212 | 212 | 0.5 | 74.1 | 71.4 | 9.5 | 49.5 | 0.69 | 3 | 0 | 3 | 0 |
| | | | 480 | 75.4 | 72.7 | 9.0 | 52.7 | 0.72 | 5 | 0 | 5 | 0 |
| | 300 | 300 | 0.5 | 67.1 | 64.9 | 8.0 | 56.8 | 0.88 | 5 | 0 | 5 | 0 |
| | | | 96 | 67.3 | 64.6 | 9.0 | 56.8 | 0.88 | 6 | 0 | 6 | 0 |
| | | | 480 | 65.1 | 62.6 | 9.5 | 61.3 | 0.98 | 5 | 0 | 5 | 0 |
| 2024-T3 | −320 | −320 | (a) | 83.2 | 60.5 | 11.5 | 82.0 | 1.36 | 16 | 22 | 38 | 363 |
| | −112 | −112 | (a) | ... | ... | ... | 77.7 | ... | 18 | 32 | 50 | 520 |
| | RT | RT | ... | 67.8 | 48.2 | 18.5 | 75.1 | 1.56 | 19 | 33 | 52 | 540 |
| | RT | 212 | 0.5 | 70.2 | 48.2 | 19.0 | ... | ... | ... | ... | ... | ... |
| | | | 96 | 70.6 | 48.2 | 20.5 | 75.6 | 1.57 | 14 | 35 | 49 | 565 |
| | | | 960 | 70.0 | 48.2 | 19.0 | 72.0 | 1.50 | 17 | 34 | 51 | 552 |
| | RT | 300 | 0.5 | 67.8 | 46.3 | 18.5 | 75.1 | 1.62 | 17 | 35 | 52 | 565 |
| | | | 96 | 70.9 | 53.4 | 16.3 | 75.5 | 1.41 | 13 | 27 | 40 | 443 |
| | RT | 400 | 0.5 | 69.4 | 49.5 | 12.5 | 76.1 | 1.28 | 17 | 32 | 49 | 526 |
| | 212 | 212 | 0.5 | 65.6 | 46.5 | 17.0 | 72.5 | 1.55 | 11 | 31 | 42 | 507 |
| | | | 96 | 65.2 | 46.4 | 17.0 | 73.5 | 1.58 | 14 | 28 | 42 | 457 |
| | 300 | 300 | 0.5 | 60.2 | 43.0 | 19.5 | 71.1 | 1.66 | 15 | 32 | 47 | 522 |
| | | | 96 | 63.0 | 50.1 | 15.5 | 73.5 | 1.47 | 10 | 29 | 39 | 465 |
| | | | 480 | 64.9 | 61.7 | 8.5 | 65.0 | 1.05 | 8 | 19 | 27 | 308 |
| | 400 | 400 | 0.5 | 54.0 | 49.1 | 10.5 | 64.5 | 1.31 | 10 | 31 | 41 | 497 |
| 2024-T81 | −320 | −320 | (a) | 87.1 | 76.5 | 8.0 | 62.6 | 0.76 | 6 | 10 | 16 | 155 |
| | −112 | −112 | (a) | 77.5 | 73.0 | 6.0 | 57.8 | 0.79 | 6 | 9 | 15 | 150 |
| | RT | RT | ... | 72.7 | 67.5 | 5.5 | 53.4 | 0.79 | 6 | 8 | 14 | 135 |
| | RT | 212 | 0.5 | 72.2 | 67.2 | 5.5 | 55.6 | 0.83 | 5 | 10 | 15 | 165 |
| | RT | 300 | 0.5 | 72.0 | 67.2 | 6.0 | 58.3 | 0.87 | 6 | 12 | 18 | 195 |
| | | | 1000 | 70.8 | 64.8 | 6.0 | 59.9 | 0.92 | 7 | 17 | 24 | 270 |
| | RT | 400 | 0.5 | 71.5 | 66.4 | 6.0 | 61.7 | 0.93 | 8 | 16 | 24 | 250 |
| | 212 | 212 | 0.5 | 66.6 | 62.2 | 7.5 | 61.2 | 0.98 | 6 | 15 | 21 | 245 |
| | | | 1000 | 66.6 | 63.3 | 7.5 | 57.9 | 0.91 | 5 | 10 | 15 | 155 |
| | 300 | 300 | 0.5 | 60.1 | 56.9 | 10.0 | 69.0 | 1.21 | 13 | 20 | 33 | 330 |
| | | | 1000 | 57.9 | 54.3 | 10.0 | 65.2 | 1.20 | 9 | 24 | 33 | 390 |
| | 400 | 400 | 0.5 | 51.4 | 48.2 | 8.0 | 64.3 | 1.33 | 14 | 33 | 47 | 525 |
| 2024-T86 | −320 | −320 | (a) | 94.5 | 85.2 | 7.5 | 59.6 | 0.70 | 5 | 11 | 16 | 175 |
| | −112 | −112 | (a) | 83.3 | 76.8 | 5.5 | 58.7 | 0.76 | 5 | 11 | 16 | 175 |
| | RT | RT | ... | 77.5 | 72.8 | 6.0 | 60.9 | 0.84 | 6 | 9 | 15 | 145 |
| | RT | 212 | 0.5 | 77.5 | 72.2 | 6.0 | 56.5 | 0.78 | 5 | 10 | 15 | 160 |
| | RT | 300 | 0.5 | 77.2 | 72.2 | 6.0 | 62.3 | 0.86 | 6 | 11 | 17 | 170 |
| | | | 1000 | 74.0 | 67.3 | 6.5 | 57.0 | 0.85 | 7 | 12 | 19 | 195 |

(continued)

Specimens per Fig. A1.8. Results of single test at each time-temperature combination. For tensile yield strengths, offset is 0.2%. All specimens from transverse direction. RT, room temperature. (a) Tested immediately upon reaching temperature; time at temperature has no known effect (b) Obsolete alloy

**Table 9.7 (continued)**

| Alloy and temper | Test temperature, °F | Exposure temperature, °F | Time at temperature, h | Ultimate tensile strength (UTS), ksi | Tensile yield strength (TYS) ksi | Elongation in 2 in., % | Tear strength, ksi | Ratio tear strength to yield | Initiate crack, in.-lb | Propagate crack, in.-lb | Total energy in.-lb | Unit propagation (UPE), in lb/in.[2] |
|---|---|---|---|---|---|---|---|---|---|---|---|---|
| | | | | | | | | | **Energy required to:** | | | |
| | RT | 400 | 0.5 | 76.2 | 70.6 | 5.5 | 62.2 | 0.88 | 8 | 11 | 19 | 180 |
| | 212 | 212 | 0.5 | 71.1 | 67.2 | 7.5 | 64.5 | 0.96 | 7 | ... | ... | ... |
| | 300 | 300 | 0.5 | 64.3 | 61.4 | 10.0 | 67.1 | 1.09 | 11 | 22 | 33 | 355 |
| | | | 1000 | 60.5 | 57.4 | 10.5 | 71.1 | 1.24 | 12 | 26 | 38 | 425 |
| | 400 | 400 | 0.5 | 55.0 | 50.2 | 6.5 | 67.5 | 1.35 | 16 | 34 | 50 | 540 |
| 2219-T87 | −320 | −320 | (a) | 91.0 | 70.7 | 11.5 | 76.4 | 1.08 | 11 | 15 | 26 | 240 |
| | −112 | −112 | (a) | 77.9 | 62.6 | 9.0 | 71.7 | 1.14 | 9 | 14 | 23 | 220 |
| | RT | RT | ... | 72.4 | 59.8 | 9.5 | 65.7 | 1.10 | 9 | 15 | 24 | 230 |
| | RT | 212 | 0.5 | 72.2 | 59.6 | 9.5 | 66.2 | 1.11 | 8 | 14 | 22 | 225 |
| | RT | 300 | 0.5 | 72.4 | 59.6 | 9.5 | 68.2 | 1.14 | 11 | 20 | 31 | 310 |
| | | | 1000 | 67.6 | 53.4 | 9.5 | 68.7 | 1.29 | 12 | 30 | 42 | 460 |
| | RT | 400 | 0.5 | 70.6 | 56.7 | 9.5 | 69.2 | 1.22 | 13 | 19 | 32 | 295 |
| | 212 | 212 | 0.5 | 63.8 | 55.0 | 13.5 | 70.0 | 1.27 | 11 | 21 | 32 | 330 |
| | | | 1000 | 63.2 | 54.7 | 13.5 | 67.5 | 1.24 | 9 | 14 | 23 | 225 |
| | 300 | 300 | 0.5 | 55.1 | 49.2 | 15.5 | 66.5 | 1.35 | 13 | 34 | 47 | 530 |
| | | | 1000 | 51.6 | 45.0 | 16.0 | 65.4 | 1.45 | 15 | 36 | 51 | 560 |
| | 400 | 400 | 0.5 | 44.4 | 39.7 | ... | 57.2 | 1.44 | 16 | 36 | 52 | 555 |
| 2618-T6 | −320 | −320 | (a) | 74.5 | 63.0 | 11.5 | 77.6 | 1.22 | 11 | 23 | 34 | 353 |
| | −112 | −112 | (a) | ... | ... | ... | 66.2 | ... | 8 | 16 | 24 | 250 |
| | RT | RT | ... | 60.6 | 54.2 | 6.0 | 58.8 | 1.09 | 8 | 15 | 23 | 234 |
| | RT | 212 | 0.5 | 59.7 | 53.1 | 7.0 | 60.8 | 1.15 | 8 | 15 | 23 | 245 |
| | | | 480 | 60.2 | 53.4 | 6.0 | 63.0 | 1.18 | 7 | 19 | 26 | 292 |
| | RT | 300 | 0.5 | 59.8 | 53.3 | 6.5 | 65.5 | 1.23 | 6 | 18 | 24 | 285 |
| | | | 96 | 60.2 | 54.0 | 7.0 | 64.3 | 1.19 | 8 | 16 | 24 | 248 |
| | | | 480 | 60.1 | 53.3 | 7.0 | 65.5 | 1.23 | 8 | 19 | 27 | 300 |
| | 212 | 212 | 0.5 | 56.2 | 51.4 | 7.5 | 62.8 | 1.22 | 10 | 20 | 30 | 311 |
| | | | 480 | 56.3 | 51.7 | 7.5 | 62.8 | 1.22 | 8 | 19 | 27 | 291 |
| | 300 | 300 | 0.5 | 50.5 | 46.9 | 12.0 | 66.6 | 1.42 | 13 | 28 | 41 | 445 |
| | | | 96 | 51.1 | 47.5 | 11.0 | 69.2 | 1.46 | 15 | 42 | 56 | 648 |
| | | | 480 | 50.6 | 46.9 | 13.5 | 68.1 | 1.45 | 14 | 36 | 50 | 570 |
| 5454-O | −320 | −320 | (a) | 53.1 | 19.2 | 28.5 | 55.3 | 2.88 | 59 | 107 | 166 | 1650 |
| | −112 | −112 | (a) | 39.3 | 18.3 | 29.0 | 49.1 | 2.68 | 54 | 113 | 167 | 1745 |
| | RT | RT | ... | 35.8 | 17.5 | 21.5 | 47.0 | 2.68 | 47 | 89 | 136 | 1375 |
| | RT | 212 | 0.5 | 36.5 | 17.0 | 22.0 | 47.5 | 2.79 | 47 | 85 | 132 | 1315 |
| | RT | 300 | 0.5 | 36.7 | 16.8 | 20.5 | 46.8 | 2.79 | 42 | 84 | 126 | 1300 |
| | RT | 400 | 0.5 | 36.7 | 16.0 | 23.5 | 46.4 | 2.81 | 48 | 83 | 131 | 1285 |
| | 212 | 212 | 0.5 | 36.9 | 16.9 | 26.5 | 45.3 | 2.68 | 43 | 87 | 130 | 1345 |
| | 300 | 300 | 0.5 | 29.1 | 16.8 | 40.5 | 43.1 | 2.56 | 56 | 122 | 178 | 1890 |
| | 400 | 400 | 0.5 | 21.2 | 14.9 | 45.5 | 36.0 | 2.42 | 60 | 175 | 235 | 2710 |
| 5454-H34 | −320 | −320 | (a) | 61.8 | 42.8 | 19.5 | 79.9 | 1.86 | 31 | 68 | 99 | 1055 |
| | −112 | −112 | (a) | 47.7 | 37.4 | 12.5 | 71.3 | 1.90 | 24 | 64 | 88 | 995 |
| | RT | RT | ... | 46.6 | 36.8 | 11.5 | 65.0 | 1.77 | 19 | 40 | 59 | 625 |
| | RT | 212 | 0.5 | 46.2 | 36.3 | 10.5 | 66.9 | 1.84 | 18 | 48 | 66 | 740 |
| | RT | 300 | 0.5 | 45.9 | 36.1 | 10.5 | 67.8 | 1.88 | 23 | 44 | 67 | 680 |
| | RT | 400 | 0.5 | 44.5 | 35.0 | 12.0 | 65.5 | 1.87 | 24 | 50 | 74 | 765 |
| | 212 | 212 | 0.5 | 46.0 | 36.5 | 11.0 | 65.0 | 1.78 | 15 | 44 | 59 | 685 |
| | 300 | 300 | 0.5 | 40.5 | 35.1 | 25.5 | 62.8 | 1.79 | 30 | 90 | 120 | 1390 |
| | 400 | 400 | 0.5 | 31.5 | 22.2 | 27.5 | 46.1 | 2.08 | 36 | 169 | 205 | 2620 |
| 5456-H343 | −320 | −320 | (a) | 71.7 | 49.7 | 10.0 | 51.8 | 1.04 | 6 | 10 | 16 | 150 |
| | −112 | −112 | (a) | 59.4 | 43.9 | 10.5 | 62.5 | 1.43 | 10 | 20 | 30 | 300 |
| | RT | RT | ... | 57.4 | 43.5 | 10.0 | 59.6 | 1.37 | 9 | 18 | 27 | 280 |
| | RT | 212 | 0.5 | 56.7 | 43.5 | 9.5 | 63.8 | 1.47 | 13 | ... | ... | ... |
| | RT | 300 | 0.5 | 56.1 | 42.9 | 10.5 | 63.8 | 1.49 | 14 | 21 | 35 | 320 |
| | | | 1000 | 50.0 | 35.3 | 10.0 | 57.8 | 1.64 | 12 | 26 | 38 | 385 |
| | RT | 400 | 0.5 | 53.6 | 39.6 | 12.0 | 62.8 | 1.58 | 14 | 31 | 45 | 465 |
| | 212 | 212 | 0.5 | 53.9 | 42.0 | 16.5 | 73.0 | 1.74 | 22 | 24 | 46 | 360 |
| | 300 | 300 | 0.5 | 44.3 | 37.5 | 27.5 | 65.9 | 1.76 | 36 | 42 | 78 | 640 |
| | | | 1000 | 38.5 | 32.8 | 28.5 | 59.2 | 1.80 | 36 | 98 | 134 | 1445 |
| | 400 | 400 | 0.5 | 32.4 | 19.8 | 23.5 | 44.1 | 2.23 | 45 | 165 | 210 | 2500 |

(continued)

Specimens per Fig. A1.8. Results of single test at each time-temperature combination. For tensile yield strengths, offset is 0.2%. All specimens from transverse direction. RT, room temperature. (a) Tested immediately upon reaching temperature; time at temperature has no known effect (b) Obsolete alloy

**Table 9.7** (continued)

| Alloy and temper | Test temperature, °F | Exposure temperature, °F | Time at temperature, h | Ultimate tensile strength (UTS), ksi | Tensile yield strength (TYS) ksi | Elongation in 2 in., % | Tear strength, ksi | Ratio tear strength to yield | Initiate crack, in.-lb | Propagate crack, in.-lb | Total energy in.-lb | Unit propagation (UPE), in lb/in.$^2$ |
|---|---|---|---|---|---|---|---|---|---|---|---|---|
| 6061-T6 | −320 | −320 | (a) | 60.1 | 47.1 | 16.0 | 82.5 | 1.75 | 34 | 89 | 123 | 1395 |
| | −112 | −112 | (a) | ... | ... | ... | 74.2 | ... | 19 | 60 | 79 | 940 |
| | RT | RT | ... | 45.6 | 40.8 | 12.5 | 66.5 | 1.63 | 23 | 49 | 72 | 772 |
| | RT | 212 | 0.5 | 45.8 | 40.8 | 12.5 | ... | ... | ... | ... | ... | ... |
| | | | 96 | 46.1 | 41.0 | 12.5 | 69.7 | 1.70 | 18 | 51 | 69 | 800 |
| | | | 960 | 46.0 | 41.0 | 12.5 | 65.0 | 1.59 | 20 | 47 | 67 | 732 |
| | RT | 300 | 0.5 | 45.7 | 40.9 | 12.5 | 70.4 | 1.72 | 19 | 51 | 70 | 796 |
| | | | 96 | 45.5 | 40.9 | 12.5 | 68.9 | 1.69 | 17 | 50 | 67 | 780 |
| | RT | 400 | 0.5 | 43.9 | 39.1 | 11.5 | 68.9 | 1.76 | 19 | 55 | 74 | 868 |
| | 212 | 212 | 0.5 | 41.9 | 38.3 | 13.0 | 64.0 | 1.67 | 14 | 53 | 67 | 835 |
| | | | 96 | 41.8 | 38.5 | 13.0 | 64.0 | 1.66 | 17 | 50 | 67 | 787 |
| | 300 | 300 | 0.5 | 37.3 | 34.6 | 15.5 | 59.5 | 1.72 | 17 | 67 | 82 | 1021 |
| | | | 396 | 38.4 | 35.9 | 16.0 | 60.6 | 1.69 | 18 | 68 | 86 | 1058 |
| | | | 480 | 37.9 | 35.1 | 15.0 | ... | ... | ... | ... | ... | ... |
| | 400 | 400 | 0.5 | ... | ... | ... | 51.2 | ... | 17 | 72 | 90 | 1400 |
| 7075-T6 | −320 | −320 | (a) | 98.7 | 87.5 | 5.0 | 39.6 | 0.45 | 4 | 0 | 4 | 0 |
| | −112 | −112 | (a) | 90.4 | 79.9 | 12.0 | 61.9 | 0.78 | 7 | 0 | 7 | 0 |
| | RT | RT | ... | 84.6 | 75.5 | 11.0 | 63.6 | 0.84 | 8 | 11 | 19 | 173 |
| | RT | 212 | 0.5 | 82.2 | 72.7 | 10.5 | 73.0 | 1.00 | 11 | 9 | 20 | 149 |
| | | | 96 | 84.3 | 75.6 | 11.0 | 72.6 | 0.96 | 10 | 11 | 21 | 179 |
| | | | 960 | 85.1 | 76.5 | 12.0 | 63.4 | 0.83 | 7 | 12 | 19 | 184 |
| | RT | 300 | 0.5 | 82.2 | 72.7 | 10.5 | 70.4 | 0.98 | 9 | 10 | 19 | 159 |
| | | | 96 | 76.2 | 64.2 | 10.0 | 68.6 | 1.07 | 8 | 12 | 20 | 192 |
| | RT | 400 | 0.5 | 71.1 | 58.7 | 10.5 | 71.3 | 1.22 | 10 | 17 | 27 | 275 |
| | 212 | 212 | 0.5 | 75.0 | 70.5 | 12.5 | 81.1 | 1.15 | 16 | 18 | 34 | 292 |
| | | | 96 | 74.9 | 70.4 | 12.5 | 77.6 | 1.10 | 9 | 18 | 27 | 285 |
| | 300 | 300 | 0.5 | 60.7 | 55.1 | 16.0 | 79.0 | 1.43 | 18 | 53 | 71 | 846 |
| | | | 96 | 56.7 | 51.0 | 19.5 | 77.1 | 1.51 | 19 | 59 | 78 | 950 |
| | 400 | 400 | 0.5 | 42.0 | 37.6 | 11.0 | 64.3 | 1.71 | 21 | 59 | 80 | 945 |
| 7079-T6(b) | −320 | −320 | (a) | 95.9 | 82.1 | 8.0 | 37.3 | 0.45 | 2 | 0 | 2 | 0 |
| | −112 | −112 | (a) | 83.6 | 74.9 | 12.0 | 62.6 | 0.84 | 8 | 2 | 10 | 30 |
| | RT | RT | ... | 78.3 | 69.8 | 10.5 | 71.3 | 1.02 | 12 | 16 | 28 | 240 |
| | RT | 212 | 0.5 | 77.6 | 68.9 | 10.0 | 71.3 | 1.04 | 11 | 17 | 28 | 250 |
| | RT | 300 | 0.5 | 76.5 | 67.3 | 10.0 | 74.4 | 1.10 | 14 | 13 | 27 | 195 |
| | | | 1000 | 59.8 | 44.4 | 9.5 | 67.3 | 1.52 | 14 | 30 | 44 | 445 |
| | RT | 400 | 0.5 | 67.5 | 54.4 | 10.5 | 71.5 | 1.32 | 16 | 25 | 41 | 375 |
| | 212 | 212 | 0.5 | 70.7 | 63.6 | 13.0 | 82.2 | 1.29 | 20 | 24 | 47 | 345 |
| | 300 | 300 | 0.5 | 59.5 | 54.5 | 16.0 | 77.8 | 1.43 | 24 | 59 | 83 | 870 |
| | 400 | 400 | 0.5 | 42.3 | 38.3 | 11.0 | 60.9 | 1.59 | 27 | 74 | 101 | 1085 |
| 7178-T6 | −320 | −320 | (a) | 10.5 | 91.6 | 5.0 | 32.7 | 0.36 | 2 | 0 | 2 | 0 |
| | −112 | −112 | (a) | 93.5 | 83.5 | 11.0 | 49.80 | 0.60 | 5 | 0 | 5 | 0 |
| | RT | RT | ... | 87.9 | 78.0 | 11.5 | 58.5 | 0.75 | 6 | 4 | 10 | 64 |
| | RT | 212 | 0.5 | 87.8 | 77.7 | 11.0 | 63.5 | 0.82 | 7 | 7 | 14 | 105 |
| | RT | 300 | 0.5 | 86.1 | 75.9 | 11.5 | 65.0 | 0.86 | 8 | 7 | 15 | 115 |
| | | | 1000 | 58.6 | 43.5 | 9.5 | 63.6 | 1.46 | 12 | 26 | 38 | 400 |
| | RT | 400 | 0.5 | 74.8 | 63.6 | 9.5 | 66.6 | 1.05 | 10 | 8 | 18 | 120 |
| | 212 | 212 | 0.5 | 79.9 | 71.7 | 15.0 | 80.4 | 1.12 | 14 | ... | ... | ... |
| | 300 | 300 | 0.5 | 67.4 | 61.4 | 18.5 | 82.5 | 1.34 | 21 | 47 | 68 | 730 |
| | | | 1000 | 39.4 | 36.8 | 26.0 | 61.3 | 1.67 | 23 | 70 | 93 | 1085 |
| | 400 | 400 | 0.5 | 46.6 | 42.7 | ... | 63.5 | 1.49 | 20 | 61 | 81 | 950 |

Specimens per Fig. A1.8. Results of single test at each time-temperature combination. For tensile yield strengths, offset is 0.2%. All specimens from transverse direction. RT, room temperature. (a) Tested immediately upon reaching temperature; time at temperature has no known effect (b) Obsolete alloy

**Table 9.8(a)  Results of tensile and tear tests of aluminum alloy plate at subzero temperatures, longitudinal**

| Alloy and temper | Thickness, in. | Test temperature, °F | Ultimate tensile strength (UTS), ksi | Tensile yield strength (TYS), ksi | Elongation in 2 in., % | Tear strength, ksi | Ratio tear strength to yield strength (TYR) | Initiate a crack, in.-lb | Propagate a crack, in.-lb | Total energy, in.-lb | Unit propagation energy (UPE), in.-lb/in.[2] |
|---|---|---|---|---|---|---|---|---|---|---|---|
| 5083-O | 0.75 | RT | 45.5 | 20.4 | 20.5 | 54.0 | 2.45 | 54 | 112 | 166 | 1120 |
| | | −320 | 62.9 | 23.8 | 33.0 | 62.6 | 2.63 | 82 | 148 | 230 | 1480 |
| 5083-H321 | 0.38 | RT | 49.9 | 34.5 | 14.5 | 66.0 | 1.91 | 45 | 112 | 157 | 1125 |
| | | −320 | 70.8 | 40.5 | 27.0 | 76.0 | 1.88 | 59 | 133 | 192 | 1330 |
| | 0.38 | RT | (a) | (a) | (a) | 63.4 | ... | 42 | 85 | 127 | 855 |
| | | −320 | (a) | (a) | (a) | 74.0 | ... | 60 | 127 | 187 | 1270 |
| | 0.75 | RT | (a) | (a) | (a) | 64.4 | ... | 38 | 82 | 120 | 820 |
| | | −320 | (a) | (a) | (a) | 74.6 | ... | 57 | 112 | 169 | 1120 |
| 5086-O | 0.75 | RT | 41.7 | 20.5 | 25.0 | 46.5 | 2.27 | 66 | 114 | 179 | 1135 |
| | | −320 | 58.5 | 23.3 | 42.0 | 53.8 | 2.31 | 101 | 175 | 276 | ... |
| 5154-O | 0.75 | RT | 35.1 | 16.1 | 30.7 | 45.1 | 2.80 | 80 | 135 | 215 | 1350 |
| | | −320 | 53.1 | 18.6 | 45.0 | 60.8 | 3.27 | 106 | 202 | 308 | 2020 |
| 5356-O | 0.75 | RT | 43.5 | 18.9 | 28.8 | 50.8 | 1.76 | 65 | 141 | 206 | 1405 |
| | | −320 | 61.9 | 25.0 | 44.0 | 54.8 | 2.19 | 84 | 190 | 274 | 1900 |
| 5356-H321 | 0.75 | RT | 53.3 | 34.7 | 16.0 | 65.6 | 1.89 | 60 | 87 | 147 | 865 |
| | | −320 | 71.6 | 39.0 | 16.0 | 82.2 | 2.11 | 86 | 170 | 256 | 1700 |
| 5454-O | 0.38 | RT | 38.8 | 22.7 | 21.2 | 51.9 | 2.29 | 69 | 107 | 176 | 1070 |
| | | −320 | 58.6 | 26.6 | 26.0 | 82.2 | 3.09 | 96 | 190 | 286 | 1900 |
| | 0.50 | RT | 39.2 | 23.6 | 20.8 | 53.9 | 2.28 | 78 | 120 | 199 | 1205 |
| | | −320 | 59.2 | 27.6 | 26.0 | 82.2 | 2.98 | 110 | 199 | 309 | 1990 |
| 5454-H34 | 0.38 | RT | 41.1 | 27.1 | 20.0 | 55.0 | 2.03 | 72 | 111 | 183 | 1110 |
| | | −320 | 63.3 | 32.0 | 35.0 | 70.6 | 2.21 | 103 | 172 | 275 | 1720 |
| | 0.50 | RT | 41.6 | 35.0 | 15.2 | 61.6 | 1.76 | 43 | 92 | 135 | 920 |
| | | −320 | 64.1 | 41.3 | 32.0 | 78.2 | 1.89 | 81 | 149 | 220 | 1490 |
| 5456-O | 0.38 | RT | 50.0 | 26.6 | 20.8 | 58.6 | 2.20 | 57 | 99 | 157 | 995 |
| | | −320 | (a) | (a) | (a) | 66.2 | ... | 56 | 127 | 183 | 1270 |
| | 0.75 | RT | 49.9 | 23.4 | 22.5 | 51.6 | 2.21 | 44 | 99 | 142 | 985 |
| | | −320 | 66.1 | 26.1 | 22.0 | 58.0 | 2.22 | 52 | 121 | 173 | 1210 |
| 5456-H321 | 0.38 | RT | 52.9 | 34.2 | 16.0 | 60.6 | 1.77 | 42 | 92 | 114 | 920 |
| | | −320 | 73.6 | 39.7 | 22.0 | 71.2 | 1.79 | 52 | 110 | 162 | 1100 |
| | 0.50 | RT | 56.3 | 34.5 | 13.5 | 68.4 | 1.98 | 54 | 104 | 158 | 1040 |
| | | −320 | 73.6 | 33.6 | 19.0 | 78.6 | 2.34 | 54 | 128 | 182 | 1280 |
| | 0.75 | RT | 57.5 | 35.6 | 14.8 | 68.8 | 1.93 | 47 | 75 | 122 | 750 |
| | | −320 | (a) | (a) | (a) | 79.6 | ... | 62 | 114 | 176 | 1140 |

Each line of data represents a separate lot of material; average of duplicate or triplicate tests. Specimens per Fig. A1.8, generally 0.100 in. thick; in a few cases, 0.063 in. thick specimens were used. For yield strengths, offset is 0.2%. RT, room temperature. (a) Not reported

**Table 9.8(b)   Results of tensile and tear tests of aluminum alloy plate at subzero temperatures, transverse**

| Alloy and temper | Thickness, in. | Test temperature, °F | Ultimate tensile strength (UTS), ksi | Tensile yield strength (TYS), ksi | Elongation in 2 in., % | Tear strength, ksi | Ratio tear strength to yield strength (TYR) | Initiate a crack, in.-lb | Propagate a crack, in.-lb | Total energy, in.-lb | Unit propagation energy (UPE), in.-lb/in.$^2$ |
|---|---|---|---|---|---|---|---|---|---|---|---|
| 5083-O | 0.75 | RT | 45.9 | 20.5 | 25.0 | 54.0 | 2.63 | 56 | 96 | 152 | 960 |
| | | −320 | 62.4 | 24.8 | 35.0 | 61.6 | 2.48 | 62 | 132 | 194 | 1320 |
| 5083-H321 | 0.38 | RT | 50.1 | 34.3 | 16.1 | 63.2 | 1.84 | 43 | 84 | 127 | 840 |
| | | −320 | 66.4 | 40.3 | 24.0 | 73.0 | 1.81 | 57 | 108 | 165 | 1080 |
| | 0.38 | RT | (a) | (a) | (a) | 61.6 | ... | 43 | 81 | 124 | 805 |
| | | −320 | (a) | (a) | (a) | 71.8 | ... | 57 | 97 | 154 | 970 |
| | 0.75 | RT | (a) | (a) | (a) | 63.0 | ... | 38 | 74 | 112 | 740 |
| | | −320 | (a) | (a) | (a) | 71.0 | ... | 36 | 82 | 118 | 820 |
| 5086-O | 0.75 | RT | 41.1 | 20.6 | 27.8 | 46.0 | 2.23 | 64 | 92 | 156 | 920 |
| | | −320 | 58.7 | 23.5 | 42.5 | 54.0 | 2.30 | 108 | 171 | 279 | 1710 |
| 5154-O | 0.75 | RT | 36.1 | 16.2 | 29.6 | 44.1 | 2.72 | 77 | 115 | 192 | 1145 |
| | | −320 | 54.3 | 18.0 | 42.5 | 60.0 | 3.33 | 90 | 186 | 276 | |
| 5356-O | 0.75 | RT | 44.7 | 21.7 | 27.7 | 49.2 | 2.27 | 58 | 114 | 172 | 1140 |
| | | −320 | 62.5 | 24.5 | 35.5 | 55.6 | 2.27 | 78 | 166 | 244 | 1660 |
| 5356-H321 | 0.75 | RT | 51.8 | 33.2 | 21.0 | 62.1 | 1.87 | 46 | 70 | 116 | 700 |
| | | −320 | 60.7 | 24.5 | 35.5 | 75.2 | 3.07 | 52 | 98 | 150 | 980 |
| 5454-O | 0.38 | RT | 38.4 | 23.2 | 24.0 | 52.8 | 2.28 | 76 | 107 | 183 | 1075 |
| | | −320 | 58.4 | 27.8 | 26.5 | 75.2 | 2.71 | 108 | 158 | 266 | 1580 |
| | 0.50 | RT | 38.6 | 23.8 | 22.8 | 55.0 | 2.31 | 79 | 120 | 198 | 1200 |
| | | −320 | 58.7 | 28.6 | 26.0 | 75.2 | 2.63 | 108 | 171 | 279 | 1710 |
| 5454-H34 | 0.38 | RT | 40.7 | 26.4 | 24.4 | 57.5 | 2.18 | 72 | 120 | 192 | 1200 |
| | | −320 | 57.8 | 30.9 | 35.0 | 68.0 | 2.20 | 94 | 154 | 249 | 1540 |
| | 0.50 | RT | 42.6 | 33.9 | 16.8 | 62.3 | 1.84 | 48 | 65 | 113 | 650 |
| | | −320 | 60.5 | 39.7 | 32.0 | 75.0 | 1.89 | 73 | 127 | 200 | 1270 |
| 5456-O | 0.38 | RT | 50.4 | 27.4 | 21.2 | 58.1 | 2.12 | 50 | 91 | 142 | 910 |
| | | −320 | (a) | (a) | (a) | 66.0 | ... | 60 | 110 | 170 | 1100 |
| | 0.75 | RT | 50.4 | 23.7 | 22.3 | 49.2 | 2.08 | 42 | 82 | 124 | 820 |
| | | −320 | 65.8 | 26.7 | 22.0 | 55.6 | 2.08 | 49 | 104 | 153 | 1040 |
| 5456-H321 | 0.38 | RT | 52.7 | 34.4 | 17.2 | 61.0 | 1.77 | 46 | 79 | 125 | 785 |
| | | −320 | 68.9 | 39.6 | 24.5 | 68.6 | 1.73 | 49 | 91 | 140 | 910 |
| | 0.50 | RT | 55.9 | 33.6 | 19.0 | 66.0 | 1.96 | 48 | 81 | 129 | 810 |
| | | −320 | 69.9 | 39.2 | 25.5 | 76.7 | 1.96 | 52 | 92 | 144 | 920 |
| | 0.75 | RT | 55.8 | 34.4 | 19.3 | 62.0 | 1.80 | 35 | 62 | 97 | 620 |
| | | −320 | (a) | (a) | (a) | 72.4 | ... | 42 | 66 | 108 | 660 |

Each line of data represents a separate lot of material; average of duplicate or triplicate tests. Specimens per Fig. A1.8, generally 0.100 in. thick; in a few cases, 0.063 in. thick specimens were used. For yield strengths, offset is 0.2%. RT, room temperature. (a) Not reported

**Table 9.9(a)   Tensile tests of groove welds in wrought aluminum alloy sheet and plate at subzero temperatures**

| Alloy and temper combination | Sheet, plate thickness,.in | Specimen orientation | Filler alloy | Postweld thermal treatment | Test temperature, °F | Ultimate tensile strength (UTS), ksi | Joint yield strength (TYS), ksi | Elongation in 2 in., % |
|---|---|---|---|---|---|---|---|---|
| 1100-H112 As-welded | 1.00 | Cross weld | 1100 | None | RT | 11.6 | 6.1 | 26.5 |
|  |  |  |  |  | −320 | 22.8 | 8.0 | 31.0 |
| 3303-H112 As-welded | 1.00 | Cross weld | 1100 | None | RT | 16.1 | 7.6 | 24.0 |
|  |  |  |  |  | −320 | 33.7 | 10.8 | 31.0 |
| 2219-T62 Postweld heat treated | 0.063 | Cross weld | 2319 | HTA | RT | 61.4 | 42.8 | 8.8 |
|  |  |  |  |  | −320 | 79.4 | 53.9 | 10.8 |
| 2219-T81 As-welded | 0.063 | Cross weld | 2319 | None | RT | 46.6 | 33.2 | 1.8 |
|  |  |  |  |  | −320 | 67.8 | 39.0 | 3.2 |
| 2219-T81 Postweld aged | 0.063 | Cross weld | 2319 | Aged | RT | 48.4 | 40.2 | 1.5 |
|  |  |  |  |  | −320 | 67.2 | 49.9 | 2.2 |
| 2219-T81 As-welded | 0.063 | Cross weld | 2319 | None | RT | 46.2 | 31.8 | 2.2 |
|  |  |  |  |  | −320 | 67.2 | 37.2 | 3.5 |
| 2219-T81 Postweld aged | 0.063 | Cross weld | 2319 | Aged | RT | 52.6 | 40.4 | 2.0 |
|  |  |  |  |  | −320 | 69.9 | 45.0 | 2.0 |
| 5052-H112 As-welded | 1.00 | Cross weld | 5052 | None | RT | 29.1 | 13.9 | 18.0 |
|  |  |  |  |  | −320 | 45.8 | 16.3 | 25.0 |
| 5154-H112 As-welded | 1.00 | Cross weld | 5154 | None | RT | 32.6 | 14.5 | 17.0 |
|  |  |  |  |  | −320 | 48.7 | 16.9 | 27.5 |
| 5083-O As-welded | 0.38 | Cross weld | 5183 | None | RT | 42.4 | (a) | ... |
|  |  |  |  |  | −320 | ... | (a) | ... |
| 5083-H113 As-welded | 1.00 | Cross weld | 5183 | None | RT | 43.1 | (a) | ... |
|  |  |  |  |  | −320 | ... | (a) | ... |
| 5083-H113 As-welded | 0.88 | Cross weld | 5556 | None | RT | 41.2 | 21.2 | 12.5 |
|  |  |  |  |  | −320 | 60.1 | 24.7 | 20.0 |
| 5456-O As-welded | 0.38 | Cross weld | 5556 | None | RT | 46.8 | (a) | ... |
|  |  |  |  |  | −320 | ... | (a) | ... |
| 5456-H321 As-welded | 1.00 | Cross weld | 5456 | None | RT | 46.8 | 30.4 | 6.8 |
|  |  |  |  |  | −320 | ... | (a) | ... |
| 7005-T63 As-welded | 1.25 | Cross weld | 5039 | None | RT | 48.4 | 32.3 | 11.5 |
|  |  |  |  |  | −112 | 55.2 | 35.4 | 11.0 |
|  |  |  |  |  | −320 | 55.8 | 40.8 | 3.8 |
| 7005-T6351 As-welded | 1.25 | Cross weld | 5356 | None | RT | 42.1 | 28.2 | 6.8 |
|  |  |  |  |  | −112 | 47.4 | 30.2 | 8.3 |
|  |  |  |  |  | −320 | 60.3 | 34.0 | 6.7 |

Each line represents average results of tests of duplicate specimens at each temperature. All specimens from welds were cross weld, with crack moving along weld centerline. RT, room temperature. (a) Joint yield strength not determined. Matching tear test data are presented in Table 9.9(b).

**Table 9.9(b)   Tear tests of groove welds in wrought aluminum alloy sheet and plate at subzero temperatures**

| Alloy and temper combination | Sheet, plate thickness, in. | Specimen orientation | Filler alloy | Postweld thermal treatment | Test temperature, °F | Tear strength, ksi | Ratio tear strength yield strength (TYR) | Initiate a crack, in.-lb | Propagate a crack, in.-lb | Total energy, in.-lb | Unit propagation energy (UPE), in.-lb/in.[2] |
|---|---|---|---|---|---|---|---|---|---|---|---|
| 1100-H112 As-welded | 1.00 | Cross weld | 1100 | None | RT | 19.5 | 3.20 | 48 | 76 | 124 | 755 |
|  |  |  |  |  | −320 | 32.2 | 4.02 | 82 | 122 | 204 | 1220 |
| 3303-H112 As-welded | 1.00 | Cross weld | 1100 | None | RT | 24.0 | 3.16 | 40 | 78 | 118 | 785 |
|  |  |  |  |  | −320 | 43.5 | 4.02 | 93 | 126 | 219 | 1260 |
| 2219-T62 Postweld heat treated | 0.063 | Cross weld | 2319 | HTA | RT | 87.2 | 2.04 | 33 | 44 | 77 | 705 |
|  |  |  |  |  | −320 | 104.8 | 1.94 | 35 | 39 | 74 | 624 |
| 2219-T81 As-welded | 0.063 | Cross weld | 2319 | None | RT | 70.8 | 2.13 | 31 | 20 | 51 | 324 |
|  |  |  |  |  | −320 | 82.9 | 2.12 | 28 | 60 | 88 | 948 |
| 2219-T81 Postweld aged | 0.063 | Cross weld | 2319 | Aged | RT | 74.1 | 1.84 | 20 | 23 | 43 | 363 |
|  |  |  |  |  | −320 | 91.0 | 1.82 | 32 | 74 | 106 | 1180 |
| 2219-T81 As-welded | 0.063 | Cross weld | 2319 | None | RT | 67.0 | 2.10 | 24 | 22 | 46 | 352 |
|  |  |  |  |  | −320 | 89.6 | 2.41 | 33 | 80 | 113 | 1780 |
| 2219-T81 Postweld aged | 0.063 | Cross weld | 2319 | Aged | RT | 72.4 | 1.79 | 17 | 26 | 43 | 419 |
|  |  |  |  |  | −320 | 87.9 | 1.95 | 25 | 64 | 89 | 1032 |
| 5052-H112 As-welded | 1.00 | Cross weld | 5052 | None | RT | 37.0 | 2.66 | 45 | 108 | 153 | 1085 |
|  |  |  |  |  | −320 | 50.7 | 3.11 | 85 | 178 | 263 | 1780 |
| 5154-H112 As-welded | 1.00 | Cross weld | 5154 | None | RT | 36.2 | 2.50 | 50 | 104 | 154 | 1040 |
|  |  |  |  |  | −320 | 43.9 | 2.60 | 50 | 107 | 157 | 1070 |
| 5083-O As-welded | 0.38 | Cross weld | 5183 | None | RT | 50.2 | (a) | 38 | 97 | 135 | 970 |
|  |  |  |  |  | −320 | 54.6 | (a) | 33 | 74 | 107 | 740 |
| 5083-H113 As-welded | 1.00 | Cross weld | 5183 | None | RT | 51.6 | (a) | 33 | 99 | 132 | 990 |
|  |  |  |  |  | −320 | 57.2 | (a) | 34 | 70 | 114 | 700 |
| 5083-H113 As-welded | 0.88 | Cross weld | 5556 | None | RT | 48.2 | 2.25 | 36 | 85 | 121 | 850 |
|  |  |  |  |  | −320 | 55.5 | 2.27 | 33 | 83 | 116 | 830 |
| 5456-O As-welded | 0.38 | Cross weld | 5556 | None | RT | 51.7 | (a) | 38 | 91 | 129 | 910 |
|  |  |  |  |  | −320 | 56.8 | (a) | 36 | 76 | 112 | 760 |
| 5456-H321 As-welded | 1.00 | Cross weld | 5456 | None | RT | 51.7 | 1.70 | 46 | 92 | 138 | 920 |
|  |  |  |  |  | −320 | 58.2 | (a) | 40 | 116 | 156 | 1160 |
| 7005-T63 As-welded | 1.25 | Cross weld | 5039 | None | RT | 60.0 | 1.86 | 30 | 95 | 125 | 950 |
|  |  |  |  |  | −112 | 65.2 | 1.84 | 40 | 93 | 131 | 930 |
|  |  |  |  |  | −320 | 51.6 | 1.26 | 11 | 24 | 34 | 335 |
| 7005-T6351 As-welded | 1.25 | Cross weld | 5356 | None | RT | 51.4 | 1.82 | 28 | 94 | 122 | 945 |
|  |  |  |  |  | −112 | 53.4 | 1.77 | 36 | 112 | 146 | 1125 |
|  |  |  |  |  | −320 | 58.4 | 1.72 | 33 | 85 | 118 | 855 |

Specimens per Fig. A1.8. Each line represents average results of tests of duplicate specimens of each temperature. All specimens from welds were cross weld, with crack moving along weld centerline. RT, room temperature; HTA, heat treated and artificially aged after welding. (a) Joint yield strength not determined; ratio of test strength to yield strength not available. Matching tensile test data are presented in Table 9.9(a).

**Table 9.10(a)    Results of tensile tests of aluminum alloy plate at subzero temperatures**

| Alloy and temper | Filler alloy | Thickness, in. | Test temperature, °F | Ultimate tensile strength (UTS), ksi | Tensile yield strength (TYS), ksi | Elongation in 2 in. or 4D, % |
|---|---|---|---|---|---|---|
| **Unwelded plate** | | | | | | |
| 2014-T651 | None | 1.000 | RT | 72.0 | 65.8 | 9.2 |
| | | | −112 | ... | ... | ... |
| | | | −320 | 86.0 | 75.0 | 10.0 |
| 2024-T651 | None | 1.375 | RT | 70.8 | 64.4 | 7.2 |
| | | | −112 | ... | ... | ... |
| | | | −320 | 72.0 | 65.8 | 9.2 |
| 5083-O | None | 7.000 | RT | 45.0 | 20.8 | 18.8 |
| | | | −260 | 56.4 | 23.8 | 24.0 |
| | | | −320 | 60.0 | 23.6 | 24.5 |
| 5083-O | None | 7.700 | RT | 38.0 | 17.5 | 24.0 |
| | | | −320 | 50.9 | ... | 15.5 |
| | | | RT | 38.0 | 17.5 | 24.0 |
| | | | −320 | 50.9 | ... | 15.5 |
| | | | RT | 41.1 | 18.7 | 15.0 |
| | | | −320 | 54.3 | 21.1 | 15.0 |
| | | | RT | 38.0 | 17.5 | 14.0 |
| | | | −320 | 50.9 | ... | 15.5 |
| | | | RT | 35.6 | 16.8 | 10.0 |
| | | | −320 | 45.8 | 19.6 | 11.7 |
| 6061-T6 | None | 1.500 | RT | 51.0 | 43.4 | 12.0 |
| | | | −112 | 50.1 | 45.5 | 12.0 |
| | | | −320 | 57.9 | 47.2 | 16.8 |
| 7075-T651 | None | 1.375 | RT | 86.1 | 77.7 | 10.8 |
| | | | −112 | 91.4 | 82.8 | 9.2 |
| | | | −320 | 104.0 | 92.0 | 5.8 |
| 7075-T7351 | None | 1.375 | RT | 68.2 | 56.8 | 12.0 |
| | | | −112 | 73.8 | 59.1 | 11.0 |
| | | | −320 | 87.4 | 66.0 | 10.8 |
| 7079-T651(b) | None | 1.375 | RT | 82.5 | 72.08 | 11.2 |
| | | | −112 | 89.9 | 81.2 | 10.2 |
| | | | −320 | 100.6 | 90.6 | 4.5 |
| **Welded plate** | | | | | | |
| 5083-O | 5183 | 7.000 | RT | 43.7 | 25.0 | 16.2 |
| | | | −320 | 60.3 | 30.1 | 19.0 |
| | | | RT | 38.4 | 22.8 | 12.7 |
| | | | −320 | 49.4 | 26.2 | 9.8 |
| 5083-O | 5183 | 7.700 | RT | 35.8 | 22.5 | 6.5 |
| | | | −320 | 57.0 | 27.3 | 15.0 |
| | | | RT | 35.8 | 22.5 | 6.5 |
| | | | −320 | 57.0 | 27.3 | 15.0 |
| | | | RT | 43.8 | 24.5 | 23.5 |
| | | | −320 | 58.0 | 28.1 | 15.5 |
| | | | RT | 41.1 | 24.4 | 15.5 |
| | | | −320 | 55.6 | 27.7 | 15.0 |
| | | | RT | 39.9 | 19.7 | 14.0 |
| | | | −320 | 51.5 | 25.8 | 10.5 |

Matching fracture toughness data are presented in Table 9.10(b).

**Table 9.10(b)   Results of notched bend and compact tension fracture-toughness tests of aluminum alloy sheet and plate at subzero temperatures**

| Alloy and temper | Filler alloy | Thickness, in. | Type of specimen | Specimen orientation | Test temperature,°F | Specimen width W, in. | Initial crack length, 2a, in. | $K_Q$, ksi $\sqrt{in.}$ | $K_{max}$, ksi $\sqrt{in.}$ | Valid $K_{Ic}$, ksi $\sqrt{in.}$ |
|---|---|---|---|---|---|---|---|---|---|---|
| **Unwelded plate** | | | | | | | | | | |
| 2014-T651 | None | 1.000 | NB | T-L | RT | 2.00 | 0.99 | 21.2 | ... | Yes |
| | | | | | −112 | ... | ... | ... | ... | ... |
| | | | | | −320 | 2.00 | 1.02 | 26.1 | ... | Yes |
| 2024-T651 | None | 1.375 | NB | T-L | RT | 3.00 | 1.51 | 20.3 | ... | Yes |
| | | | | | −112 | 3.00 | 1.54 | 22.0 | ... | Yes |
| | | | | | −320 | 3.00 | 1.48 | 22.2 | ... | Yes |
| 5083-O | None | 7.000 | NB | T-S | RT | 3.00 | 3.71 | ... | 53.3 | No |
| | | | | | −260 | 3.00 | 3.72 | ... | 67.2 | No |
| | | | | | −320 | 3.00 | 3.73 | ... | 67.1 | No |
| 5083-O | None | 7.700 | NB | T-L | RT | 7.70 | 4.12 | ... | 44.3 | No |
| | | | | | −320 | 7.70 | 3.97 | ... | 55.8 | No |
| | | | NB | T-S | RT | 7.70 | 4.20 | ... | 48.0 | No |
| | | | | | −320 | 7.70 | 4.11 | ... | 59.0 | No |
| | | | CT | L-S | RT | 6.00 | 3.14 | ... | 48.6 | No |
| | | | | | −320 | 6.00 | 3.24 | ... | 56.0 | Yes |
| | | | | T-S | RT | 6.00 | 3.26 | ... | 41.2 | No |
| | | | | | −320 | 6.00 | 3.26 | ... | 47.9 | Yes |
| | | | | S-L | RT | 6.00 | 3.21 | ... | 36.2 | No |
| | | | | | −320 | 6.00 | 3.23 | ... | 41.8 | Yes |
| 6061-T6 | None | 1.500 | NB | L-T | RT | 3.00 | 1.47 | 26.5 | ... | Yes |
| | | | | | −112 | 3.00 | 1.49 | 30.1 | ... | Yes |
| | | | | | −320 | 3.00 | 0.99 | 21.2 | ... | Yes |
| 7075-T651 | None | 1.375 | NB | T-L | RT | 3.00 | 1.54 | 20.5 | ... | Yes |
| | | | | | −112 | 3.00 | 1.52 | 22.6 | ... | Yes |
| | | | | | −320 | 3.00 | 1.58 | 25.1 | ... | Yes |
| 7075-T7351 | None | 1.375 | NB | T-L | RT | 3.00 | 1.53 | 28.2 | ... | Yes |
| | | | | | −112 | 3.00 | 1.39 | 28.2 | ... | Yes |
| | | | | | −320 | 3.00 | 1.58 | 29.2 | ... | Yes |
| 7079-T651(b) | None | 1.375 | NB | T-L | RT | 3.00 | 1.64 | 23.6 | ... | Yes |
| | | | | | −112 | 3.00 | 1.67 | 26.1 | ... | Yes |
| | | | | | −320 | 3.00 | 1.51 | 26.7 | ... | Yes |
| **Welded plate** | | | | | | | | | | |
| 5083-O | 5183 | 7.000 | NB | CNT | RT | 7.00 | 3.50 | ... | 46.6 | No |
| | | | | | −320 | 7.00 | 3.50 | ... | 57.9 | ... |
| | | | NB | FNT | RT | 7.00 | 3.48 | ... | 50.3 | No |
| | | | | | −320 | 7.00 | 3.48 | ... | 62.7 | No |
| 5083-O | 5183 | 7.700 | NB | CNT | RT | 7.70 | 3.64 | ... | 49.2 | No |
| | | | | | −320 | 7.70 | 3.78 | ... | 62.5 | No |
| | | | NB | FNT | RT | 7.70 | 3.77 | ... | 49.8 | No |
| | | | | | −320 | 7.70 | 3.64 | ... | 113.6 | No |
| | | | CT | CPT | RT | 6.00 | 3.55 | ... | 58.0 | No |
| | | | | | −320 | 6.00 | 3.67 | ... | 64.4 | No |
| | | | CT | CTP | RT | 6.00 | 2.92 | ... | 35.8 | No |
| | | | | | −320 | 6.00 | 3.08 | ... | 45.6 | No |
| | | | CT | FNT | RT | 6.00 | 3.04 | ... | 22.2 | No |
| | | | | | −320 | 6.00 | 3.44 | ... | 26.7 | No |

Specimens per Fig. A1.11(a) or (b) and A1.12(a) or (b). Each line of data represents the average of four tests of one lot of material. For tensile yield strengths, offset is 0.2%. $R_{sc}$ or $R_{sb} = s_N/s_{ys}$, which is ratio of maximum net-section stress to tensile yield strength. NB, notched bend; RT, room temperature; CT, compact tension.(a) Not valid by present criteria; excessive plasticity and/or insufficient thickness for plane-strain conditions. (b) Obsolete alloy. Matching tensile test data are presented in Table 9.10(a).

**Table 9.11  Summary of toughness parameters for thick 5083-O plate and 5183 welds in 5083-O plate**

| Alloy and temper | Specimen orientation (Fig. A.1.2) | Notch-yield ratio (NYR)(a) | | Unit propagation energy (UPE), in.-lb/in.$^2$(b) | | Estimated fracture toughness(c) | | | |
|---|---|---|---|---|---|---|---|---|---|
| | | | | | | Plane strain ($K_{Ic}$), ksi-in.$^2$ | | Plane stress/mixed mode ($K_c$), ksi-in.$^2$ | |
| | | RT | −320 °F | RT | −320 °F | RT | −320 °F | RT | −320 °F |
| 5083-O | L-T, L-S | 2.41 | 2.51 | 850 | 1280 | 50 | 60 | >100 | >100 |
| | T-L, T-S | 2.35 | (d) | 730 | 1020 | 45 | 50 | 100 | >100 |
| | S-L, S-T | (d) | (d) | 500 | 590 | 35 | 40 | (e) | (e) |
| 5183 welds | Cross or through weld | 2.09 | 2.21 | 1090 | 1065 | 50 | 45 | >100 | >100 |
| | Along root pass | (d) | (d) | 865 | 985 | 40 | 45 | >100 | >100 |
| | Heat-affected zone | (d) | (d) | 830 | 1065 | 50 | 60 | >100 | >100 |

RT, room temperature. (a) Specimens per Fig. A1.4(a, b). (b) Specimens per A1.7(a, b). (c) Estimated utilizing correlations in Fig. 8.3 and 8.4. (d) Not determined. (e) $K_c$ not applicable to S-L, S-T orientations

CHAPTER **10**

# Subcritical Crack Growth

IN MOST APPLICATIONS, structures do not experience complete fracture from the initial design discontinuities or internal flaws (metallurgical discontinuities) that are present when some component of the structure goes into service. It is likely that with whatever discontinuity or localized stress raiser is present, the structure will perform for some time in service without change. After more time in service, the structure is likely to experience some time-dependent or temperature-dependent growth of whatever discontinuity was originally present. Eventually, if not discovered and repaired, the original discontinuity may grow to a "critical" length, that is, a length as predicted by fracture-toughness testing that is likely to cause unstable crack growth to complete fracture.

Unstable crack growth generally occurs by one of three mechanisms:

- Fatigue crack propagation (see section 10.1)
- Creep crack propagation (see section 10.2)
- Stress-corrosion cracking (see section 10.3)

It is beyond the scope of this book to provide a summary of data on subcritical crack growth rates for aluminum alloys; it is appropriate, however, to provide some representative data and amplify further on their relationship of subcritical crack growth data to fracture toughness data. Readers are referred to the excellent discussion by Bucci, Nordmark, and Starke in Volume 19 of *ASM Handbook* (Ref 2).

## 10.1 Fatigue Crack Growth

As noted previously, in designing fracture-critical structures, it is important to consider the case when a fatigue crack may have been initiated and is growing from an internal discontinuity of some type in the stress field. Discontinuities may be metallurgical in nature (e.g., forging defect, porosity) or design based (e.g., rivet hole or window). For fracture-mechanics analyses of such situations, it is appropriate to consider that whatever size of flaw or discontinuity cannot be ruled out reliably by nondestructive testing may well be present somewhere in the structure and may serve as the initiation site of fatigue crack growth that could lead to complete fracture. Data from fracture-mechanics-based presentations such as those in Fig. 10.1 and 10.2 (Ref 2) would be used to estimate how fast that crack might grow and if/when that crack might grow to a length predicted by fracture toughness parameters to initiate unstable crack growth to failure.

In fracture-mechanics-based presentations of fatigue crack growth, such as those in Fig. 10.1 and 10.2, the data are presented in terms of the rate of crack growth as a function of the stress-intensity factor, $K$, and so, as the crack grows, it may be tracked in the same terms as those used to define the conditions for unstable fracture, $K_c$ or $K_{Ic}$, depending upon the material thickness and stress state. As the crack grows longer, the stress intensity increases, and, at some point, potentially approaches the limiting critical conditions predicted from the fracture toughness tests at which complete fracture must be expected.

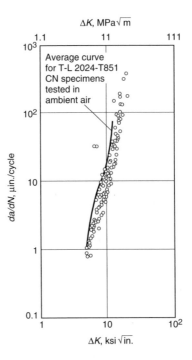

**Fig. 10.1** Fatigue crack growth rate data for 2124-T851 plate and comparison to data for 2024-T851 plate. Aluminum alloy 4.5 in. 2124–T851 plate, T-L direction, center-notched (CN) specimen; $W = 3$ in., $B = 0.75$ in.; $R = 1.3$; $t = 5.2$ Hz; RT constant load tests. $\Delta K$ is the stress intensity range during fatigue cycling; $da/dN$ is the increment of crack growth per cycle of loading.

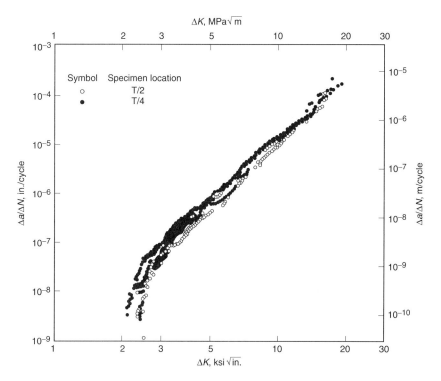

**Fig. 10.2** Fatigue crack growth rates for 7050-T7451 plate (5.67 and 5.90 in. thick). Long transverse, T/2 and T/4 test locations, $R = 0.33$, humid air (relative humidity > 90%)

Thus, fatigue crack growth data and fracture toughness data represent two components of the continuum of analysis of the life of a structure by fracture-mechanics methods. There are two additional potential modes of subcritical crack growth that should be considered in such life analyses: sustained load or creep crack growth, and stress-corrosion crack (SCC) growth, as covered in sections 10.2 and 10.3, respectively.

## 10.2 Creep Crack Growth

Evaluations of notched tensile and compact tension specimens under sustained loads have shown that some aluminum alloys widely used in high-temperature applications may experience some time-dependent crack growth at certain temperatures, referred to as *creep crack growth*. This phenomenon has been observed in at least one of the alloys of the 2*xxx* (aluminum-copper) series, namely 2219, which is highly recommended for elevated temperature service (Ref 67).

Data in Fig. 10.3 present creep crack growth rates, *da/dt*, for 2124-T851 and 2219-T851 at 300 °F in terms of the applied stress intensity factor, $K_I$. As in the case of fatigue crack growth rates, presentation in this format

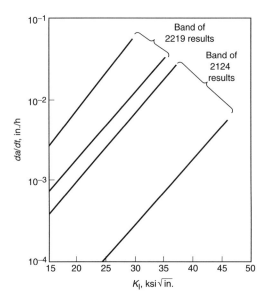

**Fig. 10.3** Crack growth rates (*da/dt*) for 2124-T851 and 2219-T851 plate at 300 °F. $K_I$ is the instantaneous stress intensity

permits tracking of the crack growth in fracture-mechanics terms, relatable to the critical fracture conditions defined by fracture toughness tests.

Of the two alloys shown in Fig. 10.3, 2219-T851 exhibited considerably faster crack growth at 300 °F than did 2124-T851, even though 2219-T851 has the higher fracture toughness over the whole temperature range, as shown in Fig. 10.4.

Parallel to the case with fatigue crack growth, total life of a structure under sustained loading may be estimated by assuming that some initial

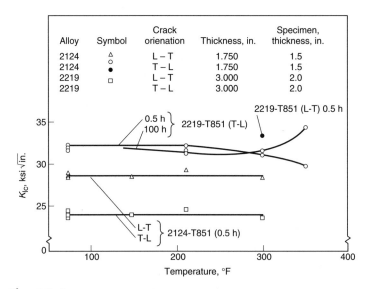

**Fig. 10.4** $K_{Ic}$ vs. temperature for 2124-T851 and 2219-T851 plate

metallurgical or design flaw may be present and that it may grow at the rate predicted by data of the type in Fig. 10.3 from the creep crack growth tests. The possibility of fracture must be assumed when the time-dependent stress-intensity value approaches that determined in the fracture toughness tests (Fig. 10.4) at that temperature. In the case of the two alloys for which data are presented, it is clear that the apparent advantage suggested for 2219-T851 by its higher fracture toughness values may not always be borne out when the potential for a higher rate of time-dependent crack growth is considered.

It is interesting to note that the results of stress-rupture tests of smooth and notched tensile specimens may provide a clue to those alloys for which the previously mentioned behavior might be expected. The results of such tests of 2219-T851 are presented in Fig. 10.5; after about 25 h, the stress-rupture lives of notched specimens are shorter than those of smooth specimens. For some other alloys, such as 5454-O and 5454-H32 for which data are shown in Fig. 10.6, the rupture lives of notched specimens remain about equal to or greater than those of smooth specimens. It is important to note that the specific relationships of notched-to-smooth specimen lives will depend upon the notch geometry, and that it is relative performance that is important in such cases. Regrettably, stress-rupture test data for notched specimens of 2124-T851 are not available to complete the comparisons referred to previously.

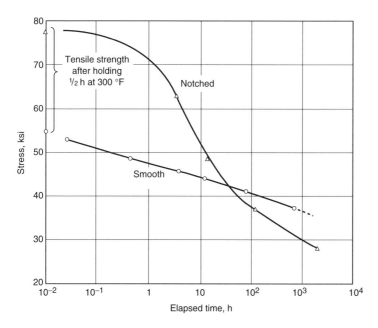

**Fig. 10.5** Effects of notches on stress-rupture strengths of 2219-T851 plate (1 in. thick) at 300 °F. Specimens were 0.5 in. diameter smooth and notched (Fig. A1.7a) and taken in the longitudinal direction of rolling.

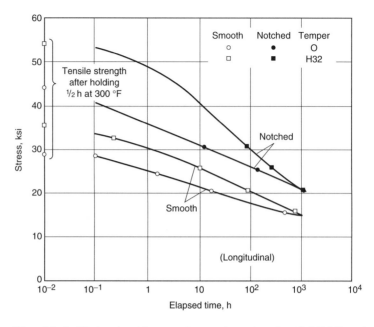

**Fig. 10.6** Effects of notches on stress-rupture strengths of 5454-O and 5454-H32 plate (0.750 in.) at 300 °F. Specimens were 0.5 in. diameter smooth and notched (Fig. A1.7a), taken in the longitudinal direction.

## 10.3 Stress-Corrosion Cracking

For certain 2xxx and 7xxx aluminum alloys, especially when subjected to stresses in the short-transverse (through-the-thickness) direction of thick plate, forgings, and extrusions, the potential for intergranular SCC growth must be considered (Ref 2). While this phenomenon has long been studied with tensile loading of smooth specimen subjected to exposure in potentially troublesome environments, it also can be examined in fracture-mechanics terms of the rate of crack growth, $da/dt$, as a function of the applied stress-intensity factor, $K_I$.

Representative data of this type are shown in Fig. 10.7 for several aluminum alloys (Ref 2). Such presentations are similar to those for fatigue and creep crack growth, except that a more pronounced upper limit to the rate of crack growth is apparent; at stress intensities beyond the bend in the curve, crack growth continues, but at a rate no longer greatly dependent on the instantaneous applied stress intensity.

Once again, it should be assumed when designing with these alloys under short-transverse stresses that the largest crack that cannot be detected reliably may be present in the stress field; the crack growth rate data can be used to determine how rapidly that crack may grow to the critical size indicated by the fracture toughness tests. Thus, presentation of SCC growth data, like fatigue and creep crack growth data, provides a means of estimating life expectancy of structures potentially susceptible to such phenomena.

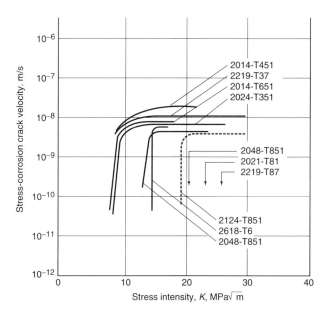

**Fig. 10.7(a)** Crack propagation rates in stress-corrosion tests using precracked thick, double-cantilever beam specimens of high-strength 2xxx series aluminum alloy plate, TL (S-L) orientation. Specimens were wet twice a day with an aqueous solution of 3.5% NaCl, 23 °C.

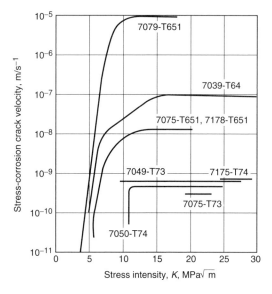

**Fig. 10.7(b)** Crack propagation rates in stress-corrosion tests using precracked specimens of 7xxx series aluminum alloys; 25 mm thick, double-cantilever beam, short-transverse orientation of die forging, long transverse orientation of hand forgings and plate. Specimens were subject to alternate immersion tests, 3.5% NaCl solution, 23 °C. Source: M.O. Speidel, *Met. Trans.,* Vol 6A, 1975, p 631

For non-fracture-mechanicians, there is a particularly useful way of dealing with design against SCC growth that combines the results of conventional smooth-specimen and pre-cracked specimen SCC testing, as illustrated in Fig. 10.8 (Ref 2, 60). It has been the experience of investigators in stress-corrosion testing of smooth tensile specimens that there are "thresholds" of applied stress below which SCC growth and failure are not likely to occur. Combining such results with the "safe" stress-flaw size results from fracture-mechanics types of SCC tests leads to the dual treatment in Fig. 10.8. On the left side of the chart in Fig. 10.8, where flaw size is quite small, SCC growth is governed by stress, and levels above line *A-B* are to be avoided. On the right side of the chart, for larger flaw sizes, SCC growth is governed by stress-intensity factor, and stresses above line *D-B* are to be avoided. Representative presentations of this type for aluminum alloys 2219-T87 and 7075-T651 are presented in Fig. 10.9.

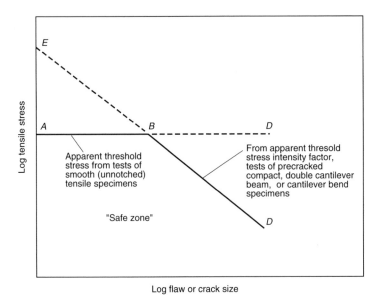

**Fig. 10.8** Stress-corrosion safe-zone plot. Apparent threshold stress is maximum stress at which tensile specimens do not fail by stress-corrosion cracking when stressed in environment of interest. Apparent threshold stress intensity factor is maximum stress intensity at which no significant stress-corrosion crack growth takes place in precracked fracture specimens, environment of interest.

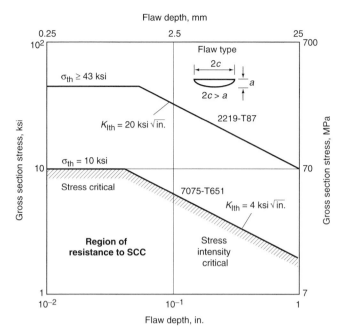

**Fig. 10.9** Composite stress-stress intensity-SCC threshold safe-zone plot for two aluminum alloys exposed in a salt-dichromate-acetate solution. $\sigma_{th}$ is threshold of applied tensile stress for SCC in smooth specimens. $K_{th}$ is threshold of applied stress intensity for SCC in notched or precracked specimens.

# **11**

# Metallurgical Considerations in Fracture Resistance

## 11.1 Alloy Enhancement

THE APPLICATION OF toughness testing to alloy development has led to a number of high-strength aluminum alloys and special tempers of some alloys with outstanding combinations of strength and toughness. An underlying basis of such work arose from the findings of Staley et al. (Ref 2, 37, 51–54) that the presence of large amounts of impurity elements such as iron and silicon, in high-strength alloys provides sites for potential crack initiation and growth as well as paths for more rapid crack growth than would otherwise be expected. The elimination of these sites would be expected to improve the toughness of the nominal composition, a concept borne out by many experiments. The combination of this principle with other optimization of compositions and thermomechanical treatments has led to the development of high-toughness alloys 2124, 2324, and 2524, all superior to 2024, and of high-toughness alloys 7175 and 7475, both substantial improvements on 7075. Similar principles have been applied to the development of newer alloys such as 7050 and 7055.

The advantages these high-toughness alloys hold over the older, conventional compositions may be seen from the following illustrations from Ref 2 and 52:

- *2124-T851 versus 2024-T851:* Fig. 11.1 illustrates a comparison of $K_{Ic}$ values for 2124-T851 plate with data for 2024-T851 plate from a consistent series of tests; $K_{Ic}$ is 3 to 5 ksi $\sqrt{\text{in.}}$ higher for the 2124-T851

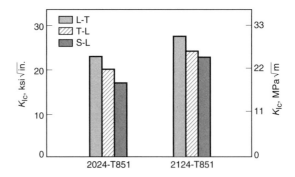

**Fig. 11.1** Average plane-strain fracture toughness data for production lots of 4 to 5.5 in. thick 2024 plate

in all test orientations included, and the difference is greatest in the often-critical short-transverse (S-L) orientation.

- *2524-T3 versus 2024-T3:* A comparison of the crack resistance curves for these two alloys is presented in Fig. 7.6, demonstrating the advantages of the composition and processing controls for 2524-T3.
- *2419-T851 versus 2219-T851:* Fig. 11.2 illustrates a comparison of $K_{Ic}$ values for 2419-T851 plate with data for 2219-T851 plate. $K_{Ic}$ is about 3 to 5 ksi $\sqrt{\text{in}}$. higher for the 2419-T851 in all test orientations included, and once again, the percentage difference is greatest in the short-transverse (S-L/S-T) orientations.
- *7050-T73651 (now T7451) versus conventional high-strength alloys:* Fig. 11.3 illustrates the range of $K_{Ic}$ data for production lots of 7050-T73651 plate in the L-T orientation compared with a band of data for conventional high-strength aluminum alloys. The amount of

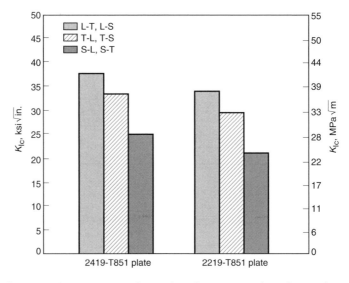

**Fig. 11.2** Comparisons of $K_{Ic}$ values for commercial production lots of 2419-T851 and 2219-T851 plate

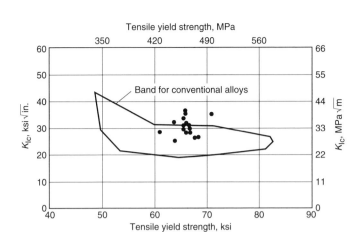

**Fig. 11.3** Plane-strain fracture toughness, $K_{Ic}$, for production lots of 7075-T73651 plate in L-T orientation

Al$_2$CuMg content present in 7050 has a significant effect on the strength-toughness combination.

• *7175-T66 and T736 (now T74) versus 7075-T6 and T73:* Fig. 11.4 shows the results of comparison tests of die forgings of exactly the same configuration of 7175 and conventional alloy 7075. The 7175 data in both the T66 and T736 (T74) tempers consistently exhibit a superior combination of strength and fracture toughness.

• *7475 versus 7075:* Fig. 11.5 through 11.8 illustrate the advantages of 7475 sheet and plate in various tempers compared with 7075 and other alloys in comparable tempers. Figure 11.5 compares representative $K_{Ic}$ data for production lots of 7475-T651 and T7651 with the range

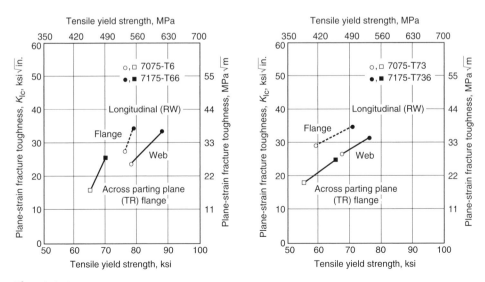

**Fig. 11.4** Plane-strain fracture toughness of 7075 and 7175 die forgings of the same configuration

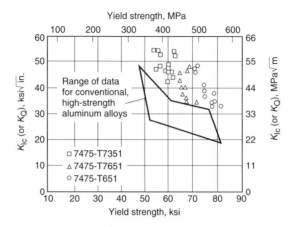

**Fig. 11.5** Plane-strain fracture toughness, $K_{Ic}$, of 7475 plate compared to band of data for conventional high-strength aluminum alloys

of data for 7075 in comparable tempers. Fig. 11.6 shows a similar comparison for 7475 sheet, where the combination of toughness and strength of 7475 is greatly superior to those of a variety of aluminum alloys, including 2024-T3, long renowned for its high toughness. The significance of this comparison is seen in the stress-flaw-size graphs in Fig. 11.7; at any stress, 7475 will tolerate cracks three to four times

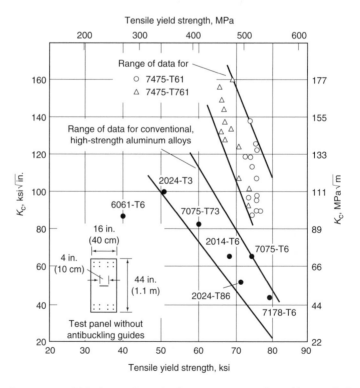

**Fig. 11.6** Critical stress-intensity factor, $K_c$, vs. tensile yield strength for 0.040 to 0.188 in. aluminum alloy sheet

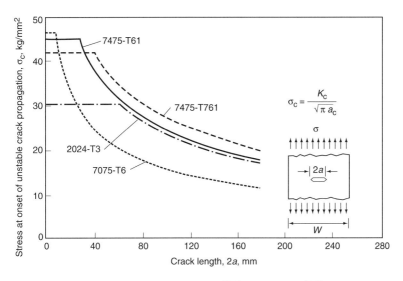

| Alloy | $K_c$, ksi$\sqrt{\text{in.}}$ | $K_c$, MPa$\sqrt{\text{m}}$ |
|---|---|---|
| 7075-T6 | 55 | 195 |
| 7475-T61 | 85 | 300 |
| 7475-T761 | 95 | 340 |
| 2024-T3 | 85 | 300 |

**Fig. 11.7** Gross section stress at initiation of unstable crack propagation vs. crack length for wide sheet panels of four aluminum alloy/temper combinations. $W$ is total panel width; $\sigma$ is uniform applied stress.

longer than 7075-T6, and at a given flaw size, 7475 will safely tolerate almost twice the stress. The advantages shown in the crack resistance curves in Fig. 7.7 for 7475 are borne out in totally independent crack growth-resistance curve tests carried out by other investigators, shown in Fig. 11.8.

Several more general metallurgical trends regarding toughness have been confirmed by extensive fracture testing, including:

- Finer, recrystallized grain size leads to higher toughness in comparable products.

- As noted earlier, total iron + silicon content is directly related to the toughness of 2xxx and 7xxx alloys; the same effect leads to the toughness advantage that A356.0 sand and permanent-mold castings hold over 356.0 castings in corresponding tempers.

- While artificial aging 7xxx alloys past peak strength (i.e., "overaging") leads to higher toughness, the strength-toughness relationship suffers; the strength of T73-type tempers is reduced to a greater extent than toughness is enhanced.

- Warm-water quenching of 7075-type alloys leads to an inferior combination of strength and toughness than cold-water or room temperature water quenching.

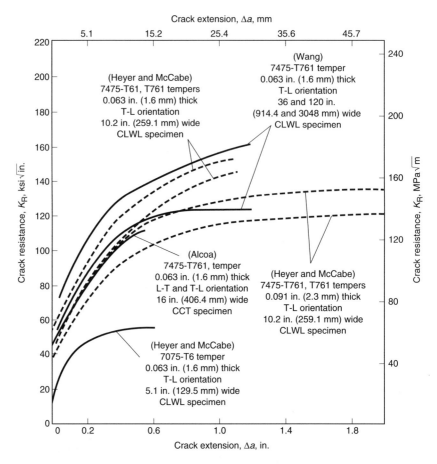

**Fig. 11.8** Crack resistance curves for 7475 sheet. Specimen type: CLWL is crack line wedge loaded; CCT is center crack tension.

## 11.2 Enhancing Toughness with Laminates

The early recognition of the limitations of the toughness of traditional high-strength aluminum alloys for aerospace applications led to studies of the effect of interleaving layers of high-strength aluminum alloy sheet with polymers (Ref 68). Center-notched panels of 0.063 in., 0.125 in., 0.250 in., and 0.500 in. thickness 7075-T6 sheet and plate were tested in full thickness. Then panels of the various thicknesses were produced by laminating the sheets and plates together to produce comparable thicknesses to the monolithic samples and tested using identical procedures as for the monolithic panels. A two-part epoxy was used to produce the multilayered panels.

Center-slotted specimens of the type in Fig. A1.9(a) with very sharp notch-tip radii, and from each monolithic layer and each composite were tested. The specimens were instrumented, and both $K_{Ic}$ and $K_c$ values were measured. The $K_{Ic}$ values were obtained using the loads observed at "pop-in" type of behavior; even with the thinnest sheet specimens, the pop-in and/or the initial deviation from elastic behavior was clear enough with high-strength alloy 7075-T6, T651 to permit comparative measurements of relative plane-strain behavior. The $K_c$ values were generated using the crack lengths and loads at fracture instability.

The results of the tests of these center-slotted panels are summarized in Table 11.1 and are plotted in Fig. 11.9. The tests of the monolithic panels reflected the thickness insensitivity of the plane-strain $K_{Ic}$ toughness level as well as the gradual decrease in stress/mixed mode toughness $K_c$ values with increasing thickness, approaching the $K_{Ic}$ values at the 0.500 in. thickness. These represent classic behavior for 7075-T6, T651. Most importantly, the tests of the laminated panels indicated clearly that the higher toughness of the individual thinner layers is retained in the multi-layered panels, even when four layers of 0.063 in. material was used to produce 0.500 in. thick panels. The $K_c$ values for the 0.500 in. thick, multilayered panel were about twice those of the monolithic panels of the same total thickness.

It is clear that for high-strength aluminum alloys, the metallurgical advantages of thin sheets of high-strength aluminum alloys may be retained in relatively thick panels by producing the required thicknesses of multilayered panels of the thinner sheet. The higher-level plane-stress or mixed mode toughness levels of the thinner sheet are retained in the thicker panel, provided that the layers are built up by a means (such as epoxy bonding) that permits the individual layers to deform plastically locally rather than acting monolithically in the thick panel. While the type of specimen design used in this study would not meet the desired rigor of the standard methods of today, the findings are unambiguous and meaningful.

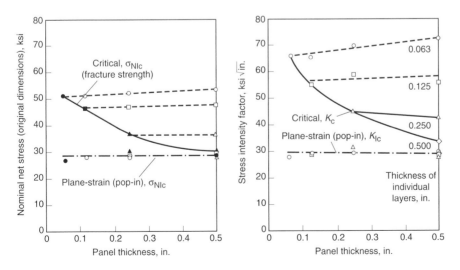

**Fig. 11.9** Results of fracture toughness tests of plain and laminated panels of 7075-T6 and 7075-T651 sheet and plate (transverse). Solid symbols, single thickness; open symbols, multilayered

**Table 11.1(a)   Results of fracture toughness tests of 7075-T6 and 7075-T651 sheet, plate, and multilayered adhesive-bonded panels bonded with two-part epoxy, transverse direction (at initial pop-in instability)**

| Total nominal thickness, in. | Represented by | Width, W, in. | Thickness, t, in. Includes adhesive | Thickness, t, in. Net | Total crack length, in. Original, $2a_o$ | Total crack length, in. Critical, $2a_c$ | At pop-in instability Load, $P_{Ic}$, lb | At pop-in instability Stress, ksi Gross, $\sigma_{Ic}$ | At pop-in instability Stress, ksi Net(a), $\sigma_{NIc}$ | Plane-strain stress-intensity factor, ksi $\sqrt{in.}$, $K_{Ic}$ | Plane-strain strain-energy release rate, in.-lb/in.,[2] $G_{Ic}$ |
|---|---|---|---|---|---|---|---|---|---|---|---|
| 0.063 | 0.063 in. sheet | 3.99 | ... | 0.062 | 1.72 | 2.05 | 3,800 | 15.4 | 27.1 | 28.1 | 68 |
| | | 4.00 | ... | 0.062 | 1.72 | 2.08 | 3,800 | 15.3 | 27.0 | 28.0 | 68 |
| | | Average | | | | | | | | **28.0** | **68** |
| 0.125 | 0.125 in. sheet | 4.00 | ... | 0.122 | 1.72 | 1.96 | 7,950 | 16.3 | 28.6 | 29.8 | 77 |
| | | 4.00 | ... | 0.122 | 1.71 | 1.89 | 7,700 | 15.8 | 27.6 | 28.7 | 71 |
| | | Average | | | | | | | | **29.2** | **74** |
| | Two layers of 0.063 in. sheet | 4.00 | 0.131 | 0.124 | 1.71 | 1.96 | 7,500 | 15.1 | 26.4 | 27.5 | 65 |
| | | 3.99 | 0.132 | 0.124 | 1.70 | 2.08 | 8,500 | 17.2 | 29.9 | 31.3 | 85 |
| | | Average | | | | | | | | **29.4** | **75** |
| 0.250 | 0.250 in. plate | 3.99 | ... | 0.253 | 1.70 | 2.06 | 18,220 | 18.0 | 31.5 | 32.8 | 94 |
| | | 4.00 | ... | 0.253 | 1.71 | 2.11 | 16,710 | 16.5 | 28.9 | 30.1 | 78 |
| | | Average | | | | | | | | **31.4** | **86** |
| | Two layers of 0.125 in. sheet | 4.00 | 0.254 | 0.244 | 1.70 | 1.96 | 15,900 | 16.3 | 28.3 | 29.5 | 76 |
| | | 4.00 | 0.254 | 0.244 | 1.71 | 2.10 | 16,050 | 16.4 | 28.7 | 29.9 | 78 |
| | | Average | | | | | | | | **29.7** | **77** |
| | Four layers of 0.063 in. sheet | 4.00 | 0.268 | 0.248 | 1.71 | 2.07 | 15,050 | 15.2 | 26.5 | 27.6 | 66 |
| | | 4.00 | 0.273 | 0.248 | 1.70 | 2.09 | 16,400 | 16.5 | 28.8 | 30.0 | 78 |
| | | Average | | | | | | | | **29.8** | **72** |
| 0.500 | 0.500 in. plate | 4.00 | ... | 0.500 | 1.71 | 1.86 | 31,700 | 15.8 | 27.7 | 28.8 | 72 |
| | | 4.00 | ... | 0.500 | 1.72 | 2.00 | 33,600 | 16.8 | 29.5 | 30.6 | 81 |
| | | Average | | | | | | | | **29.7** | **76** |
| | Two layers of 0.250 in. plate | 3.99 | 0.520 | 0.506 | 1.70 | 2.05 | 31,600 | 15.7 | 27.3 | 28.3 | 70 |
| | | 4.00 | 0.517 | 0.506 | 1.71 | 2.00 | 31,200 | 15.4 | 26.9 | 28.0 | 68 |
| | | Average | | | | | | | | **28.2** | **69** |
| | Four layers of 0.125 in. plate | 4.00 | 0.526 | 0.488 | 1.71 | 1.94 | 31,300 | 16.0 | 28.0 | 29.2 | 73 |
| | | 3.99 | 0.522 | 0.488 | 1.71 | 1.92 | 31,300 | 16.1 | 28.1 | 29.3 | 74 |
| | | Average | | | | | | | | **29.2** | **74** |
| | Eight layers of 0.063 in. sheet | 4.00 | 0.562 | 0.496 | 1.70 | 2.10 | 35,200 | 17.7 | 30.8 | 32.4 | 91 |
| | | 4.00 | 0.562 | 0.496 | 1.72 | 2.11 | 31,200 | 15.7 | 27.6 | 28.8 | 71 |
| | | Average | | | | | | | | **30.6** | **81** |

Specimens per Fig. A1.9. (a) Based on original cross section, $(W-2a_o)t$; nominal net fracture strength

**Table 11.1(b)  Results of fracture toughness tests of 7075-T6 and 7075-T651 sheet, plate, and multilayered adhesive-bonded panels bonded with two-part epoxy, transverse direction (measurements at fracture instability)**

| Total nominal thickness, in. | Represented by | Width, W, in. | Thickness, t, in. Includes adhesive | Thickness, t, in. Net | Total crack length, in. Original $2a_0$ | Total crack length, in. Critical $2a_c$ | Load, $P_c$, lb | Stress, ksi Gross $\sigma_c$ | Stress, ksi Net(a) (nominal) $\sigma_{Nc}$ | Stress, ksi Net(b) (actual) $\sigma_N$ | Critical stress-intensity factor, $K_c$, ksi $\sqrt{in.}$ | Critical energy release rate, in.-lb/in.,$^2$ $G_c$ | $\sigma_N/\sigma_{ys}$ |
|---|---|---|---|---|---|---|---|---|---|---|---|---|---|
| 0.063 | 0.063 in. sheet | 3.99 | ... | 0.062 | 1.72 | 2.05 | 7,250 | 29.4 | 51.8 | 60.3 | 67.1 | 437 | |
| | | 4.00 | ... | 0.062 | 1.72 | 2.08 | 7,250 | 29.2 | 51.4 | 60.5 | 67.3 | 440 | |
| | Average | | | | | | | | | **60.4** | **67.2** | **438** | 0.86 |
| 0.125 | 0.125 in. sheet | 4.00 | ... | 0.122 | 1.72 | 1.96 | 13,625 | 27.9 | 49.0 | 54.7 | 59.6 | 345 | |
| | | 4.00 | ... | 0.122 | 1.71 | 1.89 | 12,325 | 25.3 | 44.2 | 47.9 | 51.4 | 256 | |
| | Average | | | | | | | | | **51.3** | **55.6** | **300** | 0.69 |
| | Two layers of 0.063 in. sheet | 4.00 | 0.131 | 0.124 | 1.71 | 1.96 | 14,500 | 29.2 | 51.1 | 57.3 | 63.8 | 395 | |
| | | 3.99 | 0.132 | 0.124 | 1.70 | 2.08 | 14,675 | 29.6 | 51.7 | 62.0 | 69.3 | 467 | |
| | Average | | | | | | | | | **60.1** | **66.6** | **431** | 0.85 |
| 0.250 | 0.250 in. plate | 3.99 | ... | 0.253 | 1.70 | 2.06 | 21,250 | 21.1 | 36.7 | 43.5 | 45.2 | 198 | |
| | | 4.00 | ... | 0.253 | 1.71 | 2.11 | 21,300 | 21.0 | 36.8 | 44.5 | 46.2 | 207 | |
| | Average | | | | | | | | | **44.0** | **45.7** | **202** | 0.59 |
| | Two layers of 0.125 in. sheet | 4.00 | 0.254 | 0.244 | 1.70 | 1.96 | 26,125 | 26.8 | 46.6 | 52.5 | 56.7 | 312 | |
| | | 4.00 | 0.254 | 0.244 | 1.71 | 2.10 | 27,025 | 27.7 | 48.3 | 58.3 | 63.2 | 387 | |
| | Average | | | | | | | | | **55.4** | **60.0** | **350** | 0.75 |
| | Four layers of 0.063 in. sheet | 4.00 | 0.268 | 0.248 | 1.71 | 2.07 | 29,750 | 30.0 | 52.4 | 62.2 | 69.9 | 474 | |
| | | 4.00 | 0.273 | 0.248 | 1.70 | 2.09 | 30,150 | 30.4 | 52.9 | 63.6 | 72.0 | 503 | |
| | Average | | | | | | | | | **62.9** | **71.0** | **488** | 0.88 |
| 0.500 | 0.500 in. plate | 4.00 | ... | 0.500 | 1.71 | 1.86 | 34,300 | 17.1 | 30.0 | 32.1 | 33.3 | 108 | |
| | | 4.00 | ... | 0.500 | 1.72 | 2.00 | 34,300 | 17.1 | 30.0 | 34.3 | 35.3 | 121 | |
| | Average | | | | | | | | | **33.2** | **34.3** | **114** | 0.45 |
| | Two layers of 0.250 in. plate | 3.99 | 0.520 | 0.506 | 1.70 | 2.05 | 41,550 | 20.6 | 35.8 | 42.4 | 43.9 | 187 | |
| | | 4.00 | 0.517 | 0.506 | 1.71 | 2.00 | 41,200 | 20.4 | 35.5 | 40.7 | 42.4 | 175 | |
| | Average | | | | | | | | | **41.6** | **43.2** | **181** | 0.56 |
| | Four layers of 0.125 in. plate | 4.00 | 0.526 | 0.488 | 1.71 | 1.94 | 53,650 | 27.5 | 48.0 | 53.3 | 57.9 | 326 | |
| | | 3.99 | 0.522 | 0.488 | 1.71 | 1.92 | 52,950 | 27.2 | 47.6 | 52.4 | 56.8 | 312 | |
| | Average | | | | | | | | | **52.8** | **57.4** | **319** | 0.71 |
| | Eight layers of 0.063 in. sheet | 4.00 | 0.562 | 0.496 | 1.70 | 2.10 | 61,550 | 31.0 | 53.9 | 65.3 | 74.5 | 539 | |
| | | 4.00 | 0.562 | 0.496 | 1.72 | 2.11 | 61,250 | 30.8 | 54.2 | 65.3 | 74.4 | 537 | |
| | Average | | | | | | | | | **65.3** | **74.4** | **538** | 0.93 |

Specimens per Fig. A1.9. (a) Based on original cross section, $(W-2a_0)t$; nominal net fracture strength. (b) Based on cross section at onset of rapid fracture, $(W-2a_c)t$

# CHAPTER 12

# Summary

NOTCH-TENSILE, tear, and fracture toughness tests have been most widely used to evaluate the resistance of aluminum alloys to unstable crack growth. These tests and the parameters determined from them and a representative set of each type of data for a broad range of aluminum alloys, tempers, and products have been covered herein. Relative ratings of the various alloys and tempers are provided based upon the key parameters from these tests, the effects of temperature are described, and the role of alloy development and process control are discussed.

The specific types of tests may be summarized and categorized as follows:

**Tests providing relative toughness indicators**

- *ASTM E* 338: Sharp-notch tensile test—sheet-type specimens
- *ASTM E* 602: Sharp-notch tensile test—cylindrical specimens
- *ASTM B* 871: Kahn-type tear test

**Tests providing fracture toughness parameters**

- *ASTM E* 399: Plain-strain fracture toughness test (thick sections) as augmented by ASTM B 645 and B 646 for aluminum alloys
- *ASTM E* 561: Crack-resistance curve test (thin sections) as augmented by ASTM E 646, Section 7

Other ASTM standard methods are available for the measurement of fracture characteristics of metals, such as E 23, Notched Bar Impact Testing; E 436, Drop-Weight Tear Testing for Indicator Purposes; and E 813, *J*-Integral for Direct Measurements. However, these tests are not widely used for aluminum alloys and therefore, are not covered herein.

Notch-yield ratio (notch tensile strength/tensile yield strength) from the notch-tensile test and unit propagation energy from the tear tests provide the most useful and consistent relative indications of the overall levels of toughness of aluminum alloys. These indices generally correlate well with direct measures of toughness, such as $K_c$ and $K_{Ic}$, from the fracture toughness tests. Fracture toughness parameters, $K_c$ and $K_{Ic}$, and complete crack-resistance curves are the most useful measures of fracture toughness because they allow designers to directly relate existing or potential discontinuities in the stress fields to safe, applied stresses and to consider the effects of repeated loading (fatigue), environmental exposure (stress-corrosion cracking), or long, sustained loading (creep cracking) on component or structure life expectancy.

While many aluminum alloys are too tough for fully valid measurements of fracture toughness parameters, $K_c$ or $K_{Ic}$ values of such parameters may often be estimated from the results of notch-tensile and tear tests, and such estimates conservatively applied can provide useful projections to designers of the conditions under which unstable fracture might be experienced.

Representative data from all of these types of tests are presented herein, in some cases as a function of temperatures as low as –452 °F and in a few cases at temperatures up to 500 °F.

Among the most important trends illustrated by the data are:

- For fracture-critical structural, tankage, and transportation applications, high-strength aluminum-magnesium (5xxx) alloys such as 5083-O provide exceptional toughness at a moderate strength level. The choice of this alloy and temper for shipboard liquefied natural gas tankage is a good illustration of this advantage.
- For fracture-critical aerospace applications, alloys 2124, 2524, 2419, 7050, 7150, 7175, and 7475, providing both composition control and thermomechanical practices to achieve superior combinations of strength and toughness, are highly recommended.
- Among alloys for cast components, premium quality sand and permanent-mold castings of alloys such as A356.0 and A357.0 consistently exhibit superior combinations of strength and toughness to those of conventional sand castings; if strength is not an issue, casting alloys A444.0-F and B535.0-F offer exceptional toughness.
- Welds made with 5xxx filler alloys consistently provide superior combinations of strength and toughness to those in most other filler alloys, the only exception being when they are used to weld high-silicon-bearing castings, in which case the lower toughness of the high-silicon composition dilutes the positive effect of the high-magnesium alloys.

# CHAPTER 13

# References

## CITED REFERENCES

1. J.G. Kaufman and M. Holt, *Fracture Characteristics of Aluminum Alloys*, Aluminum Company of America, Pittsburgh, PA, 1960
2. R.J. Bucci, G. Nordmark, and E.A. Starke, Jr., Selecting Aluminum Alloys to Resist Failure by Fracture Mechanisms, *Fatigue and Fracture*, Vol 19, *ASM Handbook*, ASM International, 1996, p 771–812
3. D.G. Altenpohl, *Aluminum: Technology, Applications, and Environment: A Profile of a Modern Metal Aluminum from Within*, The Aluminum Association and TMS, 1998
4. *Annual Book of ASTM Standards*, ASTM, published annually
5. "Notched Bar Impact Testing of Metallic Materials," E23, *Annual Book of ASTM Standards*, ASTM, published annually
6. "Sharp-Notch Tensile Testing of High-Strength Sheet Materials," E338, *Annual Book of ASTM Standards*, ASTM, published annually
7. "Sharp-Notch Testing with Cylindrical Specimens," E602, *Annual Book of ASTM Standards*, ASTM, published annually
8. "Tear Testing of Aluminum Products," B871, *Annual Book of ASTM Standards*, ASTM, published annually
9. "Plane-Strain Fracture Toughness Testing of Metallic Materials," E399; "Practice for R-Curve Determination," E561; "Practice for Plane Strain Fracture Toughness Testing of Aluminum Alloys," B645; and "Practice for Fracture Toughness Testing of Aluminum Alloys," B646, *Annual Book of ASTM Standards*, ASTM, published annually
10. "Tension Testing of Wrought and Cast Aluminum Alloys" (English/Engineering and Metric Versions), B557 and B557M, *Annual Book of ASTM Standards*, published annually
11. "SI Quick Reference Guide: International System of Units (SI) the Modern Metric System" IEEE/ASTM Standard SI-10, *Annual Book of ASTM Standards*, ASTM, published annually

12. *Aluminum Standards and Data,* Standard and Metric ed., The Aluminum Association, published periodically

13. *The Aluminum Association Alloy and Temper Registrations Records*: *International Alloy Designations and Chemical Composition Limits for Wrought Aluminum and Aluminum Alloys*, The Aluminum Association, July 1998; *Designations and Chemical Composition Limits for Aluminum Alloys in the Form of Castings and Ingot*, The Aluminum Association, Jan 1996; *Tempers for Aluminum and Aluminum Alloy Products*, The Aluminum Association, Feb 1995

14. D. Zalenas, Ed., *Aluminum Casting Technology*, 2nd ed., The American Foundrymen's Society, Inc., 1993

15. *Standards for Aluminum Sand and Permanent Mold Castings*, The Aluminum Association, published periodically

16. *The NFFS Guide to Aluminum Casting Design: Sand and Permanent Mold*, Non-Ferrous Founder's Society, 1994

17. J.G. Kaufman, *Introduction to Aluminum Alloys and Tempers*, ASM International, 2000

18. J.G. Kaufman, *Properties of Aluminum Alloys: Tensile, Creep and Fatigue Data at High and Low Temperatures*, The Aluminum Association and ASM International, 1999

19. M. Holt and J.G. Kaufman, Indices of Fracture Characteristics of Aluminum Alloys Under Different Types of Loading, *Curr. Eng. Pract.*, Vol 16 (No. 3), July–Aug 1973

20. N.A. Kahn and E.A. Imbembo, A Method of Evaluating the Transition from Shear-to-Cleavage-Type Failure in Ship Plate, *Weld. J.,* Vol 27, 1948

21. W.S. Pellini, Notch Ductility of Weld Metal, *Welding Research Supplement*, May 1956, p 217s

22. T.W. Crooker et al., "Metallurgical Characteristics of High Strength Structural Materials," NRL Report 6196, Sept 1964 (also related NRL reports)

23. H. Neuber, *Theory of Notch Stresses*, McGraw-Hill, 1946; R.E. Peterson, *Stress-Concentration Design Factors*, John Wiley & Sons, Inc., 1953

24. J.G. Kaufman and E.W. Johnson, *The Use of Notch-Yield Ratio to Evaluate the Notch Sensitivity of Aluminum Alloy Sheet*, ASTM Proc., Vol 62, ASTM, 1962, p 778–791

25. J.G. Kaufman, Sharp-Notch Tension Testing of Thick Aluminum Alloy Plate with Cylindrical Specimens, ASTM STP 514, ASTM, 1972, p 82–97

26. J.G. Kaufman and E.W. Johnson, Notch Sensitivity of Aluminum Alloy Sheet and Plate at –320 °F Based Upon Notch-Yield Ratio, *Advances in Cryogenic Engineering*, Vol 8, 1963, p 678–685

27. M.P. Hanson, G.W. Stickley, and H.T. Richards, Sharp-Notch Behavior of Some High-Strength Sheet Aluminum Alloys and Welded Joints at 75°, –320°, and –423 °F, *Low-Temperature Properties of*

*High-Strength Aircraft and Missile Materials*, ASTM STP 287, ASTM, 1960, p 3–15

28. J.G. Kaufman and G.W. Stickley, Notch Toughness of Aluminum Alloy Sheet and Welded Joints at Room and Subzero Temperatures, *Cryogenic Technol.*, July/Aug 1967

29. J.G. Kaufman, F.G. Nelson, and E.W. Johnson, The Properties of Aluminum Alloy 2219 Sheet, Plate, and Welded Joints at Low Temperatures, *Advances in Cryogenic Engineering*, Vol 8, 1963, p 661–670

30. W.A. Anderson, J.G. Kaufman, and J.E. Kane, Notch Sensitivity of Aluminum-Zinc-Magnesium Alloys at Cryogenic Temperatures, *Advances in Cryogenic Engineering*, Vol 9, 1964, p 104–111

31. F.G. Nelson, J.G. Kaufman, and E.T. Wanderer, Tensile Properties and Notch Toughness of Groove Welds in Wrought and Cast Aluminum Alloys at Cryogenic Temperatures, *Advances in Cryogenic Engineering*, Vol 14, 1969, p 71–82

32. J.W. Coursen, J.G. Kaufman, and W.E. Sicha, "Notch Toughness of Some Aluminum Alloy Castings at Cryogenic Temperatures," *Advances in Cryogenic Engineering*, Vol 12, 1967, p 473–483

33. F.G. Nelson, J.G. Kaufman, and E.T. Wanderer, Tensile, Notch-Tensile, and Tear Properties of Groove Welds in X7005-T63 Plate and 7005-T53 Extrusions at Room and Cryogenic Temperatures, *Proceedings of the XIIIth International Conference of Refrigeration* (Washington, DC), Vol 1, 1971, p 655–672

34. F.G. Nelson, J.G. Kaufman, and M. Holt, Fracture Characteristics of Welds in Aluminum Alloys, *Weld. J.*, July 1966, p 3–11

35. J.G. Kaufman, K.O. Bogardus, and E.T. Wanderer, Tensile Properties and Notch Toughness of Aluminum Alloys at –452 °F in Liquid Helium, *Advances in Cryogenic Engineering*, Vol 13, 1968, p 294–308

36. J.G. Kaufman and A.H. Knoll, Tear Resistance of Aluminum Alloy Sheet as Determined from Kahn-Type Tear Tests, *Materials Research and Standards*, Vol 4 (No. 4), April 1964, p 151–155

37. J.G. Kaufman and H.Y. Hunsicher, Fracture-Toughness Testing at Alcoa Research Laboratories, *Fracture-Toughness Testing and Its Applications*, ASTM STP 381, April 1965, p 290–309

38. A.A. Griffith, "The Phenomenon of Rupture and Flow in Solids," *Philos. Trans. R. Soc. (London)*, A221, 1920

39. G.R. Irwin, "Fracturing and Fracture Mechanics," T and AM Report Number 202, Department of Applied Mechanics, University of Illinois, Oct 1961

40. "Fracture Testing of High-Strength Sheet Materials: A Report of a Special ASTM Committee," *ASTM Bulletin*, No. 243, Jan 1960, p 29–40; No. 244, Feb 1960, p 18–28

41. *Fracture-Toughness Testing and Its Applications*, ASTM STP 381, ASTM, April 1965

42. W.F. Brown and J.E. Srawley, Fracture-Toughness Testing, *Fracture Toughness Testing and Its Applications*, ASTM STP 381, ASTM, April 1965, p 133–196

43. Progress in Measuring Fracture Toughness and Using Fracture Mechanics, *Materials Research and Standards*, Vol 4 (No. 3), March 1964, p 107–118

44. H.Y. Hunsicher and J.A. Nock, Jr., High-Strength Aluminum Alloys, *J. Met.*, Vol 15 (No. 3), March 1963, p 216–224

45. J.G. Kaufman, F.G. Nelson, Jr., and M. Holt, Fracture Toughness of Aluminum Alloy Plate Determined with Center-Notch Tension, Single-Edge-Notch Tension, and Notch-Bend Tests, *Eng. Fract. Mech.*, Vol 1, 1968, p 259–274

46. J.G. Kaufman, F.G. Nelson, Jr., and E.T. Wanderer, Mechanical Properties and Fracture Characteristics of 5083-O Products and 5183 Welds in 5083 Products, *Proc. of the XIIIth International Congress of Refrigeration* (Washington, DC), Vol 1, 1971, p 651–658

47. F.G. Nelson, Jr., P.E. Schilling, and J.G. Kaufman, "The Effect of Specimen Size on the Results of Plane Strain Fracture Toughness Tests," *Eng. Fract. Mech.*, Vol 4, 1972, p 33–50

48. J.G. Kaufman, Fracture Toughness of Aluminum Alloy Plate from Tension Tests of Large Center-Slotted Panels, ASTM STP 601, ASTM, July 1967, p 889–914

49. R.J. Bucci, R.W. Bush, and G.W. Kuhlman, "Damage Tolerance Characterization of Thick Wrought Aluminum Products With and Without Stress Relief," paper presented at the 1997 USAF Aircraft Structural Integrity Program Conference, 2–4 Dec 1997 (San Antonio), U.S. Air Force

50. R.J. Bucci and R.W. Bush, "Purging Residual Stress Effects from Fracture Toughness Measurements," 94th MIL-HDBK-5 Coordination Meeting (Williamsburg, VA) 15 Oct 1997

51. J.G. Kaufman and S.F. Collis, A Fracture Toughness Data Bank, *ASTM J. Test. Eval.*, March 1981, p 121–126; and S.F. Collis, D.J. Brownhill, and R.H. Wygonik, "Fracture Toughness Data Bank for Aluminum Alloy Mill Products," Final Report, Alcoa Laboratories for the Metals Properties Council (New York), 8 Aug 1979. J.G. Kaufman and S.F. Collis, A Fracture Toughness Data Bank, *ASTM J. Test. Eval.*, March 1961, p 121–126

52. *Metallic Materials and Elements for Aerospace Vehicle Structures*, MIL-HDBK-5H, CD version, Battelle, 1 Dec 1998

53. "2124 Plate; 6013-T6; 7050 Plate; 7075 Plate; 7475 Plate," Alcoa Aerospace Technical Fact Sheets, Aluminum Company of America (Bettendorf, IA), issued periodically

54. J.G. Kaufman, Design of Aluminum Alloys for High Toughness and High Fatigue Strength, AGARD Conf. Proc. No. 185, *Alloy Design for Fatigue and Fracture Resistance,* Advisory Group, 1975

55. R.J. Bucci, C.J. Warren, and E.A. Starke, Jr., Need for New Materials in Aging Aircraft Structures, *J. Aircr.,* Vol 37 (No. 1), Jan–Feb 2000, p 122–129

56. J. Liu and M. Kulak, A New Paradigm in the Design of Aluminum Alloys for Aerospace Applications, *Proc. of the 7th International Conf. on Aluminum Alloys (ICAA7): Their Physical and Mechanical Properties,* 9–14 April 2000 (Charlottesville, VA)

57. R.J. Bucci et al, "New Aluminum Aircraft Alloys for Damage Tolerant Stiffened Skins," paper presented at the 1994 USAF Structural Integrity Program Conference, 6–8 Dec 1994 (San Antonio, TX)

58. G.H. Bray, R.J. Bucci, J.R. Yeh, and Y. Macheret, "Prediction of Wide-Cracked-Panel Toughness from Small Coupon Tests," paper presented at Aeromat '94 Advanced Aerospace Materials Conference, 6–9 June 1994 (Anaheim, CA), ASM International

59. "Standard Guide for Plane Strain Fracture Toughness Testing of Non-Stress Relieved Aluminum Products," Designation BXXX-XX, ASTM draft

60. J.G. Kaufman, G.T. Sha, R.F. Kohm, and R.J. Bucci, Notch-Yield Ratio as a Quality Control Index for Plane Strain Fracture Toughness, ASTM STP 601, ASTM, July 1976, p 169–190

61. J.G. Kaufman and M. Holt, Evaluation of Fracture Characteristics of Aluminum Alloys at Cryogenic Temperatures, *Advances in Cryogenic Engineering,* Vol 10, 1965

62. J.G. Kaufman, F.G. Nelson, and R.H. Wygonik, Large-Scale Fracture Toughness Tests of Thick 5083-O Plate and 5183 Welded Panels at Room Temperature, –260 °F, and –320 °F, ASTM STP 556, ASTM, 1974

63. J.G. Kaufman, F.G. Nelson, and R.H. Wygonik, Mechanical Properties and Fracture Characteristics of 5083-O and 5183 Welds in 5083 Products, *Proc. 13th International Conf. on Refrigeration,* Vol 1, International Institute of Refrigeration, 1971

64. R.A. Kelsey, R.H. Wygonik, and P. Tenge, Crack Growth and Fracture of Thick 5083-O Plate Under Liquified Natural Gas Ship Spectrum Loading, ASTM STP 556, ASTM, 1974

65. R.W. Judy, Jr., R.J. Goode, and C.N. Freed, "Fracture Toughness Characterization Procedures and Interpretations to Fracture-Safe Design for Structural Aluminum Alloys," Naval Research Laboratory Report 6871, 31 March 1969

66. R.L. Lake, F.W. DeMoney, and R.J. Eiber, Burst Tests of Pre-Flawed Welded Aluminum Alloy Pressure Vessels at –220 °F, *Advances in Cryogenic Engineering,* Vol 13, 278–293, 1967

67. J.G. Kaufman and J.R. Low, Jr., "The Micro Mechanism of Sustained Load Crack Growth (Creep Cracking) in Al-Cu Alloys 2124 and 2219 at 300 °F," *Proc. Second International Conf. on Mechanical Behavior of Materials*, 16–20 Aug 1976, ASM International, 1978, p 415–472
68. J.G. Kaufman, Fracture Toughness of 7075-T6 and T651 Sheet, Plate, and Multilayered Adhesive-Bonded Panels, *J. Basic Eng. (Trans. ASME)*, No. 67-Met-4, 19 Jan 1967

## SELECTED REFERENCES

- *The Aluminum Design Manual*, The Aluminum Association, Washington D.C., 1999
- J.R. Kissell and R.L. Ferry, *Aluminum Structures, A Guide to Their Specifications and Design*, John Wiley & Sons, Inc., 1995
- M.L. Sharp, *Behavior and Design of Aluminum Structures*, McGraw-Hill, Inc., 1993
- M.L. Sharp, G.E. Nordmark, and C.C. Menzemer, *Fatigue Design of Aluminum Components and Structures*, McGraw-Hill, Inc., 1996
- *Fatigue Design Handbook*, SAE AE-10, 2nd ed., Society of Automotive Engineers, 1988
- H.E. Boyer, Ed., *Atlas of Fatigue Curves*, ASM International, 1986
- H.E. Boyer, Ed., *Atlas of Stress-Strain Curves*, ASM International, 1990
- H. Chandler, Ed., *Heat Treater's Guide: Practices and Procedures for Nonferrous Alloys*, ASM International, 1996
- B.D. Craig, Ed., *Handbook of Corrosion Data*, ASM International, 1990
- J.R. Davis, Ed., *Aluminum and Aluminum Alloys, ASM Specialty Handbook*, ASM International, 1994
- R.H. Jones, Ed., *Stress-Corrosion Cracking: Materials Performance and Evaluation*, ASM International, 1992
- K. Laue and H. Stenger, *Extrusion,* ASM International, 1981
- J.D. Minford, *Handbook of Aluminum Bonding Technology and Data*, Marcel Dekker, Inc., 1993

# Notch-Tensile, Tear, and Fracture Toughness Specimen Drawings

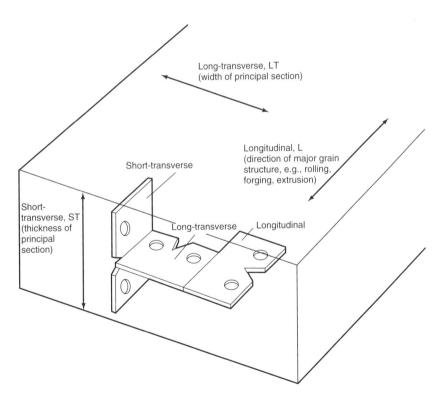

**Fig. A1.1** Orientations of tear specimens in aluminum alloy products

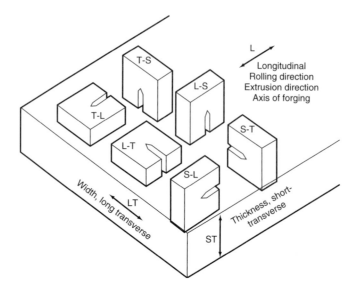

**Fig. A1.2(a)** Crack plane orientation code for fracture toughness specimens from rectangular sections

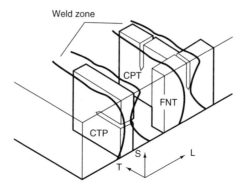

**Fig. A1.2(b)** Crack plane orientation code for fracture toughness specimens from welded plate. First letter designates crack tip location. Second letter designates direction of principal stress at crack tip with respect to weld. Third letter designates direction of crack growth. C, center of weld; H, heat-affected zone; F, fusion zone; P, parallel; N, normal; T, through

Notch-tip radius, ≤0.001 in.

**Fig. A1.3** Sheet-type notch-tensile specimen, ½ in. wide test section. Notch-tip radius ≤ 0.001 in.

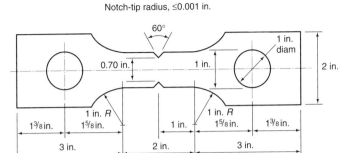

**Fig. A1.4(a)** Sheet-type notch-tensile specimen, 1 in. wide test section. Notch-tip radius ≤0.001 in.

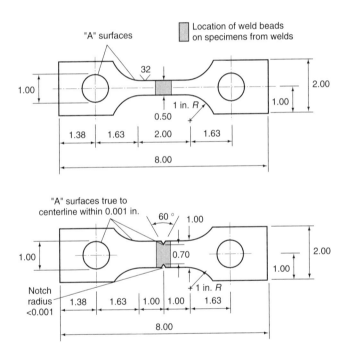

**Fig. A1.4(b)** Sheet-type notch-tensile specimen, 1 in. wide test section, from welded panels. Notch-tip radius ≤0.0005

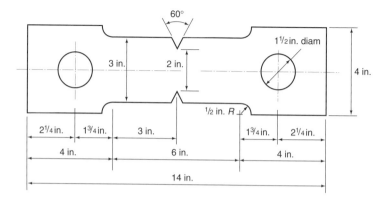

**Fig. A1.5**  Sheet-type notch-tensile specimen, 3 in. wide test section. Notches to be symmetrical about centerline within ±0.002 in. and notch-tip radii ≤0.0005 in.

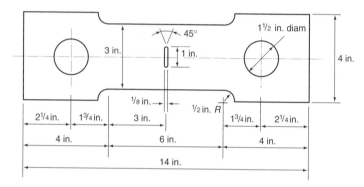

**Fig. A1.6**  Center-slotted sheet-type notch-tensile specimen, 3 in. test section. Fatigue-cracked

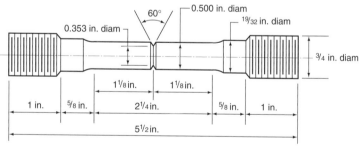

**Fig. A1.7(a)**  Cylindrical notch-tensile specimen, ½ in. test section. Notch-tip radius ≤0.0005 in.

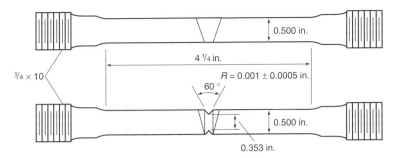

**Fig. A1.7(b)** Cylindrical notch-tensile specimen, ¹/₂ in. test section, from welded panels. Notch-tip radius ≤0.0005 in.

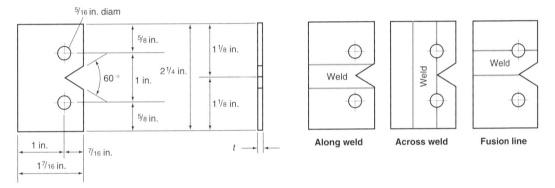

**Fig. A1.8** Tear specimen from unwelded and welded panels. Notch-tip radius ≤0.001 in.

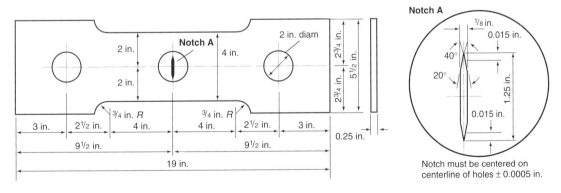

**Fig. A1.9(a)** Small center-notched fracture toughness specimen. Specimen was subsequently fatigue-precracked to about 1.33 in. total center-slot length.

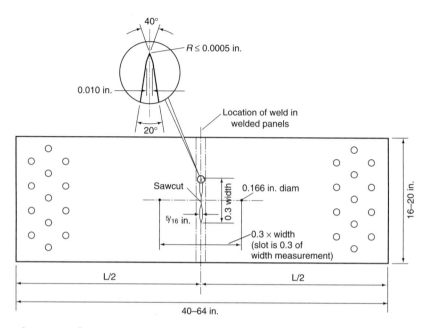

**Fig. A1.9(b)** Large center-slotted fracture toughness specimen

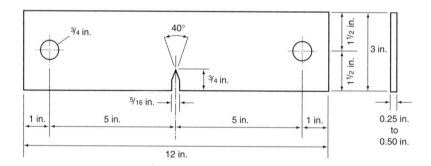

**Fig. A1.10** Single-edge-notched fracture toughness specimen. Subsequently fatigue precracked to length of approximately 1.0 to 1.5 in.

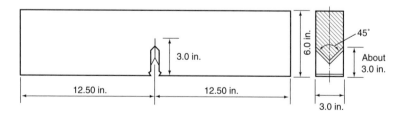

**Fig. A1.11(a)** Notched-bend fracture toughness specimen. Subsequently fatigue precracked to length of approximately 3 in.

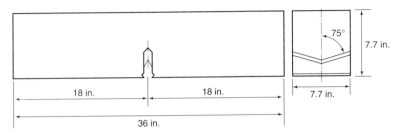

**Fig. A1.11(b)** Large notched-bend fracture toughness specimen used for 5083-O plate. Subsequently fatigue precracked to length of approximately 3.5 to 4.0 in.

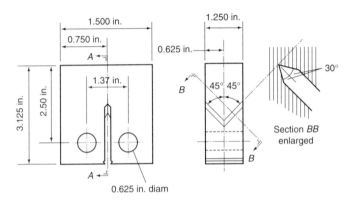

**Fig. A1.12(a)** Compact tension fracture toughness specimen. Subsequently fatigue precracked to length of approximately 1.3 in.

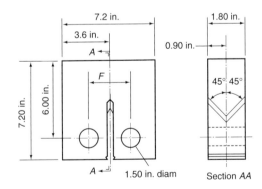

**Fig. A1.12(b)** Small compact tension fracture toughness specimen used for 5083-O plate. Subsequently fatigue precracked to length of approximately 3.3 in.

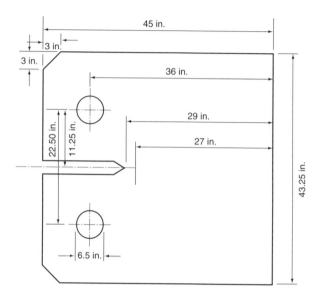

**Fig. A1.12(c)** Large-plate 4 in. thick, compact tension specimen used for 5083-O plate. Subsequently fatigue precracked to length of approximately 10 to 12 in.

# Metric (SI)
# Conversion Guidelines

Because the majority of the data presented herein were generated in an environment of the usage of English/engineering units, and because of the mass of data involved, almost the entire book is presented in those units. While the customary ASM International and Aluminum Association, Inc. policies are to present engineering and scientific data in both Standard International (SI) and English/engineering units, in this case it would have involved a considerable amount of time, effort, and expense to perform the conversions, to expand, reformat, and reset the tables, and to add the substantial number of pages to the book. In addition, foregoing conversion avoids the inevitable compromises surrounding rounding techniques for relatively complex conversions with a multitude of units.

For those interested in the properties in metric/SI units and who would like to make their own conversions, the following applicable conversion factors from English/engineering units to SI units are presented from the ASTM standard on such conversions (Ref 11):

$$1 \text{ ksi} = 6.897 \text{ MPa}$$

$$1 \text{ lbf} = 4.45 \text{ N}$$

$$1 \text{ in.-lb} = 113 \text{ mN-m}$$

$$1 \text{ ksi } \sqrt{\text{in.}} = 1.1 \text{ MPa } \sqrt{\text{m}}$$

$$1 \text{ in.} = 25.4 \text{ mm}$$

For additional information on such conversions, readers are referred to that ASTM standard.

# Alloy Index

## Cast Alloys

# Wrought Alloys